BROADBAND CABLE TV ACCESS NETWORKS:

FROM TECHNOLOGIES TO APPLICATIONS

SHLOMO OVADIA

Intel Corporation

Prentice Hall PTR
Upper Saddle River, NJ 07458

ISBN 0-13-086421-8

90000

9 780130 864215

Library of Congress Cataloging-in-Publication Data

Ovadia, Shlomo
    Broadband Cable TV Access Networks—From Technologies to Applications/ Shlomo Ovadia.
    Includes bibliographical references and index.
    ISBN 0-13-086421-8
    1. Cable TV Access. I. Title

QA9.64 .M46 2000 511.1--dc21   00-054756

Production Editor: Rose Kernan
Acquisitions Editor: Bernard Goodwin
Editorial Assistant: Michelle Vincenti
Marketing Manager: Dan DePasquale
Manufacturing Manager: Alexis R. Heydt
Cover Designer: Nina Scuderi
Cover Design Direction: Jerry Votta
Composition: Shlomo Ovadia
Cover: TV image is courtesy of DIVA

 © 2001 Prentice Hall PTR
Prentice-Hall, Inc.
Upper Saddle River, NJ 07458

The publisher offers discounts on this book when ordered in bulk quantities. For more information contact: Corporate Sales Department, Prentice Hall PTR, One Lake Street, Upper Saddle River, NJ 07458. Phone: 800-382-3419; Fax: 201-236-7141; E-mail: corpsales@prenhall.com
Intel is not affiliated with this book

Printed in the United States of America

10 9 8 7 6 5 4 3 2 1

ISBN 0-13-086421-8

Prentice-Hall International (UK) Limited, *London*
Prentice-Hall of Australia Pty. Limited, *Sydney*
Prentice-Hall Canada Inc., *Toronto*
Prentice-Hall Hispanoamericana, S.A., *Mexico*
Prentice-Hall of India Private Limited, *New Delhi*
Prentice-Hall of Japan, Inc., *Tokyo*
Pearson Education Asia Pte. Ltd.
Editora Prentice-Hall do Brasil, Ltda., *Rio de Janeiro*

To my parents, Mazal and Moshe,
My wife, Nina, and our children,
Yafit, Shlomit, Binyamin, and Leora

תם ונשלם שבח לאל בורא עולם

CONTENTS

Preface   **xiii**

1 Broadband Hybrid Fiber/Coax Access Networks Overview   **1**

1.1 Introduction   *2*
1.2 Traditional Cable TV Networks   *5*
1.3 Two Way HFC Access Networks   *6*
    1.3.1 Subscriber's Home Terminal   6
    1.3.2 Cable Modems   9
    1.3.3 IP Telephony   10
1.4 Competing Access Technologies   *12*
    1.4.1 Asymmetric Digital Subscriber Line (ADSL)   12
    1.4.2 Fiber-In-The-Loop (FITL)   15
    1.4.3 Direct Broadcast Satellite (DBS)   17
    1.4.4 Multichannel Multipoint Distribution Service (MMDS) 18
References   19

2 Basic Cable TV Background: Modulation, Signal Formats and Coaxial Cable Systems   **21**

2.1 Analog Modulated Video Signal Formats   *21*
    2.1.1 NTSC and AM-VSB Video Signals   21
    2.1.2 NTSC Signal Test Parameters   25
    2.1.3 PAL and SECAM Video Signals   27
2.2 Digital Video and Audio Signals   *29*
    2.2.1 MPEG-1 Standard   29
    2.2.2 MPEG-2 Standard   33
    2.2.3 MPEG and AC-3 Audio   35
    2.2.4 MPEG-4 Standard   36
    2.2.5 Other Digital Video Standards   37
2.3 Cable TV Frequency Plans   *37*
2.4 Coaxial Cable TV Components and Systems   *42*
    2.4.1 Coaxial Cable   42
    2.4.2 RF Amplifiers   44

      2.4.3 Taps   47
  2.5 Multichannel Coaxial Cable TV Systems   *47*
      2.5.1 CNR of a Single and Cascaded Amplifiers   47
      2.5.2 Nonlinear Distortions: CSO, CTB, and XMOD   49
      2.5.3 Multipath Reflections (Echoes) and Group Delay   50
      2.5.4 AM Hum Modulation   55
  2.6 Cable TV Return-Path Transmission Characteristics   *56*
      2.6.1 Return-Path Noise Sources   56
      2.6.2 Return-Path Noise Filtering   60
  References   61

3 Directly Modulated Cable TV Lightwave Laser Transmitters   **63**

  3.1 Semiconductor Laser Diodes   *63*
      3.1.1 Basic Laser Physics Concepts   63
      3.1.2 Semiconductor Laser Structures   68
      3.1.3 Light-Current Characteristic   70
  3.2 DFB and Multiple-Quantum-Well (MQW) Laser Diodes   *72*
      3.2.1 DFB Laser Diodes   72
      3.2.2 Multiple-Quantum-Well (MQW) Lasers   73
  3.3 Laser Dynamic Characteristics   *74*
      3.3.1 Small-Signal Response   75
      3.3.2 Large-Signal Response Circuit Model   79
  3.4 Noise in Laser Diodes   *82*
      3.4.1 Relative Intensity Noise (RIN)   82
      3.4.2 Laser Phase Noise   84
  3.5 DFB Laser Transmitter   *85*
      3.5.1 DFB Laser Transmitter System Design   85
      3.5.2 Optical Isolators   87
      3.5.3 Thermoelectric Cooler (TEC) Design and Operation   88
      3.5.4 Linearization Methods   90
      3.5.5 DFB Laser Transmitter Performance Requirements   91
  3.6 Return-Path Laser Transmitters   *92*
      3.6.1 Mode-Partition Noise and Mode-Hopping Noise   93
      3.6.2 Performance Requirements   94
  References   96

4 Externally Modulated Cable TV Lightwave Laser Transmitters   **99**

4.1 LiNbO₃ Optical Modulators   *100*
   4.1.1 Basic Operation of LiNbO3 Intensity Modulators   100
   4.1.2 Distortion Characteristics of MZI and BBI Modulators   104
4.2 Linearization Methods of Optical Modulators   *106*
   4.2.1 Feedforward Linearization Method   107
   4.2.2 Predistortion Linearization Method   108
   4.2.3 Linearizer Circuits   110
4.3 Optical Linearization Methods   *112*
   4.3.1 Optical Dual Parallel Linearization Method   112
   4.3.2 Optical Dual Cascade Linearization Method   114
4.4 Externally Modulated Laser Transmitter Design   *116*
   4.4.1 Externally Modulated YAG Laser Transmitter   116
   4.4.2 Externally Modulated DFB Laser Transmitter   119
References   120

5 Lightwave Receivers for Cable TV Networks   **123**

5.1 p-i-n Photodiode   *123*
5.2 Noise Sources in Lightwave Receivers   *129*
   5.2.1 Shot Noise   129
   5.2.2 Thermal Noise   130
   5.2.3 Laser RIN Noise   132
5.3 Carrier-To-Noise Ratio at the Receiver   *133*
5.4 Nonlinear Behavior of p-i-n Photodetectors   *135*
5.5 Basic Cable TV Receiver Design Configurations   *138*
   5.5.1 Low- and High-Impedance Front-End Receiver Design   140
   5.5.2 Transimpedance Front-End Receiver Design   141
   5.5.3 High Performance Receiver Design for Cable TV   143
References   145

6 Optical Fiber Amplifiers for Cable TV Networks   **147**

6.1 Optical Fiber Amplifier Components   *147*
   6.1.1 Wavelength-Division Multiplexers (WDMs)   147
   6.1.2 Erbium-Doped Fibers (EDFs)   154
   6.1.3 Pump Lasers   161
6.2 Basic EDFA System Configurations   *163*
6.3 Amplifier Noise and CNR Calculation   *165*
   6.3.1 Optical Fiber Amplifier Noise   166

            6.3.2 CNR and Noise-Figure Calculation   167
            6.3.3 Noise-Figure Measurement   169
        6.4 EDFA Requirements for Cable TV Networks   *170*
            6.4.1 EDFA's Noise-Figure Requirement   170
            6.4.2 EDFA's CNR Requirement   175
            6.4.3 Gain-Flattened EDFAs   176
    References   178

7 RF Digital QAM Modems   **183**

        7.1 RF QAM Modem Building Blocks   *183*
        7.2 MPEG Transport Framing   *185*
        7.3 Reed-Solomon Codes   *188*
        7.4 Interleaver/Deinterleaver   *191*
        7.5 Trellis-Coded Modulation (TCM)   *193*
            7.5.1 Punctured Convolutional Coding   193
            7.5.2 Viterbi Decoding   195
            7.5.3 TCM for 64/256-QAM Modulation   198
            7.5.4 Differential Precoder   200
        7.6 Randomizer/Derandomizer   *201*
        7.7 M-ary QAM Modulator Design and Operation   *202*
            7.7.1 Baseband Shaping Filter   202
            7.7.2 Building Blocks of M-ary Modulator   204
        7.8 M-ary QAM Receiver Design and Operation   *206*
        7.9 Adaptive Equalizer   *208*
        7.10 Carrier and Timing Recovery   *214*
            7.10.1 Carrier Recovery in QAM Receivers   214
            7.10.2 Timing Recovery in QAM Receivers   217
        7.11 MER and EVM   *219*
            7.11.1 MER and EVM Definition   219
            7.11.2 MER and EVM Test Procedure   220
        7.12 BER of M-ary QAM Signals in AWGN Channel   *223*
    References   228

8 Subscriber Home Terminals   **231**

        8.1 Digital Set-Top Box Building Blocks   *231*
        8.2 Cable TV RF Tuner   *233*
        8.3 Out-of-Band (OOB) Receiver   *235*
            8.3.1 OOB Randomizer   237
            8.3.2 OOB Reed-Solomon Coding   238

8.3.3 OOB Interleaver   238
8.3.4 OOB QPSK Mapping   239
8.4 RF QAM Transceiver   *239*
8.4.1 RF Upstream FEC   242
8.4.1.1 Upstream R-S Coding   242
8.4.1.2 Upstream Randomizer   242
8.5 MPEG Video/Audio Demultiplexer and Decoder   *243*
8.5.1 VBI Retriever and Decoder   245
8.6 Conditional Access and Control   *246*
8.6.1 Digital Encryption/Decryption Basics   246
8.6.2 Access Control Basics   248
8.6.3 Renewable Security   249
8.7 Graphics Processor   *250*
8.7.1 Basic 3D Graphics Concepts and Techniques   251
8.7.2 On-Screen Video and 2D/3D Graphics Rendering Requirements   255
8.8 Set-Top Box CPU and Memory   *257*
8.8.1 Set-Top Box CPU   257
8.8.2 Set-Top Box Memory   258
8.9 Advanced Set-Top Box with Built-in DOCSIS Cable Modem   *260*
8.10 M-QAM Transmission Impairments in HFC Networks   *261*
8.10.1 Set-Top Box Front-End Losses   262
8.10.2 QAM Transmitter Losses   266
8.10.3 Additive White Gaussian Noise (AWGN)   266
8.10.4 Multipath Reflections   267
8.10.5 Amplitude and Group Delay Variations   270
8.10.6 Burst and Impulse Noise   271
8.10.7 AM Hum Modulation   273
8.10.8 64/256-QAM System Budget Link   273
References   276

9 Transmission Impairments in Multichannel AM/QAM Lightwave Systems   **279**

9.1 Clipping-Induced Nonlinear Distortions   *280*
9.1.1 Asymptotic Statistical Properties of Clipping Noise   281
9.1.2 BER of M-ary QAM Channels Due to Clipping Noise   283
9.1.3 Comparison to Experimental Results   287
9.1.4 BER of QAM Channel Due to "Dynamic Clipping" Noise   290
9.1.5 Clipping Noise Reduction Methods   292
9.2 Bursty Nonlinear Distortions   *294*
9.3 Multiple Optical Reflections   *299*
9.3.1 Interferometric Noise due to DRB   300

9.3.2 Laser RIN Due to Multiple Discrete Optical Reflections   302

9.3.3 Multiple Reflections Effect on AM-VSB Channels   303

9.3.4 Multiple Reflections Effect on QAM Channels   307

9.4 Dispersion-Induced Nonlinear Distortions   *309*

9.5 Optical Fiber Nonlinear Effects   *312*

9.5.1 Stimulated Brillouin Scattering Effect   312

9.5.2 Self and External Phase Modulation Effects   316

9.5.3 Stimulated Raman Scattering Effect   319

9.5.4 Cross-Phase Modulation Effect   322

9.6 Polarization-Dependent Distortion Effects   *323*

References   326

10 EDFA-Based WDM Multichannel AM/QAM Video Lightwave Access Networks   **331**

10.1 Architecture and Performance of Multichannel AM-VSB/QAM Video Lightwave Trunking Networks   *331*

10.1.1 Multichannel AM/QAM Video Lightwave Trunking Systems   333

10.1.2 Multichannel AM-VSB Video Lightwave Trunking Results   334

10.1.3 Multichannel AM-VSB/QAM Video Lightwave Trunking Results   336

10.1.4 Differential Detection in AM-VSB Video Trunking Systems   340

10.1.5 SPM and EPM Effects in AM Video Trunking Systems   342

10.1.6 Polarization Effects in AM-VSB Video Trunking Systems   344

10.1.7 Gain Tilt Distortion in AM Video Trunking Systems   346

10.2 The Problem with the Current HFC Networks   *348*

10.3 DWDM Downstream Access Network Architecture   *349*

10.4 DWDM Upstream Access Network Architecture   *352*

10.4.1 Frequency-Stacking Scheme   352

10.4.2 Digitized Return-Path Transport   354

References   357

11 Data-Over-Cable Interface Specifications (DOCSIS) Protocol   **359**

11.1 DOCSIS Communication Protocol   *359*

11.2 Downstream PHY Layer   *361*

11.2.1 Downstream PMD Sublayer   362

11.3 Upstream PHY Layer   *363*

11.3.1 Upstream Channel Parameters and Requirements   365

11.3.2 Burst Profiles   365

11.3.3 Burst Timing   368

11.3.4 Upstream Spurious Power Outputs   370

11.3.5 Burst Frame Structure 371
11.4 Downstream Transmission Convergence Sublayer 372
11.5 Media Access Control (MAC) Layer 373
11.5.1 MAC Frame Format 374
11.5.2 MAC Management Messages 375
11.6 Random Access and Contention Resolution Methods 380
11.6.1 Random Access Methods 380
11.6.2 p-Persistence with Binary Exponential Backoff Algorithm 382
11.7 MAC Layer Protocol Operation 383
11.8 Quality of Service (QoS) and Fragmentation 384
11.8.1 Basic Concepts and Operation 385
11.8.2 Upstream Service Flow Scheduling 386
11.8.3 Upstream Fragmentation 388
11.9 CM and CMTS Interaction 390
References 392

12 Digital Set-Top Box Software Architecture and Applications **395**

12.1 Digital Set-Top Box Software Architecture 395
12.1.1 Real-Time Operating System (RTOS) 397
12.1.2 Set-Top Box Middleware 398
12.2 Native Applications 402
12.2.1 Electronic Program Guides (EPGs) 402
12.2.2 Parental Control 404
12.3 TV-Based Interactive Applications 405
12.3.1 Pay-Per-View (PPV) and Impulse PPV (IPPV) 406
12.3.2 Enhanced Broadcast, Interactive, and Targeted Advertisements 407
12.3.3 Video-on-Demand (VOD) and Near VOD (NVOD) 407
12.4 Internet-Based Applications 409
12.4.1 Set-Top Box Web Browsers 409
12.4.2 E-mail 411
12.4.3 E-Commerce 412
12.4.4 Home Banking, Education, and Gaming 413
12.5 Integrated Set-Top Box Applications 414
References 416

Appendix A Comparison of DAVIC and DOCSIS Specifications **419**

Appendix B International Cable TV Frequency Plans **423**

    B.1 CCIR System B/G Frequency Plan   *423*

    B.2 CCIR System I Frequency Plan   *425*

    B.3 CCIR System D Frequency Plan   *427*

Appendix C Satellite Transponder Parameters for Cable TV Networks   **429**

Acronyms   **431**

Index   **437**

Biography   **447**

PREFACE

Broadband cable TV access networks have been going through a dramatic transformation worldwide since the late 1980s. The technology vision, which was articulated by several visionary industry leaders such as Bill Gates of Microsoft, of "information at your finger tips" has started to become a reality. With the invention and emergence of the Internet and Intranet as the information superhighway, cable subscribers and small businesses did not want to be left behind and were eager for high-speed access.

Traditionally, cable TV networks were broadband coaxial networks that offered one-way broadcast of analog video channels. The fundamental understanding of the physics of various opto-electronics devices and components led to the invention and development of key fiber-optics transmission technologies such as high-power directly and externally modulated DFB laser transmitters operating at 1310 nm and 1550 nm, optical fiber amplifiers, and optical receivers. These fiber-optics technologies transformed cable TV network architecture to 750-MHz and greater bandwidth hybrid fiber/coax (HFC) networks. Furthermore, the development of low-cost highly integrated communication chips, modules, and systems such as quadrature amplitude modulation (QAM) modulators, transceivers, and MPEG encoders and decoders, enabled cable operators to introduce many digital video programs to the home using a digital set-top box. Instead of low-speed Internet access using a dial-up modem, high-speed Internet access through cable TV networks is possible using a cable modem either as a stand-alone unit or built into a digital set-top box.

The continuing insatiable desire at sometimes accelerating pace for more "bandwidth" and more services on demand has forced the cable TV operators to rethink their two-way HFC network architecture, paving the road toward two-way dense wavelength division multiplexed (DWDM) cable TV networks.

From 1992 to 1996 as a research scientist at Bell Communications Research (Bellcore), I had the privilege to work with many world-class scientists and participate in the emerging fiber-optics and QAM receiver technology studies. With a built-in HFC test-bed, my colleagues and I were able to provide valuable technical analysis and auditing services to the Regional Bell Operating Companies such as Pacific Bell and Ameritech, as well as cable TV equipment manufacturers. My exposure to the hardware and software development of digital set-top boxes and cable modems came during my work as a principal scientist at Digital

Network Systems, General Instrument from 1996 to 2000. In particular, I was exposed to the various discussions and debates on the different cable TV network and set-top box requirements while participating at the various cable TV standards meetings such as DOCSIS, IEEE802.14, and OpenCable.

The purpose of this book is to provide the reader with the basic understanding of today's two-way HFC cable TV network technologies and their evolution toward DWDM network architectures. This book, which can be used as the basis for a graduate-level material, is intended for engineers, scientists, cable TV professionals, and students interested in learning more about the existing and emerging cable TV technologies and applications.

The book is organized into five main sections:

- The first section, which consists of Chapters 1 and 2, provides an overview of the two-way HFC network with the competing access technologies, including digital subscriber line (DSL), fiber-in-the-loop (FITL), direct broadcast satellite (DBS), and multichannel multipoint distribution service (MMDS). Chapter 2 provides basic cable TV systems background, including different analog and digital video modulation formats, cable TV frequency plans, coaxial cable components and systems, and return-path transmission characteristics.

- The second section of the book, which consists of Chapters 3, 4, 5, and 6, introduces the key fiber-optics transmission technologies, including directly and externally modulated laser transmitters such as distributed feedback (DFB) and YAG laser transmitters operating at both 1310 nm and 1550 nm, optical receivers, and Erbium-doped optical fiber amplifiers (EDFAs).

- The third section of the book, which consists of Chapters 7 and 8, discusses the hardware architecture with the key components, features, and functionality of end-user home terminals, including cable modems and digital set-top boxes. Chapter 8 also includes a budget link analysis for digital set-top box in the presence of various HFC network impairments.

- The fourth section, which consists of Chapters 9 and 10, provides an in-depth analysis of single and multiple-wavelength fiber-optics transmission impairments over HFC and DWDM networks carrying both AM-VSB and QAM channels. The emerging two-way DWDM network architecture is also discussed in Chapter 10.

- The last section, which consists of Chapters 11 and 12, discusses the transmission protocol, the software architecture, and the various applications for the end-user home terminals. This includes the DOCSIS protocol for cable modems, native applications such as electronic program guides (EPGs), TV-based and Internet-based applications such as video-on-demand (VOD), e-mail, and e-commerce for digital set-top boxes.

I would like also to acknowledge many people who have contributed to my professional career. These include Professors Hyatt Gibbs and Murray Sargent III from the Optical Sciences Center, University of Arizona, who supervised my Ph.D. dissertation on optical bistable devices and non-linear optics, and Professor Chi Lee, University of Maryland. I also indebted to Dr. Chinlon Lin, William T. Anderson, and Burt Unger from Bellcore, who not only introduced me to the various technical issues of broadband fiber-optics transmission systems and networks, but also supported my work in the HFC test-bed lab at Bellcore. At General Instrument/Motorola corporation, I had the pleasure to work with various individuals including Mark Kolber, Victor Hou, Reem Safadi, Qin Zhang, Robert Mack, Steve Kimble, and Matt Waight, who helped me further understand the various technical aspects of the operation of cable TV networks as well as the OpenCable standard. In particular, I would like to acknowledge the efforts of Mark Kolber, Victor Hou, Petr Peterka, and Qin Zhang, along with Keith Williams, Philip Antoney, and Ron Hranac who graciously agreed to review portions of this book. Also, I would like to express my sincere thanks to the management in the Cable Network Operation business unit of Network Communication Group at Intel for their encouragement and support of my efforts to publish this book.

Finally, I am grateful to my parents, Moshe and Mazal, for their love and encouragement over the years. I am also exceedingly thankful to my wife Nina and our children, Yafit, Shlomit, Binyamin, and Leora, for their love and patience while I was working many nights and weekends over the last two years to write this book.

Shlomo Ovadia
San Jose, California

BIOGRAPHY

Shlomo Ovadia has earned his B.Sc. in Physics from Tel-Aviv University in 1978, and his M.Sc. and Ph.D. in Optical Sciences from the Optical Sciences Center, University of Arizona in 1982 and 1984, respectively. He spent two years as a postdoctoral fellow at the Electrical Engineering Department, University of Maryland, investigating different III-V optoelectronics materials and devices. In 1987, Shlomo joined IBM at East Fishkill as an optical scientist developing various IBM optical communications and storage products. He joined Bellcore in 1992, where he developed an HFC test-bed, and studied the transmission performance of multichannel AM/QAM video transmission systems. As a project manager and a senior scientist, Shlomo provided technical analysis and consulting services to the Regional Bell Operating Companies as well as to various cable TV equipment vendors. In 1996, Shlomo joined General Instrument as a principal scientist in Digital Network Systems division, where he was developing the next-generation digital set-top boxes for both domestic and international markets. In April 2000, Shlomo joined Intel's Cable Network Operation business unit in San Jose, California, as a principal system architect developing communication products such as cable modems. He is a Senior Member of IEEE/LEOS/Comsoc with more than 60 technical publications and conference presentations. Shlomo served on the technical committees of many IEEE/LEOS conferences, and he is a regular reviewer for various IEEE publications such as Photonics Technology Letters and Journal of Lightwave Technology. Shlomo has eight pending patents, and his personal biography is included in the Millennium edition of Who's Who in Science and Engineering (2000/2001).

CHAPTER 1

BROADBAND HYBRID FIBER/COAX ACCESS
NETWORKS OVERVIEW

Broadband access networks have been dramatically evolving for more than a decade to offer consumers and businesses a variety of services such as analog and digital video, high-speed data, and telephony services at their "finger tips." Various competing enabling access technologies have also been developed since the late 1980s to address the continuously growing consumer appetite for these services. The goal of this book is threefold. The first goal is to explain to the reader the architectural evolution of cable TV access networks to fulfill the consumer requirements for a wide range of entertainment and interactive information services. The cable TV access architecture has been evolving from a one-way broadband coaxial network to a two-way hybrid fiber/coax network to a dense wavelength-division-multiplexed (DWDM) network. The second goal is to introduce the reader to the technology behind all the key building blocks of cable TV access networks, and discuss their characteristics and performance requirements. This includes fiber-optics transmission technologies such as directly and externally modulated laser transmitters, optical receivers, and optical fiber amplifiers as well as digital communication technologies such as quadrature amplitude modulation modems and digital set-top boxes. Using these technologies, the third goal of this book is to explain to the reader the high-speed data transmission protocol for cable modems as well as the software architecture and the wide range of interactive applications for digital set-top boxes. These applications include native applications such as electronic program guides, TV-based interactive applications such as video-on-demand, and Internet-based applications such as e-mail and e-commerce.

Chapter 1 provides a brief overview of cable TV access networks based on hybrid fiber/coax architecture, and introduces the reader to the key end-user home terminals such as digital set-top boxes and cable modems. To complete this discussion, it is necessary to briefly explain the emerging competing broadband access technologies with two-way cable TV networks such as digital subscriber line, fiber-in-the-loop, direct broadcast satellite, and fixed wireless services.

1.1 Introduction

During the past two decades, telecommunications networks have been evolving in separate ways. In general, telecommunications networks can be divided into three different groups as follows: (A) community antenna television (CATV) network[1], (B) local and wide-area computer network, and (C) public-switched telephone network (PSTN). These telecommunication networks were truly independent networks since they delivered unique services that otherwise where not available in the other type of network. Thus, traditional cable TV networks did not offer subscribers telephony or high-speed data services, and PSTN networks did not offer broadcast analog or digital video services. In the middle of the 1990s, two strong independent forces have played a significant role in changing the telecommunication landscape. First, easy and low-cost access to the Internet opened the information superhighway to many subscribers and businesses to electronic commerce, on-line shopping, advertisement, and information services that are easy to use, quick, and tax-free. The second major force was the passing of the U.S Telecommunications Act in 1996. The main idea was to deregulate the telecommunications industry by allowing telephone companies (local and long-distance), wireless operators, cable TV operators, and broadcasters to enter

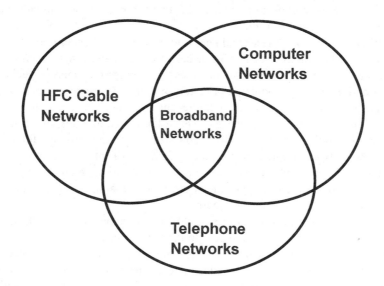

Figure 1.1 The convergence of HFC cable TV networks, computer networks and public-switched telephone networks into the emerging broadband networks carrying video, data, and voice services.

---

[1] The term "CATV" is generally accepted today to mean simply cable television or cable TV.

each other's markets. This act has opened the door to a merger mania among the different companies. Furthermore, utilities companies such as gas companies and even railway companies wanted to lay fiber optics cables along their rights-of-way and build their own telecommunications networks. Figure 1.1 shows the convergence of the three telecommunications networks into a broadband network to deliver a wide-range of entertainment and in- formation services [1]. However, as we will see later in this chapter (Section 1.4), there are many economical, business, and regulatory factors in addition to the selected technologies, which determine the feasibility of building such a telecommunications network.

Cable TV networks have been evolving from one-way broadcast of analog video channels to two-way interactive hybrid fiber/coax (HFC) networks delivering analog/digital video channels and high-speed data. The traditional tree-and-branch broadband coaxial network, which was supported by RF technology, served well the cable TV operators for the type of services it provided, namely, broadcast or point-to-multipoint services. The cascades of RF amplifiers became very long (30–40 amplifiers), which in turn reduced the quality and availability of the transmitted amplitude modulation vestigal sideband (AM-VSB) video channels and increased customer dissatisfaction. The use of terrestrial microwave links to reduce long cascades of RF amplifiers improved the transmission performance of the broadcast analog video channels. Significant improvements in fiber-optics transmission technology since the late 1980s fueled the explosive growth of the cable TV industry [2–5]. The introduction of 550 MHz directly modulated (DM) DFB laser transmitters and optical receivers operating in the 1310 nm wavelength band changed the traditional architecture of coaxial networks. HFC networks enabled robust transmission of broadcast analog video channels via standard single-mode fiber (SMF) to the optical fiber-node, resulting in a significant reduction in the number of cascaded RF amplifiers. In addition, cable TV operators were able to consolidate the deployment of headend equipment using fiber-optic rings to connect between the master headend and secondary headends or hubs in major metropolitan markets. Thus, cable TV operators were able to cut cost and further improve the quality and availability of the original broadcast services.

The development of various key systems such as quadrature amplitude modulation (QAM) modulators, low-cost QAM receivers, and moving picture-expert group (MPEG)[2] digital video encoders and decoders enabled cable operators to offer tens of new digital video services along with the traditional AM-VSB video channels using a digital set-top box (STB) at the subscriber's home [6]. The rapid deployment of 750 MHz HFC networks also allowed cable TV operators and some telecom services to provide competitive access to consumers and various businesses in major markets.

In the middle of the 1990s, the existing HFC network architecture had started to evolve in a new direction. This evolution can be attributed to the following forces in the marketplace:

(A) An explosive demand for high-speed data (e.g., Internet) access in residential areas

---

[2] Detailed overview of the various MPEG standards can be found in Chapter 2, Section 2.2.

(B) The delivery of highly targeted interactive digital services

(C) Increased competition from various telecommunication and direct-broadcast-satellite (DBS) service operators

(D) Improved cable TV network scalability with reduced life-cycle cost of various fiber and coaxial components and systems

(E) Improvements in fiber-optics technologies, particularly laser transmitters and receivers, and cable TV network management

These requirements and market forces have been forcing cable TV operators to rethink their existing HFC network architecture, and are paving the road toward DWDM cable TV access networks [7]. This book is organized in such a way as to enable the reader to understand step-by-step the technology, operation, and applications of the different key systems in HFC and DWDM networks. The architecture and services of two-way HFC networks are reviewed and contrasted with competing access technologies in the remainder of this chapter. Chapter 2 explains the different signal formats for analog and digital video and audio, and the basic characteristics of coaxial cable TV networks. The operation and design principles of directly modulated DFB laser transmitters (Chapter 3) and externally modulated laser transmitters (Chapter 4), which are used to transmit the sub-carrier multiplexed (SCM) analog and digital video signals from the cable TV headend up to the fiber node through a SMF, are discussed in detail. Chapter 5 discusses the theory and operation of optical receivers, which are located at the fiber node, and are used to convert the optical signals to electrical signals. Chapter 6 explains the operation principles and use of optical fiber amplifiers in cable TV networks. The design and use of RF digital modulators and demodulators such as M-ary QAM and 8/16-VSB, which are among the key systems in the HFC network, are explained in Chapter 7. Chapter 8 discusses the design and operation of the primary subscriber's home terminals, namely, a STB and a cable modem (CM), which are the enabling technologies to receive and transmit information over HFC networks. To understand the requirements for robust transmission on HFC and DWDM networks, the various types of transmission impairments are analyzed both theoretically and experimentally in Chapter 9. In Chapter 10, the reader is now ready for the discussion of the downstream and upstream architecture of the next generation of broadband cable TV access networks, namely, DWDM networks. To enable the operation of CMs, standard communication protocols are needed. First, CableLabs' multimedia cable network system (MCNS) data-over-cable service interface specification (DOCSIS) protocol for CMs is discussed with emphasis on the physical (PHY) layer and medium-access-control (MAC) layer in Chapter 11. Chapter 12 discusses the software architecture of a digital STB and the different types of applications, including native applications, TV-based interactive applications, and Internet-based applications.

1.2 Traditional Cable TV Networks

As the name implies, HFC networks are a hybrid combination of fiber optics and coaxial cables used for signal transmission and distribution. Fig. 1.2 shows a typical major metropolitan HFC network architecture. Analog and digital video signals from various sources such as satellite transponders, terrestrial broadcast, and video servers are multiplexed and transmitted via an SMF ring from the cable TV master headend to typically four or five primary hubs or headends. Each primary headend or hub has as many as 100,000 homes passed. Up to four or five secondary hubs and local headends, where each primary headend has up to 25,000 homes passed, are used for additional distribution of the multiplexed analog and digital video signals. To reduce the necessity to duplicate the same broadcast video channels at the various primary or secondary headends, the analog or digital channels from the master headend can be shared over the backbone network. The backbone network is typically built with a ring architecture using either synchronous optical network (SONET) with the associated SONET add-drop multiplexers (ADMs), or some proprietary technology. The SONET specification, which was proposed by Bellcore [8] and standardized by the American National Standards Institute (ANSI), defines a hierarchy of standard digital data

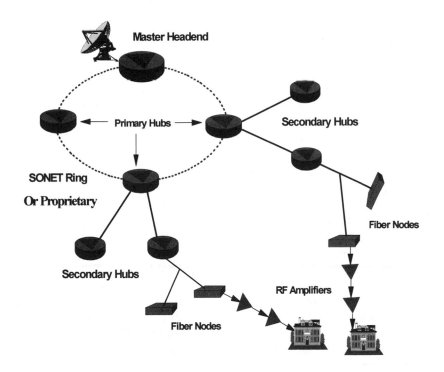

Figure 1.2 Typical HFC network architecture for a large metropolitan area.

rates from 51.84-Mb/s or OC-1 (optical carrier level 1) to integer multiples of this rate. In a typical SONET network, the analog video signals are digitized, modulated, time-division-multiplexed (TDM), and transmitted as digital baseband signals at various rates ranging from OC-12 (622-Mb/s) to OC-48 (2488-Mb/s). Statistical TDM, which can be used as an alternative to synchronous TDM, increases overall bandwidth usage by dynamically allocating time slots on demand to the different variable bit-rate services transported over SONET [9]. To further reduce deployment cost, most cable TV operators have selected to use SONET-compatible equipment, relying on proprietary network interfaces. The analog and digital video signals are further distributed to the various fiber nodes using a tree-and-branch architecture as shown in Figure 1.2. At the fiber node the optical signals are converted to electrical signals and transmitted to the subscriber through coaxial cables using different types of RF amplifiers and taps. The fiber node size, which is defined by the number of homes passed (HP), can vary from a "small node size" of about 100 HP to a "very large node size" of about 2000 HP. As we will see in both Chapter 2 and Chapter 10, the size of the fiber node has significant implications on the performance of the HFC network, particularly in the return-path or upstream spectrum. In general, a large fiber node size reduces upstream channel availability and peak user data throughput because of the time varying ingress noise as well as other impairments from all the homes passed.

1.3 Two Way HFC Access Networks

The evolution of one-way HFC cable networks into two-way broadband HFC access networks has been enabled by the introduction of three new systems as follows: (A) advanced subscriber's home terminal or STB, (B) CM, and (C) IP telephony systems operating over HFC access networks. Each of these technologies enabled the introduction of new types of services that were not possible before. In the following section, the basic functionality and the different type services are reviewed. The hardware system architectures, including the operation principles of all the primary building blocks, as well as the performance requirements of both a digital STB and a CM, are discussed in detail in Chapter 8.

1.3.1 Subscriber's Home Terminal

Subscriber home terminals, which include analog and digital STBs, are probably the most important component in the HFC access network. Digital STBs represent the connection between the present analog TVs and the advanced digital TVs of the future. With the introduction of analog STBs since the 1970s and 1980s, subscribers were able to receive broadcast analog TV channels via the HFC network instead of the terrestrially broadcast channels. Cable operators offered subscribers impulse-pay-per-view (IPPV) services of movies and special events using telephone return for interactive operation. The IPPV subscribers were able to place an order for a particular program or service directly through a cable home terminal at almost any time. An extended credit from the cable operator, which was dispensed

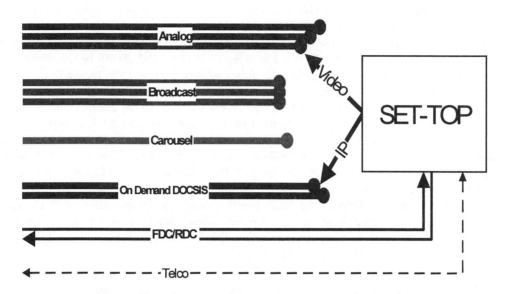

Figure 1.3 Illustration of the different types of video and data services that can be accessed through an advanced digital STB, which has a built-in cable modem and dual cable TV tuners.

by the home terminal, allowed the subscriber impulse purchases. With IPPV, the billing system calculates the credit limit of a given subscriber based upon the cable operator guidelines and payment history of the subscriber. The credit limit is transmitted from the cable headend to the subscriber's home terminal, where IPPV purchases by the subscriber are enabled up to the available credit limit. With each purchase, the available credit is reduced until the cable operator renews the credit balance.

The introduction of digital STBs in 1996 by both General Instrument and Scientific Atlanta opened the door for cable operators to offer subscribers new array of broadband services [10]. The field-installed digital STB provided the following basic functions:

- Channel tuning of digital (MPEG-2) and analog video services in the downstream bandwidth
- Demodulation of the received digital channel
- Modulation of return-path digital signals
- Encryption and decryption of the selected digital services
- Subscriber management signaling from the cable TV headend
- Subscriber user interface

The recently deployed advanced digital STBs from Motorola/General Instrument, which have dual cable TV tuners and a built-in CM, enable subscribers to "watch" TV and "surf" the web simultaneously over the HFC network. The various types of video and data services that can be accessed with the advanced digital STB are illustrated in Figure 1.3. Services such as clear and encrypted analog/digital video services, digital pay-per-view (PPV), digital IPPV, near-video-on-demand (NVOD), and video-on-demand (VOD) can be received using one of the STB tuners. The VOD service allows subscribers to order a movie from a centralized library using a downloadable electronic program guide (EPG) and navigation system. This system, which is delivered through an out-of-band (OOB) channel[3] and provides menu or icon-driven access to the resources of the network and the STB, is protected for secure access of resource areas using a personal identification number (PIN). The VOD service provides the subscriber with VCR-like functionality just as if it was a rented movie played at his home VCR. In contrast, the NVOD service allows the subscriber to order a movie from a limited selection of regularly scheduled playing movies at fixed time intervals, say every 30 minutes, without the VCR functionality. The downstream OOB channel, which carries all the specific information for the various in-band programs, is operating at 2.048-Mb/s using quadrature-phase-shift-keying (QPSK) modulation and is typically located in the 70-MHz to 130-MHz frequency range [11].

Table 1.1 Average and peak rates to the digital STB for various types of traffic. CBR is constant bit rate traffic.

Traffic type	Average rate	Peak rate
Entitlement control messages	200-kb/s	CBR
MPEG-2 compressed video and audio	3-Mb/s per active subscriber	9-Mb/s per active subscriber
Applications (e.g., Web browsing)	100-kb/s per active subscriber	5-Mb/s per active subscriber
Broadcast data carousel	0.1- 5.0-Mb/s	CBR
Analog broadcast	N/A	N/A

The second tuner in the advanced STB provides two-way high-speed Internet access using DOCSIS protocol. Additional new services such as IP telephony and IP videophone, and interactive games will be supported in the near future [12]. Thus, the advanced STB acts as a residential gateway (RGW), delivering Internet protocol (IP) packets to external devices such as IP voice or videophone. Table 1.1 shows the average and peak bit-rate to the digital STB for the various types of traffic.

The advanced STB also provides pass-through, processing, and decoding of high-definition TV (HDTV) signals. All the previously mentioned applications may utilize the 2D/3D graphics presentation capabilities of the STB when such applications are delivered to

---

[3] The downstream OOB channel is sometimes called forward data channel (FDC), while the upstream OOB channel is sometimes called reverse data channel (RDC).

a TV receiver/monitor either as baseband video or as RF modulated video. The advanced STB has a number of input/output interfaces including universal serial bus (USB), IEEE 1394 interface ("FireWire"), and Ethernet connection. The IEEE 1394 is a high-performance serial bus standard, which was developed for real-time transfer of multimedia applications at 100-Mb/s, 200-Mb/s, and 400-Mb/s rates over cable [13].[4] The IEEE 1394 standard is a scalable, flexible, easy to use, low-cost digital interface that provides the necessary integration between different consumer electronics devices such as digital camcorder, digital STB, digital TV, videophone, and personal computers (PCs).

1.3.2 Cable Modems

Cable modems (CMs) are the most exciting enabling technology that provides two-way high-speed access to the Internet over HFC networks. Since the development of the first World Wide Web software by Tim Berners-Lee and the 1993 release of the Mosaic web-browser software, the number of web sites has grown exponentially around the world, doubling about every three months. The realization of the enormous potential of these on-line global information resources available to every user at the click of his computer mouse has created the so-called global information superhighway. To access the Internet, telephone modems operating at a typical rate of 28.8 to 56-kb/s have been used. The existing telephone networks, which were not designed to handle the massive amount of data traffic, became congested, and thus accessing the multimedia-rich web pages took a relatively long time. In contrast, CM technology operating over two-way HFC networks can provide downstream data rates more than 30-Mb/s, which is more than 1000 times faster than a typical telephone modem, and upstream data rates more than 10-Mb/s. These data rates are also more than 200 times faster than the basic data rates (128-kb/s) available from the LECs using the so-called integrated services digital network (ISDN). The CM has an enormous economic advantage over ISDN services since cable TV operators and Internet service providers (ISP) charge only a small fraction of the access fees charged by the LECs. Most CMs typically in residential areas are configured for asymmetric data traffic. This is because most users are using the web to download multimedia-rich applications while using the upstream only for lower bit-rate applications such as e-mail, file transfer, and simple keystroke commands requesting downloads. CMs operating at symmetric data rates are typically needed in a corporate campus Intranet, where each CM is shared among various networked computers.

Given the enormous importance of CMs in two-way HFC networks, different aspects of their system design, operation, and protocol are discussed throughout this book. Chapter 7 reviews and explains the RF system architecture, design, and operation of the CM in HFC networks. The standard CM transmission protocols, which are known as DOCSIS 1.0 and 1.1, are discussed in detail in Chapter 11[14]. Finally, a technical proposal for the next-generation advanced physical layer specification for DOCSIS titled "Advanced TDMA

---

[4] The IEEE 1394 was conceived by Apple Computer and then developed within the IEEE 1394 Working Group.

Proposal for HFC Upstream Transmission" has been recently submitted to CableLabs for approval [15]. The goals of the proposal are (A) to increase upstream capacity with additional robustness against cable plant impairments over the existing DOCSIS modems, and (B) to achieve full backward compatibility with existing DOCSIS modems.

1.3.3 IP Telephony

The term IP telephony refers to the use of IP to carry voice over telecommunications networks [16]. It is also frequently called voice-over-IP (VoIP). It should be noted that IP networks carry data independent of the physical layer. These networks can also be categorized into *circuit-switched* networks and *packet-switched* networks. Circuit-switched networks, which are also called connection-oriented networks, operate by forming a dedicated connection (e.g., circuit) between two physical points in the network. In packet-switched networks, which are also called connectionless networks, the data is segmented into small blocks called packets and transmitted by forming virtual connections between any points in the network. Due to the Internet's explosive growth, IP has essentially become the standard packet-switched layer protocol for both local-area networks (LANs) and wide area networks (WANs).

As we learned earlier in Section 1.1, the integration of telephony services into residential broadband HFC networks carrying both video and data has enormous advantages for the information superhighway [17]. The IP telephony can be implemented using either an IP phone or a POTS phone connected to a modified CM or a digital STB. An IP phone is a relatively new device, which instead of connecting to a proprietary private branch exchange[5] (PBX) port, it connects to a standard Ethernet port in a CM, or a digital STB, or a home PC. An IP phone can operate as a standard IP device with its own IP address, and has the built-in ability to provide voice compression. In order to connect a POTS phone to a CM or a digital STB, the analog voice signals needs to be converted to short packets of digital audio. Currently, new interface modules are being developed that plug into a CM or STB and provide this functionality. These IP packets are sent through the HFC network using the DOCSIS protocol, which has built-in a guaranteed quality of service (QoS) for voice telephony [14]. It should be pointed out that DOCSIS 1.1 capable CMs have not been deployed as of this writing since CableLabs has not certified them. Some manufacturers, to enable QoS VoIP, have deployed CMs using a subset of the DOCSIS 1.1 protocol, which is called DOCSIS 1.0+.

For a mass-market deployment of IP telephony service to residential customers, seamless integration between PSTN and IP networks has to occur. Figure 1.4 shows, for example, the proposed integration between PSTN and IP networks [18]. A RGW, which is sometimes called embedded client, terminates an analog POTS connection to a phone or PBX. The RGW is responsible for supporting real-time transport protocol (RTP), which is used for the

---

[5] PBX is a telephone company's central office switch that routes many incoming and outgoing calls to their proper destinations.

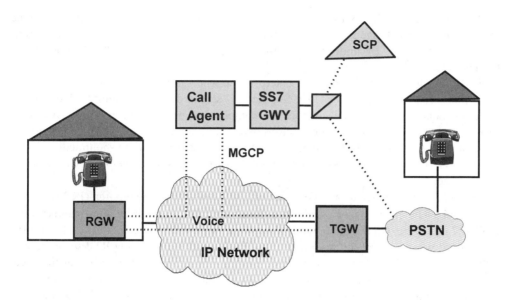

Figure 1.4 Integration of MGCP with PSTN for IP telephony services. After Ref. [18]
(© 1999 IEEE).

voice signals end-to-end communications over IP networks. The RGW also supports an interface to IP networks through the HFC access network. A trunking gateway (TGW) is responsible for bridging between the PSTN and the IP network. The TGW converts the TDM voice signals, which were originated from the PSTN, to RTP packets. Similarly, calls that originated from the IP side of the network and terminated at the PSTN are converted to the proper TDM voice format. The TGW also supports the media gateway control protocol (MGCP), which is the pending standard interface between the TGW and call agents [19]. Using MGCP messages, the call agent instructs both the RGWs and TGWs on how to proceed with the call processing. The call agent also handles the so-called signaling-system number 7 (SS7) signaling between TGWs that connect the PSTN with the IP network [20]. SS7 is a reliable open-architecture protocol that divides a telephone call into two separate components: (A) the signaling component, which contains address information for call setup and teardown, and (B) the voice and data component for communication between the originator and terminator of a call. These two components are transmitted on separate facilities. The facility for carrying signaling information is a signaling link or data link, and the facility for carrying voice and data traffic is a voice trunk. Telephone switches use the SS7 ISDN user part (ISUP) protocol for communicating with each other when setting up calls that span more than one switch. The call agent sends ISUP messages over SS7 for setting up calls between the IP network and the PSTN. A PacketCable network-based call signaling (NCS) protocol is currently being drafted by CableLabs [21]. The NCS protocol, which is part of

PacketCable, defines the control of VoIP in RGWs by the call agents. The MGCP assumes that the different call agents can synchronize with each other in order to send the proper commands to each RGW under their control. For further details on DOCSIS protocol, see Chapter 11.

The IP telephony requirements over enterprise networks are quite different from those for residential broadband networks. In 1996, the ITU standard organization introduced ITU-T Recommendation H.323, which defines the components, procedures, and protocols necessary to provide multimedia communications over LANs and IP networks that do not provide a guaranteed QoS [22]. H.323 terminals, which may carry real-time multimedia services over LANs or IP networks, may be integrated into a PC or as a standalone device such as a videophone.

The PSTN carries voice as a 64-kb/s stream using pulse-code-modulation (PCM) for both A-law and μ-law or as a compressed voice signal using ITU-T speech coding standards such as G.727 or G.729. On the other hand, IP networks set up the voice path for each call using call agents, which control the resources of the RGWs and TGWs, using RTP since there are tight constraints of the QoS that must be delivered. The H.323 gateway provides the necessary voice format conversion between the PSTN and the IP networks. In addition, the H.323 gateway terminates the call on the IP network side. However, existing H.323 gateways do not support SS7 signaling. The ITU-T standard defines the procedures and protocol by which network elements in the PSTN exchange information over a digital signaling network to effect call setup, routing, and control.

1.4 Competing Access Technologies

There are various competing access technologies that can bring 'broadband' services to the subscriber's home. This section provides a brief overview of the various competing access technologies with their advantages and disadvantages.

1.4.1 Asymmetric Digital Subscriber Line (ADSL)

Asymmetric digital subscriber line (ADSL) technology uses existing twisted-pair copper telephone wires to provide the required bandwidth for broadband services such as Internet access, video conferencing, interactive multimedia, and VOD. The ADSL technology is designed to solve the severe bottleneck that currently exists in telephone networks between the central office (CO) and the end customer [23]. ADSL can deliver data rates ranging from 64-kb/s up to 8.192-Mb/s for downstream channels and from 16-kb/s up to 768-kb/s for upstream channels, while simultaneously providing plain old telephone service (POTS) [24].

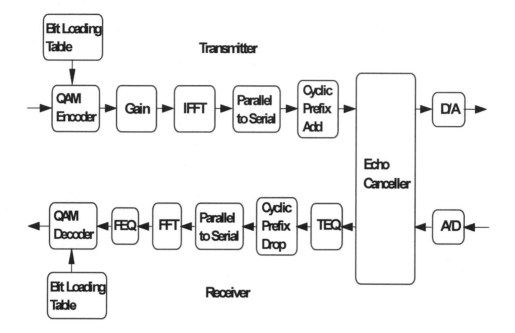

Figure 1.5 Basic block diagram of ADSL modem.

Discrete multi-tone (DMT) modulation technology has been standardized by the ANSI T1 committee for ADSL transmission, due to its unique ability to overcome the severe interference that exists in the residential twisted copper-pair environment. The DMT technique divides the input data stream into N sub-channels, which are typically referred to as "tones," each with equal bandwidth but at a different center frequency. 256 frequency tones are available for downstream channels and 32 tones for upstream channel are used, where each tone has a bandwidth of 4.3125-kHz and the same frequency separation between two successive tones. Each of the tones is QAM modulated on a separated RF carrier. The RF carrier frequencies are multiples of one basic frequency. The available spectrum band ranges from about 20-kHz to 1.104-MHz, while below 20 kHz is reserved for voice service (POTS). Noise and channel conditions are constantly measured for each tone separately, to achieve optimal transmission at any time. Figure 1.4 shows a simplified block diagram of an ADSL modem. The QAM encoding is done according to a bit-loading table, which defines the number of bits carried by each tone. Thus, using higher QAM constellation modes in those tones with high SNR provides higher bandwidth efficiency (b/s/Hz). The bit-loading table is calculated during startup according to the actual measured SNR to allow optimal use of channel capacity. Loading is limited by the ADSL standard to 2–15 bits per tone. When a given bit rate is required, the bits are allocated to the carriers in such a way that the sum of all the bits on all carriers matches the desired rate, and the probability of error on each carrier is about the same. When the maximum available bit rate is needed, each carrier is allocated with the maximum number of bits that can be transmitted without errors, based on the measured SNR of that carrier. This mode is typically referred to as the rate-adaptive mode. This mode allows higher SNR and data rates for subscribers who are living close to the CO, and lower SNR and data rates for subscribers who are far away from the CO due to additional line attenuation and noise. From a practical point of view, transmission distance for an ADSL service is typically limited to about 17,500 feet from the CO.

Another key element in the ADSL modem design is the use of fast Fourier transform (FFT) and inverse FFT (IFFT). The IFFT allows one to create the sum of N tones, where each tone is independently QAM modulated. At the DMT transmitter, the time-domain samples at the output of the IFFT (2N) are converted to a serial bit stream and fed to cyclic prefix (CP). The CP is designed to separate the transmitted symbols in time to decrease inter-symbol interference (ISI). At the DMT receiver, a time equalizer (TEQ) and a frequency equalizer (FEQ) are used to minimize ISI, cochannel interference, and reduce the signal amplitude and phase distortion.

To support two-way traffic, ADSL modems divide the available bandwidth using a frequency division-multiplexing (FDM) scheme, where different bands are assigned for the downstream and upstream data. Another method is to use echo cancellation (EC). EC is accomplished by generating an exact replica of the transmitted signal that leaks into the receiver. Upon subtraction of the near-end echo-replica the received far-end signal can be processed as if its only impairment has been the channel-induced noise sources. EC in ADSL modem must be designed with different sampling rates for upstream and downstream due to the asymmetric data transmission rates. The ADSL modem design also implements forward-error-correction (FEC) with an interleaver to combat up to 500-μs burst noise.

ADSL modems usually include a POTS splitter, which enables simultaneous access to voice telephony and high-speed data access. The POTS splitter provides a low-pass filter between the copper line node and the telephone node, and provides a high-pass filter between the copper line node and the ADSL modem node. Thus, between the ADSL node and the telephone node the splitter attenuates all signals. An active or passive POTS splitter can be implemented. The advantage of the active POTS splitter is that it enables simultaneous telephone and data access. However, the telephone service fails if there is a power outage or modem failure with an active POTS splitter. In contrast, a passive POTS splitter maintains lifeline telephone access even if the modem fails (because of a power outage, for example), and is typically used for a regular analog voice channel.

One of the key functions of the POTS splitter is to block impulsive noise coming from the home telephone or narrowband switch at the CO into the ADSL modem. Telephone companies are finding out that residential wiring is a huge source of impulsive noise on the ADSL upstream path. The impulsive noise originates from the ring trip signals, electrical switches, electrical appliances and tools, and powerline-switching transients coupled to the home telephone lines. The filter also blocks the ADSL modem signals from going into the telephone set, reducing the quality of the line. The impedance of telephone wires varies significantly between lines. This depends heavily on the length and gauge of the wire, and also on the telephone itself, when the line is short. The addition of two filters, one at the CO and one at the home may be more complicated, particularly if one is trying to fit just one design to all cases. Most POTS splitter designs are passive due to their reliability. They do not require an external power, offer better protection from lightning, and support lifeline services. However, active POTS splitter designs may be required in countries outside the United States.

To address the need of reduced ADSL modem complexity, a new ADSL standard called G.Lite or "splitterless" DSL has been developed by the ADSL working group. In October 1998, the G.Lite standard was renamed as the G.922.2 standard. The G.Lite technology is capable of providing data rates up to 1.5-Mb/s for downstream channels and up to 386-kb/s for upstream channels. One of the main drivers for the G.Lite standard was to eliminate the installation of the POTS splitter by a telephone company technician, and thus reduce deployment cost.

Very-high-bit-rate DSL (VDSL) and high-bit-rate DSL (HDSL) modems are related technologies that are expected to be used for video distribution and interactive multimedia applications for the residential consumer market [24]. There are many unresolved technical and business issues that impede the mass deployment of these modems.

1.4.2 Fiber-In-The-Loop (FITL)

Fiber-in-the-loop (FITL) access technology, which uses optical fibers in a star architecture (point-to-multipoints), consists of a family of network architectures such as fiber-to-the-node (FTTN), fiber-to-the-curb (FTTC) and fiber-to-the-home (FTTH) [25]. The FITL systems were intended to be compatible with typical local exchange carrier (LEC) service, transmission, and operation. The initial FITL architecture is the FTTC as shown in Figure 1.5. A FITL network consists of a host digital terminal (HDT) with subtending broadband network units (BNUs) in a star architecture that is managed by the HDT. The HDT provides the necessary transmission and operations interface of the FITL system to the rest of the LEC network. For example, the HDT may separate locally switched traffic and nonlocally switched traffic for routing management. Broadband services such as Internet, interactive multimedia, and telephony are transmitted to the HDT, which may be located at the CO or a remote site, as baseband digital signals. This is in contrast with HFC networks, where the broadband services are RF modulated. At the HDT, the baseband digital signals are switched and sent to various broadband network units via optical fibers in a star architecture[6]. The BNU is located in the customer neighborhood and serves multiple customers. The BNU, which terminates the optical fiber from the HDT, performs optical-to-electrical

---

[6] Broadband network unit is also sometimes called Optical Network Unit (ONU).

conversion and other essential functions. The electrical signals are then transmitted to the subscriber's home via coaxial cable or over twisted copper pair wires. A network interface unit at the side of the subscriber's home separates the video signals, data signals, and telephony signals as shown in Figure 1.6. The digital video signals are demultiplexed and decoded using a set-top box.

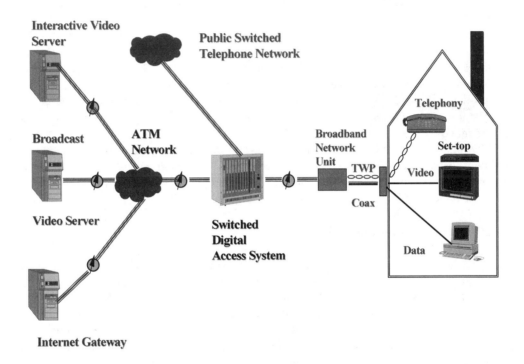

Figure 1.6 Basic fiber-to-the-curb network architecture.

Another FITL architecture is FTTH. As the name implies, the optical fiber replaces the coaxial cable or the twisted copper pair wires from the BNU to the subscriber's home. The difference between the FFTC and FTTH architecture lies in the location of the BNU.

In FTTH architecture, the BNU is located at the subscriber's home. Thus, the need to use twisted copper pair technology outside the subscriber's home is eliminated in the FTTH architecture. The portion of the network between the HDT and the BNUs becomes a passive optical fiber network, which is essential for its future upgrade. The large available bandwidth (> 2-GHz) of the optical fiber to deliver broadband services to the subscriber's home is one of the main advantages of this all-digital fiber architecture. With the use of WDM technology, FTTH systems are capable of delivering multi-gigabit, e.g. OC-48 or SONET-compatible broadband services to each subscriber. In addition, since each BNU is located at the subscriber's home, no external power or additional maintenance is needed. The main disadvantage of the FTTH architecture is the relatively high cost of each BNU, and the initial installation costs of the optical fibers. Despite this disadvantage, the FTTH architecture has a very promising potential to deliver multi-gigabit broadband services to each subscriber's home.

1.4.3 Direct Broadcast Satellite (DBS)

While ADSL and FITL access technologies use wireline networks to provide broadband services, direct-broadcast-satellite (DBS) technology is based on various geosynchronous satellites that provide multichannel digital video programming services to subscribers equipped with DBS receivers [26]. DirecTV and EchoStar (also known as the Dish network) dominate the DBS industry and offer several hundred digital video channels. DirecTV service, which was launched in the summer of 1994, consists of an 18" satellite antenna dish and a digital integrated receiver and decoder (IRD), which tune and decode the selected digital channel. The DirecTV service is distributed by three high-power HS 601 satellites (DBS-1, DBS-2, and DBS-3). Each satellite features 16 120-watt Ku-band transponders with DBS-2 and DBS-3 each configured to provide 8 transponders at 240 watts. DBS-1 delivers up to 60 channels and over 20 channels of programming from USSB®. Recently, USSB was acquired by DirecTV and merged its service offering. With DBS-1, DBS-2, and DBS-3, DirecTV service has several hundred video and audio channels of programming. All three satellites are colocated in geosynchronous orbit 22,300 miles above the earth at 101 degrees west longitude.

EchoStar Communications Corporation, which was founded in 1980, was granted a DBS orbital slot at 119 degrees west longitude, and three years later, the DISH (Digital Sky Highway) network brand name was created. On December 28, 1995, EchoStar successfully launched its first DBS, EchoStar I, and DISH network came to life on March 4, 1996. The second DBS satellite, EchoStar II was successfully launched on September 10, 1996 and is also orbiting at 119°. On October 5, 1997 EchoStar launched the satellite EchoStar III into an orbital slot at 61.5 degrees west longitude. EchoStar IV was launched from Kazakhstan on May 8, 1998. These four satellites give DISH network delivery capacity for over 250 channels of digital video, audio, and data services. The satellite-delivered programming service, which is extracted from various video sources, is first digitized, encrypted, and up-linked to the orbiting DBS satellites. Then, these DBS satellites retransmit the video signals back down to every subscriber's home with the properly oriented receiver dish. Each of the transponders on the DBS-1 satellite can transmit MPEG-2 video streams at a rate more than 23 Mb/s, while the DBS-2 and DBS-3 satellites can transmit about 30 Mb/s. For further details on the MPEG standard, please see Chapter 2, Section 2.3.

1.4.4 Multichannel Multipoint Distribution Service (MMDS)

Multichannel multipoint distribution service (MMDS) access technology is another wireless technology that is based on terrestrial broadcast of analog and digital video channels. In the early 1960s, the FCC began setting aside certain RF frequencies, which were called instructional television fixed service (ITFS), for distance learning. In the 1970s, various equipment vendors petitioned the FCC to allow commercial use of part of the ITFS spectrum. The industry began by offering one and two video channel service to consumers, which was known as multipoint distribution service (MDS). Since many of the ITFS frequencies were not used in many geographical locations, the MMDS operators petition the FCC to reallocate the ITFS frequencies for MMDS use. When additional frequencies in the ITFS spectrum were allocated for video services, MDS was renamed as MMDS.[7] In September 1998, the FCC ruled for two-way MMDS. The MMDS spectrum has been allocated worldwide in various RF frequency bands from 2-GHz to 2.7-GHz.

Table 1.2 MMDS frequency plan for downstream and upstream channels.

Downstream Center Frequency (MHz)	Downstream Channel #	Upstream Center Frequency (MHz)	Upstream Channel #
2599	E1	2675	H3-A
2605	F1	2677	H3-B
2611	E2	2679	H3-C
2617	F2	2681	G4-A
2623	E3	2683	G4-B
2629	F3	2685	G4-C
2635	E4	2687	H4-A
2641	F4	2689	H4-B
2647	G1	2151	MDS1-A
2653	H1	2153	MDS1-B
2659	G2	2155	MDS2-A
2665	H2	2157	MDS2-A
2671	G3	2159	MDS2-B

Table 1.2 shows the MMDS frequency plan in the United States for both downstream and upstream channels. In addition, the FCC has allocated additional RF spectrum in 900-MHz, 2.4-GHz, 5.8-GHz, and 24-GHz bands for unlicensed use to encourage competition. The unlicensed bands are referred sometimes as industrial scientific and medical (ISM) bands.

---

[7] MMDS is sometimes called "wireless cable."

The wireless network operators would use the unlicensed spectrum to offer broadband services such as Internet, telephony, and interactive multimedia to businesses at cost-effective rates compared with HFC cable networks.

The basic MMDS architecture consists of MMDS radio transmitters mounted on radio towers with antennas, a subscriber premises antenna, a downconverter, and an STB. Each geographical serving area is divided into slightly overlapping cells, each with a typical radius of 40-km. For robust transmission, line-of-sight between the transmitter and receiver antenna is typically required. Since line-of-sight is not always available, the primary channel impairment mechanism in MMDS systems is multipath signal fading.

A related access technology to the MMDS is called hybrid fiber wireless (HFW) or hybrid fiber radio (HFR). This architecture is similar to HFC where a central headend transmits broadband services to various RF cell sites via SMF cables, except the connection from the RF cell site to the subscriber is accomplished via two-way MMDS. There are many advantages to such architecture. First, it is a more reliable two-way transmission between the subscriber and the headend compared with a traditional MMDS architecture. Second, it reduces RF site installation and maintenance costs (less equipment at each site). Third, this architecture is well suited for large-scale deployment in metropolitan areas where fiber networks may already be installed.

References

1. Special issue on "Residential Broadband Services and Networks," *IEEE Communications Magazine*, **35**, no.6 (1997).
2. R. Olshansky, V. A. Lanzisera, and P. M. Hill, "Subcarrier Multiplexed Lightwave Systems for Broadband Distribution," *IEEE Journal of Lightwave Technology* **7**, 1329–1324 (1989).
3. W. I. Way, "Subcarrier Multiplexed Lightwave System Design Considerations for Subscriber Loop Applications," *IEEE Journal of Lightwave Technology* **7**, 1806–1818 (1989).
4. T. E. Darcie, "Subcarrier Multiplexing for Lightwave Networks and Video Distribution Systems," *IEEE Journal on Selected Areas in Communications* **8**, 1240–1248 (1990).
5. J. A. Chiddix, J. A. Vaughan, and R. W. Wolfe, "The Use of Fiber Optics in Cable Communications Networks," *IEEE Journal of Lightwave Technology* **11**, 154–166 (1993).
6. W. S. Ciciora, "Inside the Set-top Box," *IEEE Spectrum Magazine*, 70–75 (1995).
7. S. Ovadia and C. Lin, "Performance Characteristics and Applications of Hybrid Multichannel AM-VSB/M-QAM Video Lightwave Transmission Systems," *IEEE Journal of Lightwave Technology* **16**, 1171–1186 (1998).
8. TR-NWT-00253, *Synchronous Optical Network (SONET) Transport System: Common Generic Criteria*, Issue 2 (Bellcore, December 1991).

9.   U. Black and S. Waters, *SONET and T1: Architectures for Digital Transport Networks*, PTR Prentice Hall, New York (1997).

10. A. N. Nair, "Interactive Television Broadcasting and Reception," *Digital Consumer Electronics Handbook*, R. K. Jurgen, Editor-in-Chief, McGraw-Hill, New York (1997).

11. SCTE DVS/178, "Cable System Out-Of-Band Specifications," (1998).

12. T. C. Kwok, "Residential Broadband Internet Services and Applications Requirements," *IEEE Communications Magazine*, 76–83 (1997).

13. IEEE 1394-1995 High Performance Serial Bus Interface Standard (1998).

14. CableLabs, MCNS Data-Over-Cable Service Interface Specifications: Radio Frequency Interface Specification SP-RFIv1.1-I01-990311 (1999).

15. Broadcom and Texas Instruments, "Advanced TDMA Proposal for HFC Upstream Transmission," Rev. 1.0, December (1999).

16. D. Minoli and E. Minoli, *Delivering Voice over IP Networks*, John Wiley & Sons, New York (1998).

17. Jim Forster, "An integrated Architecture for Video, Voice, and IP services over SONET and HFC Cable Systems," *SCTE Proceedings Manual of Conference on Emerging Technologies*, 155–199 (1998).

18. C. Huitema, J. Cameron, P. Mouchtaris, and D. Smyk, "An Architecture for Residential Internet Telephony Service," *IEEE Network* **13**, 50–56 (1999).

19. M. Arango et al., "Media Gateway Control Protocol (MGCP)," Internet Engineering Task Force (IETF) Internet draft, October (1998).

20. F. Cuervo et al., "SS7-Internet Interworking Architectural Framework," Internet Engineering Task Force (IETF) Internet draft, July 9 (1998).

21. CableLabs, PacketCable Network-Based Call Signaling Protocol Specification, PKT-SP-EC-MGCP-I02-9901201.

22. ITU-T Recommendation H.323, *Visual Telephone Systems and Equipment for Local Area Networks which Provide a Non-Guaranteed Quality of Service* (1996).

23. W. Y. Chen and D. L. Waring, "Applications of ADSL to Support Video Dial Tone in the Copper Loop," *IEEE Communications Magazine*, 102 (1994).

24. D. Rauschmayer, *ADSL/VDSL Principles: A Practical and Precise Study of Asymmetric Digital Subscriber Lines and Very High Speed Digital Subscriber Lines*, MacMillan technical publishing (1998).

25. TR-NWT-000909, *Generic Requirements and Objectives for Fiber in the Loop*, Issue 2 (Bellcore, December 1993).

26. See for example http://www.dbsdish.com/ Web site, which has the latest news and information on the DBS industry.

27. S. B. Moghe and D. Urban, "Broadband Wireless Services Based on Combined Use of MMDS, LMDS and Unlicensed Band Spectrum," *Technical Papers of the Fifth Annual Technical Symposium Wireless Communication Association International* (1998).

28. J. C. Cornelius, "Design and Implementation of a Two Way Data System in an Existing Analog market," *Technical Papers of the Fifth Annual Technical Symposium Wireless Communication Association International* (1998).

CHAPTER 2

BASIC CABLE TV BACKGROUND: MODULATION SIGNAL FORMATS AND COAXIAL CABLE SYSTEMS

As we learned in chapter 1, HFC networks have evolved from traditionally broadcasting only multichannel analog video signals to a two-way mixture of analog and digital signals carrying both digital video and high-speed data. This chapter provides the basic technical background of cable TV systems carrying analog and digital video signals. Section 2.1 will explain the most commonly used analog video modulation formats, while Section 2.2 will discuss the different digital video and audio signal standards. The primary cable TV frequency plans that are used in the U.S. are reviewed in Section 2.3. In Section 2.4, the basic characteristics of multichannel coaxial systems are reviewed in terms of the properties of their components such as coaxial cables, taps, and RF amplifiers. The different degradation mechanisms over multichannel coaxial cable plants such as second- and third-order nonlinear distortions, multipath micro-reflections or echoes, group delay, and AM hum modulation are reviewed in Section 2.5. The last section will discuss the sources of upstream noise, including ingress noise, and their impact on the return-path transmission over cable TV networks.

2.1 Analog Modulated Video Signal Formats

2.1.1 NTSC and AM-VSB Video Signals

The National Television Systems Committee (NTSC) initiated the basic monochrome TV standard in the U.S. in 1941 by broadcasting 525 interlaced lines at 60 fields per second, which was designated as system M by the Commite′ Consulatif International Radiocommunications (CCIR). In 1953, the NTSC of the Electronic Industries Association (EIA) established color TV standards, which are now in use for terrestrial broadcasting and cable TV transmission systems in North America, Japan, and many other countries. This color TV system was designed to be compatible with the monochrome (black and white) TV systems that existed previously [1]. The specifications of a composite NTSC video signal, as described by NTSC, include a 525-line interlaced scan at horizontal frequency of 15,734.26

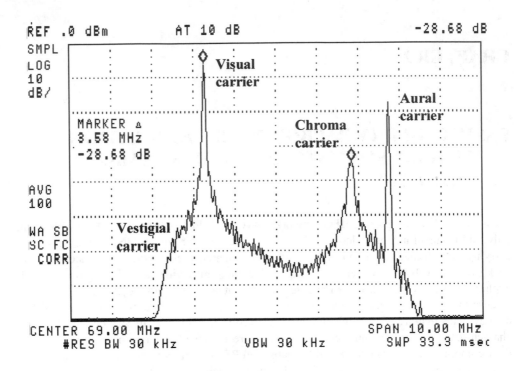

Figure 2.1 Measured frequency spectrum of NTSC video signal at channel 4, showing the luma, chroma, and aural carriers.

Hz; 15,750 Hz for monochrome TV, and a vertical scan frequency of 59.94 Hz (60 Hz for monochrome TV).

A complete video image as seen on a TV monitor is called a frame, which consists of two interlaced vertical fields with 262.5 lines each. The TV image is scanned at a vertical frequency of 59.94 Hz such that the lines of field 2 are interlaced with the lines of field 1 to create the desired 525 lines frame at a repetition rate of 29.97 Hz. Historically, the 60-Hz vertical scan frequency was selected for monochrome TV sets to match the 60-Hz power line rate such that any power-related distortions would appear stationary. For color TV, both the horizontal and vertical scan frequencies have been slightly reduced from the monochrome display case to allow for the interference beat between the chrominance carrier and the aural carrier to be synchronized to the video signal. Figure 2.1 shows the modulated RF spectrum of an NTSC video signal based on AM-VSB format at an RF visual carrier frequency of 67.25 MHz (channel 4). The brightness portion of the video signal, which contains all the information of the picture details, is commonly called the luminance or visual carrier. The color portion of the video signal, which contains information about the picture hue (or tint) and color saturation, is commonly called chrominance or *chroma* carrier. The chroma carrier is located 3.579545 MHz above the luma carrier. The hue information, which is contained in the phase angle of the 3.58-MHz chroma carrier, allows one to distinguish between the different colors such as red (R), green (G), blue (B), etc. Notice that white, gray, and black are not hues. Color saturation indicates the amount of white light dilution of the hue, and often is expressed in percentages. Thus, 100%-saturated color is a pure hue, which is undiluted by white light. The saturation information is carried by the amplitude magnitude of the chroma carrier. Since the human eye response varies from one hue to another, different amplitudes for different colors are required for 100% color saturation. For more details on the 1931 Commission Internationale de l'Eclairage (CIE) colorimetric standard, the reader is encouraged to look at these references [1–3].

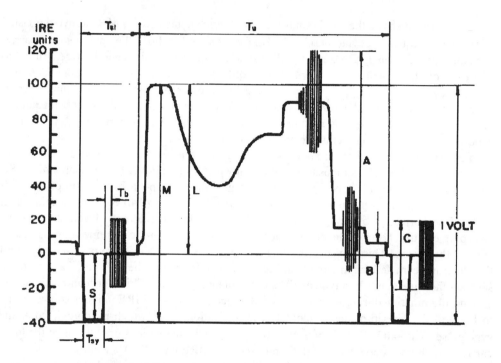

Figure 2.2 Time domain waveform of standard NTSC composite color video signal, showing its components as follows: (A) the peak-to-peak amplitude of the composite color video signal, (B) the difference between the black level and blanking level, (C) the peak-to-peak amplitude of the color burst, (L) nominal value of luminance signal, (M) peak-to-peak amplitude of monochrome video signal, (S) amplitude of synchronizing signal; T_b, duration of breezeway, T_{sl}, duration of line blanking period, T_{sy}, duration of synchronizing pulse, T_u, duration of active line period. After Ref. [2].

It is well known that the human eye is more sensitive to the color green than blue. Thus, the white color or visual carrier can be constructed from the three primary colors, namely R, G, and B, each with different weighting normalized to 1.0 as follows:

$$Y = 0.3R + 0.59G + 0.11B = 1.0 \tag{2.1}$$

Using this relationship, the color information of any point can be coded into two chosen color difference signals: $(R - Y = V)$ and $(B - Y = U)$. The NTSC system employs quadrature amplitude modulation (QAM) of the 3.58-MHz chroma carrier to convey the color difference information of the composite chroma signal. This method of color broadcast provides a compatible signal that can be reproduced on black-and-white as well as color TV sets.

Figure 2.1 also shows the aural (sound) carrier, which is located 4.5 MHz above the visual carrier with frequency deviation of ±25 kHz for monaural audio. The analog sound carrier is frequency modulated (FM), and in HFC networks must be maintained between 10 dB to 17 dB below the visual carrier amplitude according to Federal Communications Commission (FCC) requirements [4]. Notice the 0.75-MHz lower sideband from the luma carrier, which is typically called the vestigial sideband. This is the portion of the lower sideband of the composite video signal that remained as a tradeoff between minimizing the NTSC signal bandwidth and preserving picture details as much as possible.

To gain further understanding about NTSC signals, let us examine the time-domain waveform of standard NTSC composite video as shown in Figure 2.2 [2]. The standard NTSC composite video signal is specified to be 1-volt peak-to-peak (p-p) from the tip of the horizontal synchronization pulse to 100% white, and is divided into 140 equal parts called IRE units. The horizontal blanking pedestal, which is the 0 IRE reference level, starts as the sweeping electron beam of the picture tube reaches slightly over the extreme right-hand side edge of the screen. This prevents illumination of the TV screen during retrace, namely, until the electron beam deflection circuits are reset to the left edge of the TV screen and ready to start another line scan. In a typical TV set, the blanking level is "blacker than black" to assure no illumination during retrace. This difference, which is 7.5 IRE units, is shown as (B) in Figure 2.2. The tip of the horizontal synchronization pulse, which is abbreviated as the sync pulse, is at −40 IRE, and about 0.3 volt (p-p). Thus, the various gray levels of the luma carriers are divided between +7.5 IRE units and +100 IRE units (100% white).

During the vertical blanking interval, the first 21 lines are not displayed as the TV set prepares to receive a new field. This is commonly called the vertical blanking interval (VBI). The FCC requires that broadcasting TV stations include special vertical interval test signals (VITS) in each TV field [5]. The VITS are typically inserted between lines 10 and 21 to allow in-service testing of the TV broadcasting equipment. The FCC composite and the NTC-7[1] test signals have the necessary video signal components to perform the FCC re-

---

[1] NTC report number 7 (1976) was prepared by the Network Transmission Committee of the Video Transmission Engineering Advisory Committee, a joint committee of television network broadcasters and the Bell system.

quired tests for NTSC signals. Figure 2.3 shows, for example, the FCC composite test signal, which consists of a line bar, a 2T pulse, a chroma pulse, and a modulated five-riser staircase signal. The primary difference between these two test signals is the test sequence in which the video signal components are presented.

The relationship between the amplitude of the composite video signal and the percentage of amplitude modulation of the RF carrier is also unique. TV transmission uses a negative modulation such that the sync pulse produces the maximum RF peak-to-peak amplitude modulated envelope of the composite signal, while the white portion of the signal (+100 IRE units) produces the RF minimum amplitude of the modulated envelope. Negative modulation means that brightest portion of the composite video signal is represented by the minimum RF power. The depth of AM modulation of the composite video signal is specified to vary between 82.5 and 90% with a nominal value of 87.5%. The depth of modulation is less than 100% in order to allow the sound carrier to be recovered for even during the brightest portion of the composite video signal. This relationship between the amplitude and depth of modulation is advantageous, in order to minimize the visual effect of noise interference, which typically shows up as "snow" in the picture, at the weakest portion of the signal.

Figure 2.2 also shows four main time intervals in the composite video signal waveform. The horizontal blanking pedestal consists of the front porch, which lasts 1.47 µs, the -40 IRE sync pulse (T_{sy}) with duration of 4.89 µs, and a 4.4 µs back porch at the blanking level. During the back porch period, color is being generated in 8 to 10 cycles of 3.58 color burst at a specific reference phase for all the TV display lines.

2.1.2 NTSC Signal Test Parameters

There are various parameters, which are often quoted in the literature or in product specifications, to measure the quality of an NTSC video signal after its transmission. In this section a short overview of the most important test parameters of NTSC video signal is provided. The main NTSC video signal test parameters are

- Signal-to-(weighted)-noise-ratio (SNR)
- Differential gain (DG)
- Differential phase (DP)
- Chrominance-to-luminance delay inequality (CLDI)

The NTC-7 SNR is the ratio of the peak luma carrier level, 714 mV or 100 IRE, to the weighted rms noise level contained in 4-MHz bandwidth. The peak luma carrier level must be measured across a terminating impedance of 75 ohms. The weighted noise is used to compensate for the visual responses of the human eye, namely, different viewers perceive greater picture degradation at some frequencies more than other frequencies. A special luma weighted filter (NTC-7) per CCIR Recommendation 567 is primarily used in the U.S. The carrier-to-noise-ratio (CNR) is approximately equal to the SNR (within 0.5 dB) for

AM-VSB video signals. The CNR measurement on the spectrum analyzer is done with 30-kHz resolution bandwidth (RB), 100-Hz or 300-Hz video bandwidth (VB), and in automatic sweep mode.

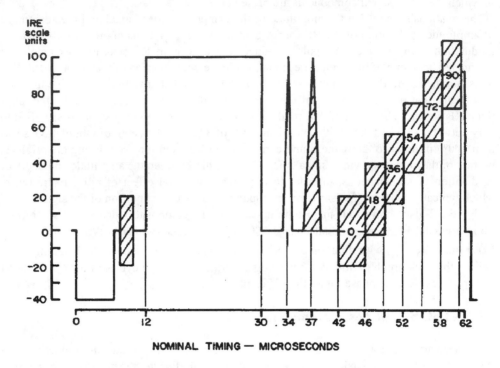

Figure 2.3 FCC Composite test signal within one horizontal scan period. After Ref. [2].

There are several noise correction factors that are best explained through an example. Suppose one measures an uncorrected CNR of 63 dB. Then, the corrected CNR is calculated as follows: CNR = 63 dB − (21.25 + 2.5 − 0.52 − 2.2) = 41.97 dB, where the 21.25 dB is the conversion factor from a RB of 30 kHz to 4 MHz, and the 2.5 dB is log detect Rayleigh noise correction factor for the spectrum analyzer. Since most spectrum analyzers have a Gaussian shaped video filter, the measured 3-dB noise bandwidth is larger than an ideal rectangular filter. Thus, the corrected noise needs to be reduced by about 0.52 dB. In addition, the 2.2 dB is the spectrum analyzer noise floor correction factor, assuming the noise floor drops only 4 dB when disconnecting the coaxial cable from the spectrum analyzer. According to the FCC Section 76.605, the measured video signal at the input to the subscriber's terminal is required to have a CNR equal to or greater than 43 dB.

The differential gain (DG) is the change in amplitude of the chroma signals (measured at 3.58 MHz) as the luminance level changes. It is measured as the difference in amplitude between the largest and smallest component of the chroma signal, divided by the largest component and expressed in % or dB of the largest component. There are two types of modulated staircase signals, namely, the previously mentioned FCC composite test and NTC-7. The NTC-7 tests are intended for baseband video measurements. The FCC composite test signal should be used when using a video test generator. However, the generator is not needed if the channel to be tested already contains the FCC VITS. In the FCC Section 76.605, the DG is required not to exceed 20% for the received NTSC signal at the subscriber's terminal.

The differential phase (DP) is the change in the phase of the chroma signal as the amplitude of the luma signal changes. The same test signals to measured DG should be used to measure DP. According to FCC Section 76.605, the DP for the color carrier is measured as the largest phase difference in degrees between each chroma component and a reference component (the component at the blanking level of 0 IRE) should not exceed ±10 degrees.

The chrominance-to-luminance delay inequality (CLDI) is the change in time delay of the chroma signal relative to the luminance signal of an NTSC video signal as measured at the output of a modulator or processing unit at the cable headend. According to the FCC Section 76.605, the CLDI shall be within 170 ns. This test is typically done inserting the VITS signal before it is received at the cable TV headend.

2.1.3 PAL and SECAM Video Signals

Two other color TV standards, which are mostly used in various countries outside of North America, are the phase alternation line (PAL) standard and the sequentiel couleur avec memoire (SECAM) standard. There are different variations of the PAL standard, which are mostly used in European countries and are designated by different letters such as PAL "I" (United Kingdom), PAL "B", PAL "G", PAL "H" (continental Europe), and PAL "M" (Brazil) [1].

Like the NTSC system, the PAL system employs QAM modulation of the chroma carrier to convey the color-difference information as a single composite chroma signal. The $(R-Y)$ color-difference carrier is phase inverted (180°) on alternate lines to improve the picture performance for different impairment conditions. For example, suppose that for a red hue, the $(R-Y)$ color component in the NTSC signal has a phase of +90°. In a PAL signal, one line is identical to the NTSC signal, but on the next line, the phase of this $(R-Y)$ component is switched to +270°, returning to +90° on the following line, etc. This phase-alternation line of the $(R-Y)$ component gives the PAL system its name. The color burst signal, which is used by the NTSC system during the horizontal synchronization time interval, is eliminated in the PAL system because of the alternating phase of line by line. The "simple" PAL system relies upon the human eye to average the line-by-line color switching process. Thus, the picture on a PAL TV can be degraded by 50-Hz line beats caused by system nonlinearities introducing visible luminance changes at the line rate. To solve this problem, an accurate and stable delay element was used to store the chroma signal for one complete line period in what is sometimes called PAL-deluxe (PAL-D).

Similar to the NTSC system, the PAL system has an equal bandwidth (1.3 MHz at 3 dB) for the two color-difference components, namely U and V. There are small variations of the PAL system among different countries, particularly where 7-MHz and 8-MHz channel bandwidths are used. Table 2.1 summarizes the key technical specifications for both NTSC and PAL video signals.

Table 2.1 Basic technical specifications for NTSC and PAL video signals

Parameter/Format	NTSC	PAL
Number of lines per picture	525	625
Field (vertical Scan Frequency	59.94 Hz	50 Hz
Horizontal Scan Frequency	15,734.26 Hz	15,625 Hz
Channel Bandwidth (MHz)	6	6 System "N" 7 System "B" 8 System "I"
Nominal Video Bandwidth (MHz)	4.2	4.2 System "N" 5 Systems "B", "G", "H" 6 Systems "D", "K", "L"
Color Subcarrier relative to the vision carrier (MHz)	3.579545	4.433618 Systems "I", "B", "G", "H" 3.575611 System "M"
Sound Sub-carrier relative to vision carrier (MHz)	4.5	4.5 System "N" 5.5 Systems "B", "G", "H" 5.9996 System "I"
Sound Modulation	FM	FM
Vestigial Sideband (MHz)	0.75	0.75 Systems "B", "G", "N" 1.25 Systems "H", "I", "L"
Gamma	2.2	2.8

The PAL system "I", which has an 8-MHz channel bandwidth, has added a new digital sound carrier, which is called *near-instantaneous companding audio multiplex* (NICAM). The NICAM carrier is located 6.552 MHz above the visual carrier (exactly nine times the bit-rate), and is differentially QPSK modulated at a bit rate of 728 kb/s and it is 20 dB below the visual carrier peak-sync power. Since the new sound carrier is located close to the analog FM sound carrier as well as the adjacent channel luminance carrier, the digital sound signal is scrambled before it is being modulated.

The SECAM system, which was adopted by France and the USSR in 1967, has several common features with the NTSC system such as the same luminance carrier given by equation 2.1 and the same color-difference U and V components. But this approach is considerably different from both the NTSC and PAL systems in the way the color information is being modulated onto the composite chroma carrier. First, the color-difference components are transmitted alternately in time sequence from one successive line to the next with the same visual carrier for every line. Since there is odd numbers of lines in a frame, any given line carries the $(R-Y)$ component on one field and the $(B-Y)$ component on the next field. Second, the color information in the $(R-Y)$ and $(B-Y)$ components is carried using FM. Similar to the PAL-D receiver, an accurate and stable delay element is used in synchronization with the switching process to have a simultaneous existence of the $(B-Y)$ and the $(R-Y)$ signals in a linear matrix to form the $(G-Y)$ color-difference component. In addition, the bandwidth of the chroma components U and V is reduced to 1-MHz. It should also be noted that the SECAM system employs AM modulation of the sound carrier as opposed to the FM modulation that is used in both NTSC and PAL systems.

2.2 Digital Video and Audio Signals

Digital coding of NTSC composite signals typically requires a raw bit rate of about 115 Mb/s, assuming analog-to-digital conversion of 8-bit and 4 times sampling of the chroma carrier. Obviously, this is a very high transmission rate to stuff in a 6-MHz channel bandwidth. To overcome this limitation, a standard for digital video (sequences of images in time) and audio compression has been developed by the MPEG (moving pictures experts group) committee under ISO (the international standards organization) [6-9]. The MPEG technology includes different patents from many companies and individuals around the world. However, the MPEG committee only deals with the technical standards and does not address the intellectual property issues. Related standards for still image compression are called joint photographic experts group (JPEG) and joint bilevel image experts Group (JBIG), which is used for binary image compression such as faxes. Other non-MPEG standards for video/audio are discussed in Section 2.2.3.

2.2.1 MPEG-1 Standard

A picture element or pixel is formed by Y, U, V sample values. Thus, if all the three components (Y, U, V) use the same sampling grid, each pixel has three samples (see Section 2.2.2). The first result of the MPEG committee was called MPEG-1. MPEG-1 compression starts with a relatively low-resolution video sequence of about 352 by 240 pixels by 29.97 frames/s (US), but with the original compact disk (CD) quality audio. The video compression method relies on the human eye's inability to resolve high frequency color changes in a picture. Thus, the color images are converted to the YUV-space, and the two-chroma components (U and V) are decimated further to 176 by 120 pixels. The 352x240x29.97 rate is derived from the CCIR-601 digital television standard, which is used by professional digital video equipment. 720 by 243 by 59.94 fields per second are used, where the fields are interlaced when displayed. Note that fields are actually acquired and displayed at a 1/59.94 second apart. The chroma components are 360 by 243 by 59.94 fields a second interlaced. This 2:1 horizontal chroma decimation is called 4:2:2 sampling. The source input format for MPEG-1, called SIF, is CCIR-601 decimated by 2:1 in the horizontal direction, 2:1 in the time direction, and an additional 2:1 in the chroma vertical direction. For PAL and SECAM video standards where the display frequency is 50-Hz, the number of lines in a field is increased from 243 or 240 to 288, and the display rate is reduced to 50 fields/s or 25 frames/s. Similarly, the number of lines in the decimated chroma components is increased from 120 to 144. Since 288.50 = 240.60, the two formats have the same source data rate. Section 2.2.2 has additional discussion on the chroma subsampling for MPEG-2 video.

MPEG frames as they arrive at the decoder

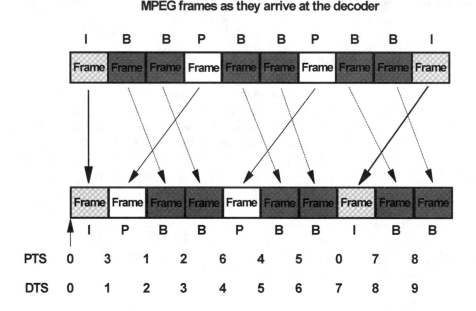

| PTS | 0 | 3 | 1 | 2 | 6 | 4 | 5 | 0 | 7 | 8 |
| DTS | 0 | 1 | 2 | 3 | 4 | 5 | 6 | 7 | 8 | 9 |

Figure 2.4 PES frames as they arrive at the decoder and as they are displayed, showing the corresponding PTS and DTS.

Macroblock is the basic building block of an MPEG picture. It consists of 16x16-luminance sample together with one 8x8 block of samples for the two-color components. The basic method is to predict image motion from frame to frame, and then to use the discrete cosine transform (DCT) to organize the redundancy in the spatial directions. The

DCTs are done on 8x8 pixel blocks, and the motion predictions are done on the luminance signals (Y) in 16x16 pixel blocks.

The second step of the encoding process is called *entropy coding*. The task of the entropy encoder is to transmit a particular stream of symbols in as few bits as possible. The technique for making optimal integer code assignments is called Huffman coding, which is used to code the result of the DCT coefficients, the motion vectors, and the quantization parameters with fixed tables. The DCT coefficients have a special Huffman table that is two-dimensional in that one code specifies a run-length of zeros and the nonzero value that ended the run. Another statistical coding model that is used to code the motion vectors and the frequency-independent DCT components is called differential pulse code modulation (DPCM). Here, the difference between each picture-element (pel) and a prediction is calculated from neighboring pel values already transmitted. Pel differences are considerably less correlated than the original pel values, and can be coded independently with reasonable efficiency.

There are three types of coded frames. Intra (I) frames are simply frames coded as still images, not using any past history, while predicted (P) frames are predicted from the most recently reconstructed I or P frames. Each macro-block in a P frame either comes with a vector and DCT coefficients for the closest match in the last I or P frame, or can be intra-coded (as with the I frames) if there was no good match. Bidirectional (B) frames are predicted from the closest two I or P frames, one in the past and one in the future. Thus, one is trying to search for the matching blocks in those frames, and see which one works best. In other words, given the 16x16 block in the current frame that one is trying to code, what is the closest match to that block in a previous or future frame? Beside the picture data, the video encoder attaches header information for each group of B, P, and I frames, known as group of pictures (GOP) to form the elementary video stream. A packetized elementary stream (PES) is formed by attaching additional header information such as program clock reference (PCR), optional encryption, packet priority levels, etc. to the elementary stream. The encoder or the MPEG multiplexer provides each PES with the so-called presentation time stamp (PTS) and decode time stamp (DTS). The DTS tells the decoder when to decode each frame, while the PTS tells the decoder when to display each frame. Before the PES is ready to be transmitted, a transport stream (TS) is formed. The packet structure for the MPEG-TS will be discussed in the next section on MPEG-2.

Figure 2.4 shows the ordering of B, P, and I frames for decoding with the corresponding PTS and DTS. The sequence of decoded frames usually goes like IBBPBBPBBPBBIBBPBBPB..., where there are 12 frames from one I frame to the next (U.S. and Japan only). This is based on a random access requirement that you need a starting point at least once every 0.4 seconds or so. The ratio of P frames to B frames may vary from one encoder to another. For the decoder to work, one would have to send that first P frame before the first two B frames, so the compressed data stream ends up looking like the example in Figure 2.4. One has to decode the I frame, then decode the P frame, keep both of those in memory, and then decode the two B frames. The I frame is probably displayed while the P frame is decoded, and the B-frames are displayed while they are decoded, and then the P frame is displayed as the next P frame is decoded, and so on. Since bidirectional macroblock predictions are an average of two macroblocks blocks, noise is reduced at low bit rates.

Table 2.2 shows the typical P-, B-, and I-frame size (in kbits) for both MPEG-1 and MPEG-2 video signals. At nominal MPEG-1 video (352x240x30, 1.15 Mbit/sec) rates, it is said that B frames improve SNR by as much as 2 dB. However, at higher bit rates, B frames become less useful since they inherently do not contribute to the progressive refinement of an image sequence (i.e., not used as prediction by subsequent coded frames).

Table 2.2 Typical frame sizes (in kilobits) for MPEG-1 and MPEG-2 compressions.

Compression Method	I frame	P frame	B frame	Average
MPEG-1 SIF @ 1.15 Mb/s	150 kb	50 kb	20 kb	38 kb
MPEG-2 @ 4.0 Mb/s	400 kb	200 kb	80 kb	200 kb

The MPEG-1 coding was originally designed only for progressive frames display (e.g., computer monitors). In order to display the MPEG-1 syntax on an analog TV set, the video frames need to be interlaced. There are two methods that can be applied to interlaced video that maintain syntactic compatibility with MPEG-1. In the field concatenation method, the encoder model can carefully construct predictions, and prediction errors that realize good compression while maintaining field integrity (distinction between adjacent fields of opposite parity). Some preprocessing techniques also can be applied to the interlaced source video that would lessen the sharp vertical frequencies. This technique is not efficient, of course. On the other hand, if the original source was progressive (e.g., film), then it is relatively more trivial to convert the interlaced source to a progressive format before encoding. Reconstructed frames are reinterlaced in the decoder display process.

An essential element in the utilization of MPEG intercompression is called motion compensation. Motion compensation is used to minimize the effect of image movement from a reference picture to the predicted picture. Motion vectors are described in terms of horizontal and vertical image displacements. The key technical issues in using motion vectors are the precision of the motion vectors, the size of the image region assigned to a single motion vector, and the selection criteria for the best motion vector values. There are various techniques that are used to estimate the motion vectors such as mean absolute distortion (MAD) and mean square error (MSE).

One of the main misconceptions about MPEG-1 is that it has fixed or limited frame size (i.e., 352x240x29.97 frame/s or 352x288x25 frame/s). In fact, MPEG-1 can use any frame size, including CCIR-601 resolutions (704x480), with frame sizes as high as 4095x4095x60 frame/s. MPEG-2 is more limited since frame sizes must be multiples of 16.

Another misunderstood issue arises from the standard profile and is known as constrained parameters bit stream (CPB). The CPB is a series of restrictions that the MPEG-1 stream must meet, including bit rate and frame sizes. Most hardware decoders accept only streams that follow the CPB profile. One can encode with any bit rate or frame size and still have standard MPEG-1. However, this encoded MPEG-1 stream cannot be decoded and displayed with some of the existing decoder chips.

CPB is a limited set of sampling and bit-rate parameters designed to normalize computational complexity, buffer size, and memory bandwidth while still addressing the widest possible range of applications. CPB limits video images to 396 macroblocks (101,376 pixels) per frame if the frame rate is less than or equal to 25 frames/s, and 330 macroblocks (84,480 pixels) per frame if the frame rate is less or equal to 30 frames/s. Therefore, MPEG-1 video is typically coded at SIF dimensions (352x240x30 frames/s or 352x288x25 frames/s). The total maximum sampling rate is 3.8 Msamples/s including chroma. The coded video rate is limited to 1.862 Mbit/sec. In industrial practice, the bit-rate is the most often waived parameter of CPB, with rates as high as 6 Mbit/s in use.

Historically, CPB was an optimum point that barely allowed cost effective VLSI implementations using 0.8 μm technology. It also implied a nominal guarantee of interoperability for decoders and encoders. MPEG decoders, which were not capable of meeting SIF rates, were not considered to be true MPEG decoders. Currently, there are several ways of getting around CPB for SIF class applications and decoder. Thus, one should remember that CPB limits frames to 396 macroblocks (as in 352x288 SIF frames). Still within the constraints are sampling rates of 416x240x24 Hz, but this only aids NTSC (240 lines/field) displays. Deviating from 352 samples/line could throw off many decoder implementations that have limited horizontal sampling-rate conversion modes. From a practical perspective, many decoders are simply doubling the sampling rate from 352 to 704 samples/line via binary taps, which are simple shift-and-add operations. Future MPEG decoders will have arbitrary sample rate converters on-chip.

2.2.2 MPEG-2 Standard

The MPEG-2 standard for compressing both video and audio signals is similar to MPEG-1, and it is targeted toward more diverse applications such as all-digital transmission of broadcast TV, digital media storage, high-definition TV (HDTV), etc. The most significant enhancement over MPEG-1 is the addition of syntax for efficient coding of interlaced video signals, such as those that originated from electronic cameras. Several other more subtle enhancements (e.g., 10-bit DCT DC precision, nonlinear quantization, VLC tables, and improved-mismatch control) are included, which have a noticeable improvement on coding efficiency, even for progressive video. Other key features of MPEG-2 are scalable extensions, which permit the division of a continuous video signal into two or more coded bit streams, representing the video at different resolutions, picture quality (i.e., SNR), or picture rates. The MPEG-2 standard defines 4:2:0, 4:2:2, and 4:4:4 chroma sampling formats relative to the luminance. Figure 2.5 illustrates the difference between 4:2:0 and 4:2:0 sampling formats. In 4:2:0 sampling, the chroma samples from two adjacent lines in a field are interpolated to produce a single chroma sample, which is spatially located half way between one of the original samples and the location of the same line but opposite field. The 4:2:0 sampling has many drawbacks, including significantly inferior vertical chroma resolution compared with a standard composite NTSC signal. In 4:2:2 sampling, the chroma is subsampled 2:1 horizontally but not vertically, with the chroma alignment with the luminance as shown in Figure 2.5. The 4:4:4 sampling has the same sampling to both chroma components, and the luminance with the same decomposition into interlaced fields, resulting in high-quality video images. MPEG-2 defines five different profiles as follows: simple profile (SP), main profile (MP), SNR scalable profile, spatially scalable profile, and high profile. The most

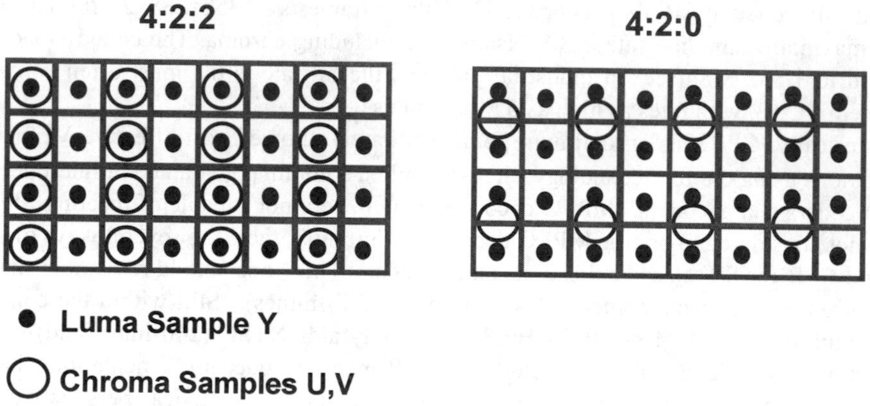

Figure 2.5 The 4:2:2 and 4:2:0 chroma components (U, V) sub-sampling relative to the luminance sample (Y).

Table 2.3 MPEG-2 level bounds for the main profile.

MPEG Parameter	MP at HL	MP at H-14	MP at ML	MP at LL
Samples/line	1920	1440	720	352
Lines/frame	1152	1152	576	352
Frame/sec	60	60	30	30
Luma Rate (samples/s)	62,668,800	47,001,600	10,368,000	3,041,280
Bit Rate (Mb/s)	80	60	15	4
VBV Buffer size (bits)	9,781,248	7,340,032	1,835,008	475,136

useful profile is the MP, which consists of four levels: main level (ML), high-1440 level (H-14), main level (ML), and low level (LL). Table 2.3 provides the level bounds for the main profile.

The video buffering verifier (VBV), which is an idealized model of the MPEG-2 decoder, is used to constrain the instantaneous bit rate of the MPEG-2 encoder such that the average bit rate target is met without overflowing the decoder data buffer. All the bounds in Table 2.3 are upper bounds except for the VBV buffer size, which is a lower bound.

The use of B frames increases the computational complexity, bandwidth, delay, and picture buffer size of the encoded MPEG video since some of the macroblock modes require averaging between two macroblocks. At the worst case, memory bandwidth is increased an extra 16 MB/s (601 rate) for this extra prediction. An extra picture buffer is needed to store the future prediction reference. In addition, extra delay is introduced in the encoding process, since the frame used for backwards prediction needs to be transmitted to the decoder before the intermediate B-pictures can be decoded and displayed. Since the extra picture buffer pushes the decoder DRAM memory requirements past the magic 1-Mbyte threshold, several companies such as AT&T and General Instrument argued against the use of the B-frames. In 1991, General Instrument introduced DigiCipher I video coding, which is similar to MPEG-2 coding but uses smaller macroblock predictions with no B frames and with Dolby AC-1 audio. In 1994, General Instrument introduced the DigiCipher II specification, which supports both the full MPEG-2 video main profile syntax and DigiCipher II with the use of the Dolby AC-3 audio algorithm [7].

The transmission of MPEG 2 compressed video streams through HFC access networks using QAM modulators to the digital STB at the subscriber's home will be discussed in detail in Chapter 8.

2.2.3 MPEG and AC-3 Audio

The MPEG compression of audio signals, which is based on MUSICAM technology, is done using high-performance perceptual schemes [10]. These schemes specify a family of three audio coding schemes called layer-1 (MP1), layer-2 (MP2), and layer-3 (MP3) with increasing complexity and performance. Since the uncompressed digital stereo music in CD quality is about 1.5 Mb/s, one can achieve 4:1 compression or 384-kb/s for stereo signals using MP1, while still maintaining the original sound quality. Compression of 8:1 to 6:1 (192 to 256 kb/s) the stereo signals can be achieved with MP2, while MP3 provides 12:1 to 10:1 compression (112 kb/s to 128 kb/s) with essentially no loss in discernible quality. The audio coding scheme can be describes as "perceptual subband transform coding." The audio coding scheme makes use of the masking properties of the human ear to reduce the amount of data. The various "spectral" components of the audio signal are analyzed by calculating a modified DCT (MDCT) and applying a psychoacoustic model to estimate the audible noise threshold. The noise caused by the quantization process is distributed to various frequency bands in such a way that it is masked by the total signal, that is, it remains inaudible.

Another perceptual coding technology, which has been adopted by the advanced television systems committee (ATSC) as the standard audio for HDTV in the United States, is called AC-3 and was developed by Dolby Laboratories [11]. AC-3 technology can support up to eight channel configurations, ranging from mono to six discrete audio channels (left, center, right, left surround, right surround, and subwoofer). As mentioned before, DigiCipher II uses the Dolby Labs AC-3 compression algorithm for the audio source. The AC-3 encoder can support data rates ranging from 32-kb/s up to 640-kb/s at sample rates of 32 kHz, 44.1 kHz, and 48 kHz. For further details, see reference [12].

2.2.4 MPEG-4 Standard

Unlike MPEG-1/2, where the scope and technology were well defined when the project started, MPEG-4 was born to answer the emerging needs of various new applications ranging from interactive audiovisual services to remote monitoring and control [13]. Thus, the MPEG-4 goal is to provide a flexible and extensible standard for the convergence of interactive multimedia applications, which currently are not addressed by the existing standards.

To enable the content-based interactive functionality, the MPEG-4 video standard introduces the concept of video object planes (VOPs). It is assumed that each frame of an input video sequence is segmented into a number of arbitrarily shaped image regions (e.g., VOPs), where each of the regions may possibly cover particular image or video content of interest (i.e., describing physical objects or content within scenes). In contrast to the video source format used for MPEG-1 and MPEG-2, the video input to be coded by the MPEG-4 verification model is no longer considered a rectangular region. The input to be coded can be a VOP image region of arbitrary shape, and the shape and location of the region can vary from frame to frame. Successive VOPs belonging to the same physical object in a scene are referred to as video objects (VOs), that is, a sequence of VOPs of possibly arbitrary shape and position. The shape, motion, and texture information of the VOPs belonging to the same VO is encoded and transmitted or coded into a separate video object layer (VOL). In addition, the bit stream must include the relevant information to identify each of the VOLs, and how the various VOLs are composed at the receiver in order to reconstruct the entire original sequence. This allows the separate decoding of each VOP and the required flexible manipulation of the video sequence. Notice that the video source input assumed for the VOL structure either already exists in terms of separate entities (i.e., is generated with chroma-key technology) or is generated by means of on-line or off-line segmentation algorithms.

It is expected that MPEG-4 video coding will eventually support all the functionalities already provided by MPEG-1 and MPEG-2, including the provision to efficiently compress standard rectangular sized image sequences at varying levels of input formats, frame rates and bit rates. In addition, content-based functionality will be assisted.

2.2.5 Other Digital Video Standards

There are several non-MPEG video standards such as ITU-T Recommendation H.261 [14] and H.263 [15], which actually were developed before the MPEG standard. The first digital video standard was developed for visual applications by the CCITT group XV for integrated services digital network (ISDN) services, operating at px64 kb/s for p = 1,..,30. The H.261 standard has many elements in common with MPEG-1. The image dimensions were restricted to two sizes, common intermediate format (CIF) with 360x288 (Y) and 180x144 (*U* and *V*), and quarter CIF (QCIF). The H.261 syntax consists of four layers as follows: (A) picture layer, which has a picture header followed by 3 or 12 group of blocks (GOBs) with each GOB consisting of 33 macroblocks and a header, (B) GOBs layer, (C) macroblock layer, and (D) block layer, which is similar to the block layer of MPEG-1. The macroblock header provides information about the position of the macroblock relative to the position of the just coded macroblock. To simplify the compression standard, macroblocks between frames can be skipped if the motion-compensated prediction is sufficiently good.

Another similar non-MPEG video teleconferencing standard is H.263, which is based on H.261 standard for low bit-rate applications. The H.263 standard has four optional modes that enhance its functionality, including unrestricted motion vector mode and arithmetic coding mode instead of variable length codes, advanced prediction mode, and coding P and B frames as one unit.

2.3 Cable TV Frequency Plans

The frequency plan of cable TV channels in the U.S. is specified by the FCC. FCC rules in Part 76 specify frequencies to in accordance with the channel allocation plan set forth by with the Electronics Industry Association (EIA) [16]. The nominal channel spacing is 6-MHz, except for the 4-MHz frequency gap between channels 4 and 5. Table 2.4 shows the standard (STD), incrementally related carrier (IRC), and harmonically related carrier (HRC) cable TV frequency plans in the U.S.

In the STD cable TV frequency plan, which is similar to the frequency plan of terrestrially broadcasted TV channels, all the visual carriers except channels 5 and 6 are located 1.25 MHz above the lower edge of the channel boundary in 6-MHz multiples (1.25 + 6N) MHz. The visual carriers for channels 5 and 6 are located 0.75 MHz below the 6-MHz multiples. Notice that the visual carrier in the following channel groups 14–15, 25–41, and 43–53 have a 12.5-kHz frequency offset relative to the rest of the channels.

In the IRC frequency plan, all the visual carriers except for channels 42, 60, and 61 are located 1.2625 MHz above the lower edge of the channel boundary in 6-MHz multiples (1.2625 + 6N) MHz. Thus, the visual carrier in the IRC frequency plan has 12.5-kHz frequency offset compared with the STD frequency plan. This frequency offset was selected to minimize interference in the 25-kHz radio channels, which are used for communications by airport control towers and aircraft navigation equipment based on FCC rulings.[2]

In the HRC frequency plan, all the visual carriers except for channels 60 and 61 are located essentially at the lower channel boundary in 6.0003-MHz multiples. As with the IRC plan, the 300-Hz incremental band increase was selected to minimize the interference in the 25-kHz radio channels used for aviation. The cable TV channel numbers are often designated according to electronic industry association (EIA) [16]. As we will see in Section 2.4,

Table 2.4 Standard, HRC, and IRC cable TV frequency plans in the US. After Ref. [16].

Channel Designation	Visual Carrier Frequency (MHz)		
EIA	STD	HRC	IRC
T7*	7.0000		
T8*	13.0000		
T9*	19.0000		
T10*	25.0000		
T11*	31.0000		
T12*	37.0000		
T13*	43.0000		
2	55.2500	54.0027	55.2625
3	61.2500	60.0030	61.2625
4	67. 2500	66.0033	67.2625
5	77. 2500	N/A	
6	83. 2500	N/A	
5	N/A	78.0039	79.2625
6	N/A	84.0042	85.2625
7	175. 2500	174.0087	175.2625
8	181. 2500	180.0090	181.2625
9	187. 2500	186.0093	187.2625
10	193. 2500	192.0096	193.2625
11	199. 2500	198.0099	199.2625
12	205. 2500	204.0102	205.2625
13	211. 2500	210.0105	211.2625
14	121. 2625	120.0060	121.2625
15	127. 2625	126.0063	127.2625
16	133. 2625	132.0066	133.2625
17	139. 2500	138.0069	139.2625
18	145. 2500	144.0072	145.2625

[2] See for example Code of Federal Regulations part 87.421 about the aviation channels and FCC rulings.
* Based on a conventional plan, which was originated in about 1960.

19	151. 2500	150.0075	151.2625
20	157. 2500	156.0078	157.2625
21	163. 2500	162.0081	163.2625
22	169. 2500	168.0084	169.2625
23	217. 2500	216.0108	217.2625
24	223. 2500	222.0111	223.2625
25	229. 2500	228.0114	229.2625
26	235.2625	234.0117	235.2625
27	241.2625	240.0120	241.2625
28	247.2625	246.0123	247.2625
29	253.2625	252.0126	253.2625
30	259. 2625	258.0129	259.2625
31	265. 2625	264.0132	265.2625
32	271. 2625	270.0135	271.2625
33	277. 2625	276.0138	277.2625
34	283. 2625	282.0141	283.2625
35	289. 2625	288.0144	289.2625
36	295. 2625	294.0147	295.2625
37	301. 2625	300.0150	301.2625
38	307. 2625	306.0153	307.2625
39	313. 2625	312.0156	313.2625
40	319. 2625	318.0159	319.2625
41	325. 2625	324.0162	325.2625
42	331. 2750	330.0165	331.2750
43	337. 2625	336.0168	337.2625
44	343. 2625	342.0171	343.2625
45	349. 2625	348.0174	349.2625
46	355. 2625	354.0177	355.2625
47	361. 2625	360.0180	361.2625
48	367. 2625	366.0183	367.2625
49	373. 2625	372.0186	373.2625
50	379. 2625	378.0189	379.2625
51	385. 2625	384.0192	385.2625
52	391. 2625	390.0195	391.2625
53	397. 2625	396.0198	397.2625
54	403.2500	402.0201	403.2625
55	409.2500	408.0204	409.2625
56	415.2500	414.0207	415.2625
57	421.2500	420.0210	421.2625
58	427.2500	426.0213	427.2625

59	433.2500	432.0216	433.2625
60	439.2500	438.0219	439.2625
61	445.2500	444.0222	445.2625
62	451.2500	450.0225	451.2625
63	457.2500	456.0228	457.2625
64	463.2500	462.0231	463.2625
65	469.2500	468.0234	469.2625
66	475.2500	474.0237	475.2625
67	481.2500	480.0240	481.2625
68	487.2500	486.0243	487.2625
69	493.2500	492.0246	493.2625
70	499.2500	498.0249	499.2625
71	505.2500	504.0252	505.2625
72	511.2500	510.0255	511.2625
73	517.2500	516.0258	517.2625
74	523.2500	522.0261	523.2625
75	529.2500	528.0264	529.2625
76	535.2500	534.0267	535.2625
77	541.2500	540.0270	541.2625
78	547.2500	546.0273	547.2625
79	553.2500	552.0276	553.2625
80	559.2500	558.0279	559.2625
81	565.2500	564.0282	565.2625
82	571.2500	570.0285	571.2625
83	577.2500	576.0288	577.2625
84	583.2500	582.0291	583.2625
85	589.2500	588.0294	589.2625
86	595.2500	594.0297	595.2625
87	601.2500	600.0300	601.2625
88	607.2500	606.0303	607.2625
89	613.2500	612.0306	613.2625
90	619.2500	618.0309	619.2625
91	625.2500	624.0312	625.2625
92	631.2500	630.0315	631.2625
93	637.2500	636.0318	637.2625
94	643.2500	642.0321	643.2625
95	91.2500	90.0045	91.2625
96	97.2500	96.0048	97.2625
97	103.2500	102.0051	103.2625
98	109.2750	108.0250	109.2750

99	115.2750	114.0250	115.2750
100	649.2500	648.0324	649.2625
101	655.2500	654.0327	655.2625
102	661.2500	660.0330	661.2625
103	667.2500	666.0333	667.2625
104	673.2500	672.0336	673.2625
105	679.2500	678.0339	679.2625
106	685.2500	684.0342	685.2625
107	691.2500	690.0345	691.2625
108	697.2500	696.0348	697.2625
109	703.2500	702.0351	703.2625
110	709.2500	708.0354	709.2625
111	715.2500	714.0357	715.2625
112	721.2500	720.0360	721.2625
113	727.2500	726.0363	727.2625
114	733.2500	732.0366	733.2625
115	739.2500	738.0369	739.2625
116	745.2500	744.0372	745.2625
117	751.2500	750.0375	751.2625
118	757.2500	756.0378	757.2625
119	763.2500	762.0381	763.2625
120	769.2500	768.0384	769.2625
121	775.2500	774.0387	775.2625
122	781.2500	780.0390	781.2625
123	787.2500	786.0393	787.2625
124	793.2500	792.0396	793.2625
125	799.2500	798.0399	799.2625
126	805.2500	804.0402	805.2625
127	811.2500	810.0405	811.2625
128	817.2500	816.0408	817.2625
129	823.2500	822.0411	823.2625
130	829.2500	828.0414	829.2625
131	835.2500	834.0417	835.2625
132	841.2500	840.0420	841.2625
133	847.2500	846.0423	847.2625
134	853.2500	852.0426	853.2625
135	859.2500	858.0429	859.2625
136	865.2500	864.0432	865.2625
137	871.2500	870.0435	871.2625
138	877.2500	876.0438	877.2625

139	883.2500	882.0441	883.2625
140	889.2500	888.0444	889.2625
141	895.2500	894.0447	895.2625
142	901.2500	900.0450	901.2625
143	907.2500	906.0453	907.2625
144	913.2500	912.0456	913.2625
145	919.2500	918.0459	919.2625
146	925.2500	924.0462	925.2625
147	931.2500	930.0465	931.2625
148	937.2500	936.0468	937.2625
149	943.2500	942.0471	943.2625
150	949.2500	948.0474	949.2625
151	955.2500	954.0477	955.2625
152	961.2500	960.0480	961.2625
153	967.2500	966.0483	967.2625
154	973.2500	972.0486	973.2625
155	979.2500	978.0489	979.2625
156	985.2500	984.0492	985.2625
157	991.2500	990.0495	991.2625
158	997.2500	996.0498	997.2625

the nonlinear distortions in both the HRC and IRC frequency plans are reduced compared with the STD plan.

2.4 Coaxial Cable TV Components and Systems

As we learned in Chapter 1, the basic architecture of HFC networks to deliver multichannel analog video/audio signals from the fiber node to the subscriber's home consists of coaxial cables, cascaded amplifiers, and taps. The transmission characteristics of these components such as transmission loss, frequency response, amplifier gain, and nonlinear distortions need to be considered when designing such a network. Furthermore, when RF digital channels are added, linear distortions such as group delay and amplitude variation within each 6-MHz channel also must be considered. The following section provides a brief overview of these properties.

2.4.1 Coaxial Cable

Network distribution coaxial cables typically consist of a copper-clad aluminum wire, which is the inner conductor, an insulating layer such as foam polyethylene, a solid aluminum shield, which serves as the outer conductor, and a PVC jacket. Subscriber drop cables are manufactured with a copper-clad steel center conductor and a combination of aluminum braid and aluminum-polypropylene-aluminum (APA) tap shield. For specialty applications such as plenum installations, the coaxial cable jacket is often made with polytretrafluoro-ethylene (PTFE) materials. The transmission characteristic of coaxial cables depends mainly on the transmission frequency and on the diameter of the cable. The primary loss mechanisms in coaxial cables are the frequency and temperature dependence of the inner conductor and the dielectric outer conductor losses. In particular, the transmission loss of the inner conductor in coaxial cables at RF frequencies is influenced by the so-called *skin effect* [17]. When DC current flows through a conductor, the current is uniformly distrib-uted throughout the cross-section of the conductor. As the RF frequency is increased, the current tends to crowd around the conductor surface, which reduces the effective current cross-section, resulting in a given conductor having higher impedance at higher RF frequencies.

There are three different types of coaxial cables that are used in the distribution system: trunk, feeder, and drop cables. The transmission loss of trunk cables, which have a typical diameter from 1/2" to 1", increases from 0.89 dB at 50 MHz to 3.97 dB at 750 MHz per 100 m for 1" cable. The feeder cable has typically a smaller diameter than the trunk cable, and is used to connect between the line-extender amplifiers to the tap. The drop cable typically has

Table 2.5 Maximum loss for drop cable (dB/100 ft at 68F) with four different cable di-ameters as a function of frequency. To obtain loss in dB/100 m, multiply by 3.281. After Ref. [16].

Frequency (MHz)	59 Series Foam	6 Series Foam	7 Series Foam	11 Series Foam
5	0.86	0.58	0.47	0.38
30	1.51	1.18	0.92	0.71
40	1.74	1.37	1.06	0.82
50	1.95	1.53	1.19	0.92
110	2.82	2.24	1.73	1.36
174	3.47	2.75	2.14	1.72
220	3.88	3.11	2.41	1.96
300	4.45	3.55	2.82	2.25
350	4.80	3.85	3.05	2.42
400	5.10	4.15	3.27	2.60
450	5.40	4.40	3.46	2.75
550	5.95	4.90	3.85	3.04
600	6.20	5.10	4.05	3.18
750	6.97	5.65	4.57	3.65
865	7.52	6.10	4.93	3.98
1000	8.12	6.55	5.32	4.35

a smaller diameter than the feeder cable, and is used between the tap and the subscriber home terminal. Table 2.5 shows the maximum drop cable loss (dB/100 ft.) at 68°F as a function of RF frequency for the following nominal cable diameters: 0.240" (59 series foam), 0.272" (6 series foam), 0.318" (7 series foam), and 0.395" (11 series foam). Notice the nearly 10 times increase in the cable loss from 5 MHz to 1 GHz for 59 series foam cable. The coaxial cable losses (in dB) are reasonably proportional to the square root of the frequency [17]. Cable plants are usually specified to operate over a wide temperature range from −40°C to +70°C. In general, the cable loss slowly increases linearly with temperature, since the cable attenuation increases with temperature.

A useful parameter to remember is the so-called cable loss ratio (CLR) given by

$$CLR = \sqrt{\frac{f_1}{f_2}} \tag{2.2}$$

where f_1 and f_2 are two different RF frequencies. For example, approximate the cable loss at 55 MHz when the cable loss at 450 MHz is 20 dB. The cable loss (CL) at 55 MHz is given by CL $= 20{\cdot}(55/450)^{1/2} = 6.99$ dB.

2.4.2 RF Amplifiers

There are also three different types of cable TV RF amplifiers, depending on their location in the coaxial portion of the HFC network. Figure 2.6 shows the respective locations of trunk, bridger, and line-extender amplifiers in the traditional tree-and-branch architecture of the coaxial portion of the HFC network. Trunk amplifiers, which are typically spaced 20–22 dB from one another, are moderate-gain amplifiers with a typical output of 30 to 36 dBmV that are used to provide high CNR with low nonlinear distortions (< −80 dBc), particularly at the high-frequency cable TV channels (> 300 MHz). Feeder amplifiers are used not only for downstream delivery of analog and digital video channels, but also to split the transmitted signals to up to four feeder cables as shown in Figure 2.7. The output power of a bridger amplifier is typically in the range of 40–50 dBmV, which is about 12 dB higher than that of trunk amplifiers. However, higher nonlinear distortions are present at the output of bridger and line extender amplifiers. To reduce the effect of nonlinear distortions on the transmitted video signals and to maintain the flatness of the entire band, a maximum of two to four line- extender amplifiers are used, depending on the number of taps between the line extender amplifiers. Line-extender amplifiers, which are typically spaced between 120 m to 350 m, are used in the vicinity of the subscriber's home. For two-way cable TV systems, the downstream or forward video channels in the United States are placed between 52 MHz and 860 MHz, while the return-path or upstream band is between 5 MHz and

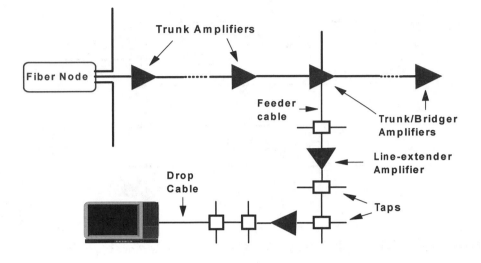

Figure 2.6 Tree-and-branch architecture of the coaxial portion of the HFC networks showing the different types of cables and amplifiers that are used.

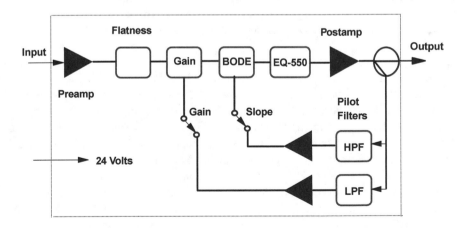

Figure 2.7 Simplified block diagram of trunk amplifier, showing the automatic gain and slope controls and equalizer up to 550 MHz.

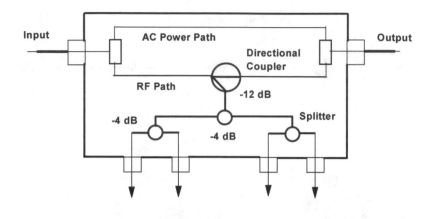

Figure 2.8 Simplified block diagram of four-way 20-dB tap.

42 MHz.[3] This partition is done using a special filter known as *diplex* filter, which has a typical isolation between these bands better than 60 dB. The diplexer filter is three-port device, with high, low, and common ports. The common-to-low port, which is marked as "L," is essentially a low-pass filter that allows the return-path signals to be transmitted. The common-to-high port, which is marked as "H," is essentially a high-pass filter that allows the forward channels to be transmitted. In a typical two-way trunk or bridger amplifier, the forward signals pass through the H ports, while the upstream signals are pass through the L ports.

A simplified block diagram of a trunk amplifier is shown in Figure 2.7. Since coaxial cables have a strong frequency-dependent loss (see Table 2.5 for example), the amplitude of the transmitted video channels must be equalized in order to maintain flatness across the transmitted RF spectrum. Forward equalizers are designed to compensate for fixed lengths of coaxial cables. By introducing additional attenuation at the lower frequencies, the equalizer allows the trunk amplifier to maintain a known slope across its transmission band. In addition, some trunk amplifiers are equipped with Bode equalizers to compensate for changes in the cable loss caused by temperature variations.

Trunk amplifiers typically use automatic gain control (AGC) and/or automatic slope control (ASC) circuitry. Typical gain and slope control ranges are 6 to 10 dB up to 750 MHz. The AGC and ASC modules in the trunk amplifier detects a sampled signal of standard pilot channels at the amplifier output, which is used to create the appropriate voltage to control the gain and/or slope of the amplifier. The standard pilot frequencies vary among the differ-

---

[3] Prior to 1994, the upstream band ended at 30 MHz. Other countries have different split frequencies such as Australia 65/85; Japan and New Zealand 55/70; India and Eastern Europe 30/48; Western Europe, Ireland, and United Kingdom 65/85.

ent manufacturers. All cable TV amplifiers use some variation of push-pull circuitry to minimize second-order distortions. Both Feedforward and power-doubling technology are used for improved distortion performance.

2.4.3 Taps

Figure 2.5 shows a simplified block diagram of a four-way 20-dB tap. Taps are used to split the transmitted signals to drop cables as shown in Figure 2.8. A typical tap consists of a RF directional coupler and power splitters. The directional coupler diverts a specific amount of the input signal power, while the power splitter splits that signal to typically two, four, or eight subscriber ports. The power loss between the input and the output ports is called is called *insertion loss*, while the power loss between different output ports is called *isolation loss*. Tap insertion loss is nominally independent on frequency or temperature. High isolation loss is essential for two-way cable systems in order to prevent the upstream signals from one customer to leak into the forward signals of another customer. Typical isolation in a tap is about 20 dB for both forward and return-path bands. Taps are marked by their tap value, which is the power ratio of the signal at the tap to the input signal. Common values are from 4 dB to 35 dB, in 3-dB steps. Taps are typically housed in a die-cast aluminum alloy housing for aerial or ground mounting with a weather seal gasket to prevent entry of moisture and dust.

2.5 Multichannel Coaxial Cable TV Systems

2.5.1 CNR of a Single and Cascaded Amplifiers

One of the most important parameters that characterize the transmission performance of a cable TV system is the carrier-to-noise ratio (CNR). The CNR at the output of a single RF amplifier is given by

$$CNR(dB) = P_{out} / k_B TB + 59.16 - F - G \qquad (2.3)$$

where P_{out} is the output power from the amplifier, k_B is Boltzmann's constant (1.38×10^{-23} joule/K), T is the effective Kelvin temperature of the amplifier, B is the signal noise bandwidth (4-MHz), and F and G are the amplifier noise figure and gain in dB. The value −59.16 dBmV is the 75-ohm impedance room temperature thermal noise in 4-MHz bandwidth. Typical noise figures for trunk amplifiers range from 7 to 10 dB at an input signal level of +10 dBmV and gain of 20 dB.

Now, consider a cable TV system with a chain of N unlike cascaded amplifiers but with the noise bandwidth, where the N_{th} amplifier has a noise figure F_N and amplification G_N as shown in Figure 2.9. Then, the overall system noise figure F (expressed as ordinary power ratio) is given by

$$F = F_1 + \frac{F_2 - 1}{G_1} + \frac{F_3 - 1}{G_1 G_2} + \ldots + \frac{F_N - 1}{G_1 G_2 \cdots G_{N-1}}$$

In the simple case of N identical RF amplifiers, Equation (2.4) is reduced to $N \cdot F$ where F is the noise figure of each amplifier. The overall noise figure in dB is given by

$$NF_N(dB) = 10 \cdot \log(N \cdot F) \tag{2.5}$$

Then, the overall system CNR is given by

$$CNR_N = CNR - 10 \cdot \log(N)$$
$$(2.6)$$

For example, if a cable TV system has four cascaded amplifiers with CNR of 56 dB for a single amplifier, then the overall CNR after the fourth amplifier is 50 dB.

The overall system CNR_N for a system of dislike amplifiers as described by Equation (2.4) is simply given by

$$CNR_N(dB) = -10 \cdot \log\left[10^{-CNR_1/10} + 10^{-CNR_2/10} + \ldots + 10^{-CNR_N/10}\right] \tag{2.7}$$

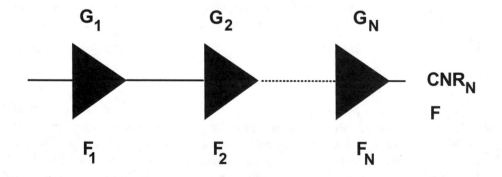

Figure 2.9 N cascaded amplifier configuration for calculating the overall system CNR and noise figure (F).

2.5.2 Nonlinear Distortions: CSO, CTB, and XMOD

When multiple band-limited signals such as AM-VSB modulated video signals are transmitted through a nonlinear component or system such as an RF amplifier or a laser transmitter, various nonlinear distortions (NLD) will be generated. The most important and strongest second-order intermodulation (IM_2) products are given by $A \pm B$, where A and B are two arbitrary RF frequencies, which are always 6-dB higher than the $2A$ or $2B$ distortion products. In a cable TV system, the summation of all possible different IM_2 products that are present in a particular channel is commonly called composite second-order (CSO) distortion. The CSO distortion beats in the standard cable TV frequency plan are 1.25 MHz and 0.75 MHz above and below the luminance carrier of a particular channel. The CSO distortion beats at ± 0.75 MHz is a result of channels 5 and 6, which are offset 2 MHz from the 6 MHz multiples, mixing with the other channels in the standard frequency plan.

In the IRC frequency plan, the CSO distortion beats are only located at ± 1.25 MHz above or below the luminance carrier of a particular channel. Due to the frequency offset of the CSO distortion from the luminance carrier, it appears as slowly moving diagonal stripes on a TV picture. In the HRC frequency plan, the CSO and CTB distortion beats are frequency-coincident with the visual carrier for coherent channel cable TV systems. As we will see later, this is one of the main advantages of the HRC frequency plan compared with the other frequency plans.

For a typical cable TV system with N dislike amplifiers, the overall CSO is given by

$$CSO_N = -15 \cdot \log_{10}\left[10^{-CSO_1/15} + 10^{-CSO_2/15} + \ldots + 10^{-CSO_N/15}\right] \tag{2.8}$$

The dominant third-order intermodulation (IM_3) products are $A+B-C$, $A-B+C$, and $A-B-C$ where $A < B < C$, and A, B, and C are three arbitrary RF frequencies. The weakest third-order intermodulation product is the $3A$ product, which is 15.56 dB weaker than the ABC distortion products. The $2A \pm B$ and $A \pm 2B$ products are 9.54 dB weaker than the ABC products. In a cable TV system, the summation of all possible different third-order intermodulation products that fall onto a particular channel is commonly called composite triple beat (CTB) distortion. Thus, the CTB distortion is deduced by measuring the two-tone IM_3 (i.e., $2A \pm B$), counting the number of IM_3 products that fall in a particular channel, and adding 6 dB.

For equally spaced luminance carriers, the estimated number of CTB beats in any channel is given by

$$N_{CTB} = \frac{(N-1)^2}{4} + \frac{(N-M) \cdot (M-1)}{2} - \frac{N}{2} \tag{2.9}$$

where N is the total number of channels and M is the number of channels being measured. For $N \gg 1$ and in the middle of the band, Equation (2.9) reduces to $3N^2/8$, while at the edge of the band, Equation (2.9) reduces to $N^2/4$. Notice that the number of CTB beats at the band edge is 2/3 of the number of CTB beats in the middle of the band, independent of the number of channels. The overall CTB distortion (in dB) is $10 \cdot \log(N_{\mathrm{CTB}})$. In Equation (2.9), it is also assumed that the luminance carriers are not phase locked. If the carriers are phase locked, then the CTB beats add coherently, and the overall CTB is $20 \cdot \log(N_{\mathrm{CTB}})$.

For a typical cable TV system with N different amplifiers, the overall CTB is given by

$$CTB_N(dB) = -20 \cdot \log\left[10^{-CTB_1/20} + 10^{-CTB_2/20} + ... + 10^{-CTB_N/20}\right] \qquad (2.10)$$

Thus, a 1 dB change in the amplifier output will change the CTB distortion ratio by 2-dB (the beat product itself changes by 3 dB). For every double in the number of amplifiers with identical CTB distortion, the overall CTB ratio degrades by 6 dB.

Cross modulation (XMOD) distortion is another type of cable TV distortion. It occurs when a group of video carriers are modulating other video carriers in a multichannel video system. The origin of XMOD distortions is the similar to the CSO/CTB distortions. These distortions usually generated when an RF coaxial amplifier or other active device is overloaded or driven beyond its compression point such that its gain becomes nonlinear. In older cable TV networks, XMOD distortions often produced picture interference that resembled a windshield wiper effect. However, in current cable TV networks with a large number of AM-VSB channels and high operating levels, the effect of XMOD distortions are typically masked by CSO/CTB distortions, and thus do not impose a serious problem to the cable TV operators.

The relative magnitude of the CSO, CTB, and XMOD distortions for a given channel is also an important consideration. Unfortunately, there is no simple answer. However, as we will learn in Chapters 3 and 4, the relative magnitude of the CSO, CTB, and XMOD distortions in a given fiber link depend primarily on the type of the laser transmitter used and the linearization techniques used to suppress these distortions.

2.5.3 Multipath Reflections (Echoes) and Group Delay

Multiple multipath reflections in the coaxial portion of the HFC network can severely degrade the propagating analog or digital signals before they reach the subscriber home. Multipath reflections occur when two or more propagation paths exist between the transmitter and receiving sites. The various reflections relative to the directly transmitted signal as measured at the receiver are called *echoes*, which are characterized by their amplitude attenuation and time delay. The echoes result from reflections off man-made or natural structures, repeaters, or the use of multiple transmitters.

The effect of multipath echoes on analog NTSC signals is quite different from that on digital signals. Let us first focus on the effect on the transmitted analog NTSC signals. For NTSC signal, a ghostlike image is horizontally displaced from the main image by an amount proportional to the time delay of the reflected signal. The effect of multipath reflections can be seen directly on an analog TV screen as ghosting, horizontally displaced from the main image. This can be used to provide an estimate of both the time delay and amplitude of the multipath echoes. The duration of each horizontal TV line is about 63.5 µs, with about 11 µs being used for the horizontal sync and blanking interval, providing about 52.5 µs for the actual video image. Because of TV set over-scan and simplicity, let us round off the video signal duration to 50 µs. Assuming a single dominant echo with a time delay less than 50 µs, its time delay can be estimated by multiplying the percentage of TV screen displaced by 50 µs. For example, if a ghost is displaced 25% from the main signal, the multipath delay is approximately 12.5 µs.

The amplitude of the multipath can also be determined by comparing the amplitude of the ghost to the amplitude of the original signal, using a video waveform monitor with the appropriate triggering to view only the VITS signals. The magnitude of the multipath echo can be determined as 20·log (% of amplitude/100). For example, if multipath reflection generates a ghost that is 25% the amplitude of the original waveform, then its amplitude is − 12 dBc, or 12 dB below the desired signal. This method can be used effectively to measure the amplitude of the echoes for one or more of the transmitted analog channels.

Multiple echo degradation is not seen in the digitally demodulated picture until some "threshold" level of the digital signal is reached, resulting in a loss of the receiver synchronization. Uncorrected multipath echoes introduce intersymbol interference (ISI) that results in a closure of the eye pattern, making the signal more susceptible to decoding errors. Using an adaptive equalizer in the digital receiver can minimize the digital degradation effect caused by multiple echoes. However, strong multiple echoes can cause the digital receiver to lose synchronization. Outside the time range of the receiver's adaptive equalizer, the effect of multiple echoes is perceived as additional noise and causes degradation to the received SNR. The topic of how an adaptive equalizer in a digital receiver works to mitigate the effect of multiple echoes will be discussed in detail in Chapter 7.

If the time delay of the multipath echoes is beyond the range of the adaptive equalizer in the digital receiver, then the adaptive equalizer is not able to combat the generated ISI. In this case, an ordinary spectrum analyzer can be used to characterize multipath on digital signals. Figure 2.10 shows, for example, the simulated spectrum of a 64-QAM channel in a 6-MHz band. In the presence of Gaussian noise, the spectrum of the 64-QAM signal is essentially flat across most of the symbol rate bandwidth. If, however, multipath reflections are present, constructive and destructive interference of the reflected paths with the direct path will cause ripples in the otherwise flat spectrum. The time delay for a single multipath echo can be estimated by taking the inverse of the measured frequency spacing of the ripples on a spectrum analyzer. The ripple spacing is measured from peak-to-peak or from null-to-null.

It turns out that for a single echo, its magnitude (in dB) with respect to the transmitted signal can be estimated according to the following approximation:

$$EM(dB) = 20 \cdot \log \left[\frac{10^{\Delta_{pp}/20} - 1}{10^{\Delta_{pp}/20} + 1} \right] \qquad (2.11)$$

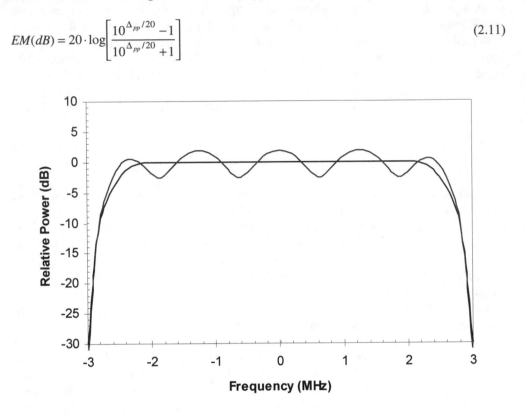

Figure 2.10 Simulated effect of a single echo with magnitude of −12 dBc and time delay of 0.8 μs relative to the main signal on the RF spectrum of a 64-QAM signal. After Ref. [9].

Table 2.6 The corresponding echo profile for Figure 2.10.

Echo #	Time Delay	Magnitude	Phase
1	0.2-μs	−11 dBc	180°
2	0.4-μs	−14 dBc	180°
3	0.8-μs	−17 dBc	180°
4	1.2-μs	−23 dBc	180°
5	2.5-μs	−32 dBc	180°

where Δ_{pp} is the measured signal peak-to-valley amplitude variation (in dB) on a spectrum analyzer. Figure 2.10 shows also, the simulated effect on the spectrum of a 64-QAM signal caused by a single multipath echo with a time delay of 0.8 µs and with a magnitude of 12 dB below the transmitted signal. Note that the frequency spacing between adjacent ripples is 1.25 MHz and the peak to valley amplitude variation of 4.5 dB.

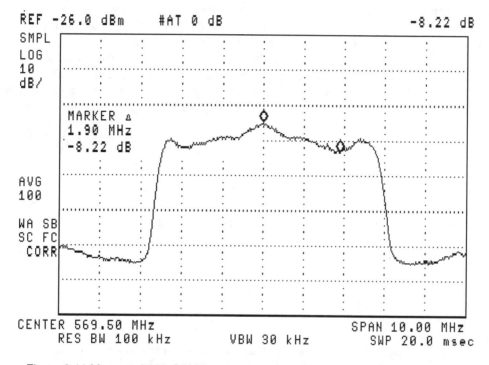

Figure 2.11 Measured 256-QAM spectrum at center frequency of 569.5 MHz in the presence of five-multipath echoes, according to Table 2.6.

When multiple multipath echoes degrade a QAM signal, the echo analysis using a spectrum analyzer becomes more difficult. This is because the ripple pattern is a complex superposition of the different echoes' time delay and magnitude, requiring a Fourier analysis. Figure 2.11 shows, for example, the measured spectrum of a 256-QAM signal at 569.5 MHz in the presence of five multipath echoes, where their time delay, amplitude, and phase are shown in Table 2.6. The QAM receiver in this measurement had an adaptive equalizer with 8 symbol-spaced feed-forward (FFE) taps and 24 symbol-spaced decision-feedback (DFE) taps. One of the FFE taps is used as the center tap, leaving 7 FFE taps for leading echoes and 24 DFE taps for lagging echoes. The total theoretical range of the adaptive equalizer is from −1.3 µs to +4.48 µs (symbol duration $\cong$ 186.5 ns). Notice the large 8.22-dB amplitude ripple (i.e., peak-to-valley variation) in the spectrum of the 256-QAM signal due to the presence of multiple echoes. In fact, the measured 256-QAM BER had increased from 10^{-8} to $1.3 \cdot 10^{-3}$ at a corrected CNR of 31.2 dB in the presence of these multipath echoes.

For multipath echoes with a short time delay (less than 0.2 µs), less than one ripple may be created in the 6-MHz spectrum. This means that multipath echoes with short-time delay, which are caused by relatively close reflections, have the effect of attenuating or amplifying entire channels, that is, it creates ripples across a wide frequency band spanning many adjacent channels. Consequently, it may actually be beneficial to those channels that are increased in level, but harmful to those channels that are attenuated.

Another important related parameter is the so-called *group delay*. The group delay is related to the frequency-dependent phase of various RF components in a cable TV plant such as diplex filters, equalizers, and impedance matching transformers. Mathematically, the complex multipath transfer function for a transmitted analog or digital channel can be described as

$$H(\omega) = |H(\omega)| \cdot \exp[i\phi(\omega)] \tag{2.12}$$

where $A(\omega) = -20 \cdot \log[\,|H(\omega)|\,]$ is the signal attenuation function (in dB), and $\phi(\omega)$ is the phase function of the propagating signal. The delay distortion or group delay is basically the instantaneous phase slope with respect to frequency, and it is given by

$$GD(\omega) = -\frac{d\phi(\omega)}{d\omega} = -\frac{1}{2\pi} \cdot \frac{d\phi}{df} \tag{2.13}$$

where f is the frequency in Hz. A critical parameter for robust transmission is the slope of the group delay across the transmission channel band. If the group delay is constant within the desired channel, there is no group-delay distortion. The term group-delay variation (GDV) is referred to simply as the maximum change (peak-to-valley) in the group delay across a given channel. For a cable TV system with N cascaded amplifiers, the total GDV across a given channel would be the sum of GDV from each amplifier in the link. In the small echo regime (for a single dominant echo), the maximum in-band GDV (in µs) can be approximated as

$$GDV(\mu s) = 2\tau \cdot 10^{-r/20} \tag{2.14}$$

where r is the echo magnitude (in dB) and τ is the echo delay (in µs) relative to the transmitted signal. Coaxial components such as amplifiers and taps typically generate GDV due to nonflat amplitude variations at different frequencies. For example, the GDV should be less than 0.2 µs per 1 MHz of channel bandwidth, which corresponds to an amplitude variation of 0.5 dB for a single echo with 0.1-µs time delay. The DOCSIS 1.1 specifications, for example, assume that the GDV of a typical cable TV network is 75 ns for a 6-MHz downstream channel, and 200 ns/MHz for an upstream channel. Even without the presence of multipath echoes, the use of various filters in the coaxial amplifiers typically generates GDV because of their nonflat frequency response.

There are various discrete multipath reflection models such as IEEE802.14 [8] and DOCSIS 1.0 [9] for both forward and return-path channels in a cable TV plant, which have been recently discussed. While the DOCSIS 1.0 model assumes a single dominant echo, the IEEE802 model breaks the echo power into multiple echoes within the time-delay range. Table 2.7 summarizes the maximum echo power and time-delay bounds for both downstream and upstream channels according to the DOCSIS model, assuming a single dominant echo.

Table 2.7 Multipath echo model according to DOCSIS 1.0 standard.

Echo Time Delay	Echo Magnitude (downstream)	Echo Magnitude (upstream)
0 to 0.5 µs	−10 dBc	−10 dBc
≤ 1.0 µs	−15 dBc	−20 dBc
≤ 1.5 µs	−20 dBc	−30 dBc
> 1.5 µs	−30 dBc	−30 dBc

Note that placing the echo power at the maximum of the delay range produces the largest GDV, but not necessarily the worst effect of the transmitted signal. Breaking the echo power into multiple echoes within the time-delay range lowers the GDV, but actually has a worse effect on the transmitted signal due to higher peak-to-rms ratio for combining multiple echoes compared with a single echo.

2.5.4 AM Hum Modulation

As we learned in Section 2.1, the vertical scan frequency of a TV image was originally selected to match the AC power frequency (60 Hz). Modulation distortion at 60 Hz or harmonics of the fundamental powerline frequency, which is called "hum," is the amplitude distortion of the transmitted signals caused by the modulation of the signal by the power source. The AM hum modulation is defined as the percentage of the peak-to-peak interference compared with the rms value of the sync peak level of the visual RF signal [2]. According to FCC regulations, Part 76, Section 76.605, the magnitude of AM hum modulation at the subscriber's terminal is required not to exceed 3% peak-to-peak of the visual signal level [4]. The major sources of power line hum (60 Hz and 120 Hz) are amplifiers with defective power supplies and overloaded system power supplies. However, other cable system components, even passive devices, can introduce hum modulation distortion under certain conditions. For additional discussion of AM hum modulation, please see Section 8.10.7.

2.6 Cable TV Return-Path Transmission Characteristics

The standard frequency plan for upstream transmission is shown in Table 2.4. The "T-channel" plan originated in about 1960 to identify the video channels that could be converted in a block to the FCC standard channel assignment T7 to T13. The current return-path or upstream transmission band is from 5 MHz to 42 MHz in the United States (5 to 65 MHz in Europe). Recently, it has been proposed by the same cable operators that the return-path band be located above the downstream channels in the 900-MHz to 1-GHz band. Although this upstream band is almost three times as large as the 5 to 42-MHz band, coaxial cable losses are almost ten times larger than at the 5–42 MHz band and exhibit strong frequency dependence. (See Table 2.5.) On the other hand, the upstream noise in these RF frequencies is significantly less than in the 5–42-MHz band. To date, cable TV operators, because of the cost to upgrade their network to 1 GHz, have mostly ignored this proposal.

2.6.1 Return-Path Noise Sources

The noise sources that impair the transmitted upstream signals in cable TV systems include ingress noise, common-path distortion, laser transmitter, and optical receiver noise. Ingress noise is the most important and dominant noise source in the return-path portion of the HFC network, and can be divided into three general types:

- Narrowband shortwave signals, primarily from radio and radar stations, that are transmitted terrestrially and coupled to the return-path band at the subscriber's home or in the cable TV distribution plant.
- Burst noise, which is generated by various manmade and naturally occurring sources, has time duration longer than the $(\text{symbol rate})^{-1}$.
- Impulse noise, which is similar to burst noise, but has time duration shorter than $(\text{symbol rate})^{-1}$ such that the receiver impulse response is effectively being measured.

The propagation of narrowband interference signals depends on the atmospheric conditions as well as the 10.7-year solar cycle. Increased solar sunspot activity has been correlated with increases in the maximum useable frequency because of ionization in the upper layers of the atmosphere. There are various methods for reducing ingress in the return-path band at the subscriber's location, which will be discussed later. Burst and impulse noise are generated from various manmade sources such as electric motors and power-switching devices. Although these sources produce burst/impulse noise events in the 60-Hz to 2-MHz portion of the spectrum, their harmonics show up in the 5–42-MHz upstream frequency band. Naturally occurring burst/impulse noise events include lightning, atmospherics, and electrostatic discharge, which typically extend from 2 kHz up to 100 MHz. Figure 2.12 shows a typical return-path spectrum as measured at a cable headend with the average signal level in dBmV. Notice the presence of various ingress peaks, particularly below 10 MHz.

As we will see later, for robust data transmission using quadrature-phase-shift-keying (QPSK) modulation, a 16–20 dB margin above the noise level is needed, as indicated by the horizontal solid line in Figure 2.12.

The time-varying ingress noise level also depends on the number of subscribers connected to the cable TV headend. Figure 2.13 shows the average ingress noise level as a function of the number of subscribers for T7, T8, T9, and T10 upstream channel frequencies (see Table 2.4), which is based on CableLabs field measurements [7]. The worst-case ingress levels can exceed +10-dBmV within a 100-kHz bandwidth. Two distinct trends emerge from Figure 2.13. First, the ingress noise levels are generally higher in the low-frequency region of the upstream band. Second, ingress noise levels increase for cable TV distribution plants with larger number of subscribers per fiber node. The solid curves in Figure 2.13 can be approximated by $A \cdot \log (N) - B$, where N is the number of subscribers, $A = 9$, and $B = 28$, 33, 37, and 40 for T7 (top curve), T8, T9, and T10 channels, respectively.

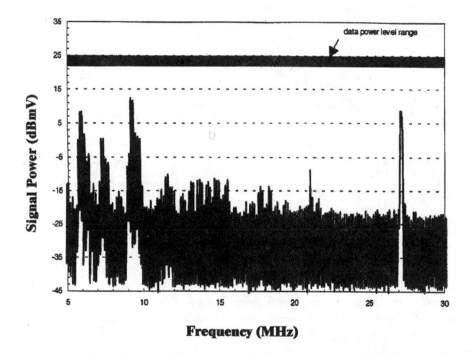

Figure 2.12 Typical return-path spectrum (5-30-MHz) as measured at the cable TV headend, with averaged signal levels (in dBmV). Notice the pronounced ingress peaks, particularly below 10 MHz. After Ref. [22] (© 1995 IEEE).

Another important noise source is common-path distortion, originating from various nonlinearities in the cable TV plant such as oxidized connectors and bad amplifiers. The common-path distortion appears as discrete noise peaks in the return-path spectrum spaced by 6 MHz (8 MHz in PAL systems). This distortion can be contained in properly maintained cable TV plants. Other types of noise sources are the upstream laser transmitter noise and optical receiver noise, which will be discussed in detail later.

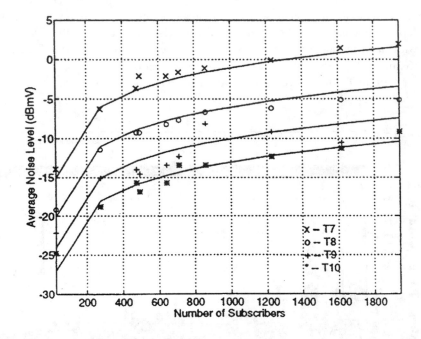

Figure 2.13 Average ingress level (in dBmV) as a function of the number of subscribers for T7, T8, T9, and T10 upstream channel frequencies. After Ref. [24].

The observed increase in the ingress noise levels as the number of homes passed is increased is due to the so-called *noise funneling* effect. This effect is based on the assumption that the unwanted noise signals are located at the subscriber's location with time-dependent amplitude. Since the unwanted signals are uncorrelated, for example, Gaussian noise, the signals sum noncoherently. If the signals from each home passed were equal, then the noise-funneling factor would be $10 \cdot \log_{10}(N)$, where N is the number of homes passed. It turns out that according to the empirically derived expression based on Figure 2.13, the noise-funneling factor is slightly smaller than one obtains from a noncoherent noise power sum.

The time-varying narrowband interferers in the return-path cable systems limits channel availability for high-speed data transmission application. To quantify this parameter, automated spectral measurements of upstream ingress noise were taken every one to five minutes over an extended period of time (48 to 72 hours). The channel availability is the percentage of the time in which a channel of a specified bandwidth is available for transmission for a given modulation format such as QPSK or 16-QAM. For example, to transmit 256-kb/s, using QPSK modulation with FEC over a 192-kHz channel bandwidth, the required carrier-to-noise-plus-interference C/(N+I) must be at least 15.8 dB at BER of 10^{-7}. Thus, if the C/N or CNR (in the presence of Gaussian noise only) is maintained at 21 dB, a single narrowband interferer as high as 10 dB below the equivalent unmodulated carrier (corresponding to C/I = 10 dB) can be tolerated.

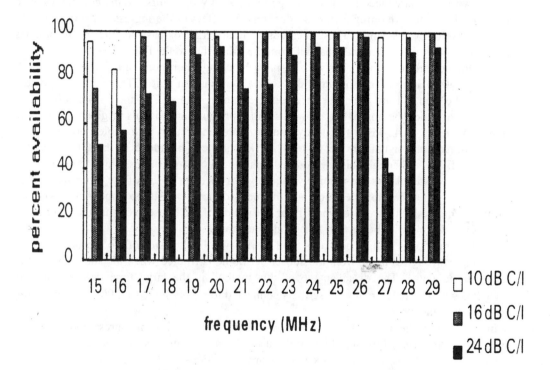

Figure 2.14 Percentage of 1-MHz channel availability in the 15 to 30-MHz frequency range for a 1500 home node. After Ref. [22] (© 1995 IEEE).

In general, the upstream channel availability depends primarily on the number of homes connected to the fiber node, the channel bandwidth, and selected upstream frequency. Based on Figure 2.10, it is clear that by decreasing the number of homes passed in the fiber node the noise funneling effect is reduced. Also, selecting an upstream frequency above, say 15 MHz, is likely to improve the upstream channel availability based on Figure 2.12. Choosing a relatively wide upstream channel band, say 1 MHz or more, is also likely to reduce the channel availability. Figure 2.14 demonstrates, for example, the availability of the 1 MHz channel in the 15- to 30-MHz frequency range of the upstream spectrum for a single fiber node with 1500 homes passed [22]. Notice the reduction in channel availability as the C/I is increased from 10 dB to 24 dB.

2.6.2 Return-Path Noise Filtering

The appearance of ingress noise and other narrowband interference on the return-path cable network has introduced a new challenge for cable TV operators. Many companies have been working on developing methods to combat the effect of the ingress noise from reaching the cable TV headend. These methods include active or passive filtering techniques anywhere along the return-path channel from the subscriber to the fiber node.

One method, for example, consists of using a blocking filter, nominally between 15 to 40 MHz, between an in-home splitter and a coaxial termination unit (CTU) at the side of the home. The signals from the subscriber home terminals such as CM and STB can be transmitted from inside the home, but a low-pass filter prevents any signals originating from the home in the 15 to 42-MHz frequency band to enter the return-path cable network. Upstream-transmitted signals in the 15 to 42-MHz band can be added in the CTU after the blocking filter. Although this method reduces the amount of ingress noise coming from each home, it still allows the relatively high ingress spectral region of the return-path (5 to 15 MHz) to enter the cable network, and the method may be costly to implement. Another proposed method is to use a low-pass filter at the side of the house such that the filter is off except when the subscriber is transmitting data upstream.

Another method is using a bandpass filter with a switch at the side of the subscriber home. The upstream ingress noise from the subscriber home is filtered when the subscriber is not transmitting data. The disadvantage of this method is the large time delays from the switches when there are many subscribers connected to a fiber node ($\approx$1200) who are trying to transmit data upstream.

In addition to the use of filters and similar approaches, upstream ingress and other impairments can largely be controlled by the use of quality materials, proper subscriber drop installation practices, good network maintenance programs, and aggressive signal leakage monitoring and repair [26]. This latter item has the benefit of reduced ingress where a leak exists and outside signals can also enter the network. To make all of these efforts successful requires effective training and quality control programs.

References

1. G. H. Hutson, P. J. Shepherd, and W. S. J. Brice, *Color Television System Principles, Engineering Practice and Applied Technology*, McGraw-Hill, Europe (1990).

2. *NCTA Recommended Practices for Measurements on Cable Television Systems*, 2nd Ed., National Cable Television Association, Washington, D.C. (1993).

3. John D. Lenk, *Lenk's Video Handbook: Operation and Troubleshooting*, McGraw-Hill (1991).

4. *Code of Federal Regulations, Title 47, Telecommunications, Part 76, Section 76.605, Cable Television Service*. Federal Communications Commission Rules and Regulations, Washington, D.C. (1991).

5. *National Association of Broadcasters Engineering Handbook*, 7th Ed., Washington, D.C. (1985).

6. L. Chariglione, "MPEG: A Technological Basis for Multimedia Applications," *IEEE Multimedia* **2**, 85–89 (1995).

7. *MPEG Video Compression Standard*, Edited by J. L. Mitchell, W. B. Pennebaker, C. E. Fogg, and D. J. LeGall, Chapman and Hill, N.Y. (1997).

8. B. G. Haskell, A. Puri, and A. N. Netravili, *Digital Video: An Introduction to MPEG-2*, Chapman & Hall, New York (1996).

9. See http://www.mpeg.org/ Web site for many technical information and news on the different MPEG standards.

10. L.Chiariglione, "MPEG and Multimedia Communications," *IEEE Transaction on CSVT* **7**, 1, (Feb. 1997).

11. U.S. Advanced Television Systems Committee, *Digital Audio Compression (AC-3), draft ATSC standard* (1994).

12. S. Vernon, "Design and Implementation of AC-3 Coders," *IEEE Transaction on Consumer Electronics* **41**, no.3 (1995).

13. ITU-T Recommendation H.261: Video Codec for Audiovisual Services at px64 kbits (1993).

14. ITU-T Recommendation H.263: Video Coding for Low Bit-rate Communication (1995).

15. See http://drogo.cselt.stet.it/mpeg/, which is the official MPEG committee Web site, for information on MPEG-4 and MPEG-7 standards.

16. Electronic Industries Association, "Cable Television Channel Identification Plan," EIA IS-132, May (1994).

17. A. S. Taylor, "Characterization of Cable TV Networks as the Transmission Media for Data," *IEEE Journal on Selected Areas in Communications* **SAC-3**, 255–265 (1985).

18. S. Ovadia and Q. Zhang, "Robust Adaptive Equalization in DVB 64/256-QAM Receiver for CATV Applications," *ICSPAT Proceedings*, **1**, 332–336 (1998).

19. M. Kolber and M. Ryba, "Measuring Multipath in Wireless Cable Environment," *RF Design Magazine* **2**, 52–74 (1999).

20. T. J. Kolze, "Proposed HFC Channel Model," General Instrument IEEE802.14 contribution, IEEE802.14-96/196; Enschede, Netherlands (1996).

21. CableLabs, "Data-Over-Cable Service Interface Specifications (DOCSIS): Radio Frequency Interface Specification," version 1.0 (1997).

22. C. A. Eldering, N. Himayat, and F. C. Gardner, "CATV Return Path Characterization for Reliable Communications," *IEEE Communications Magazine* **8**, 62–69 (1995).

23. CableLabs, "Two Way Cable Television System Characterization," (1995).

24. M. D. Carangi, W. Y. Chen, K. Kerpez, and C. F. Valenti, "Coaxial Cable Distribution Plant Performance Simulation," *SPIE* vol. 2609, 215–226 (1995).

25. F. Edgington, "Preparing for in-service Video Measurements," *Communications Engineering and Design* **6**, 94–100 (1994).

26. R. Hranac, "Mystified by Return Path Activation? Get Your Upstream Fiber Links Aligned," *Communications Technology Magazine*, May (2000).

CHAPTER 3

DIRECTLY MODULATED CABLE TV LIGHTWAVE LASER TRANSMITTERS

Semiconductor laser diodes are the primary light sources in both downstream and upstream transmission over hybrid HFC and the emerging DWDM cable TV networks. However, as the cable TV networks are evolving from the traditional HFC architecture toward DWDM architecture, the laser transmitters must have different performance specifications, depending on the type of traffic being transmitted. Thus, to understand the laser transmitter design and its performance requirements, we shall start with a review of the basic laser physics concepts, including the Fabry-Perot (FP) laser, the distributed-feedback (DFB) laser, and multiple-quantum-well laser diodes. In Section 3.3, we will review the laser modulation characteristics, including the laser rate equations, small-signal frequency response, and equivalent circuit model. The two fundamental laser noise mechanisms, namely, the laser intensity and phase noise are reviewed in Section 3.4. The DFB laser transmitter design, which includes the packaged laser diode module with an optical isolator and a thermoelectric cooler, pre-distortion circuit, as well as its performance requirements for downstream transmission are reviewed in Section 3.5. Finally, Section 3.6 provides an overview on the performance requirements of return-path laser transmitters and their limitation due to their mode-partition noise and mode-hopping noise.

3.1 Semiconductor Laser Diodes

To understand the operation of semiconductor lasers, particularly DFB laser diodes, let us review first the basic laser physics concepts. Then, the different laser diode structures, including gain-guided and index-guided laser diodes as well as their light-versus-current characteristics are discussed.

3.1.1 Basic Laser Physics Concepts

A semiconductor laser diode is essentially a forward-biased p-n junction, which is formed by bringing into physical contact n-type and p-type semiconductor material. An n-type or p-type semiconductor is made by doping it with some impurities whose atoms have an excess valence electron (i.e., donors) or one less electron (i.e., acceptors) compared with the semiconductor atoms. In an n-type semiconductor, the excess electrons occupy the conductor atoms. In an n-type semiconductor, the excess electrons occupy the conduction band states, which are typically empty for an undoped or an intrinsic semiconductor. As the doping level is increased, the quasi-Fermi level[1], which lies in the middle of the bandgap for an intrinsic semiconductor, moves toward the conduction band. Similarly, the quasi-Fermi level moves toward the valence band for the p-type semiconductor. Since the Fermi level must be continuous in thermal equilibrium, electrons and holes must diffuse across the p-n junction, and the ionized impurities create a built-in potential [1]. The space-charged region across the p-n junction, where the number of the mobile carriers is essentially 0, is called the *depletion region*. When the p-n junction is forward biased by an external source, the built-in potential is reduced, resulting in an electric current flow. The current density of a forward-bias p-n junction with voltage V is given by [1]

$$J = J_s \left[\exp(qV / k_B T) - 1 \right]$$

(3.1)

where J_s is the saturation current density, which depends on the diffusion coefficients of both the electrons and the holes. The electrons and the holes, which are simultaneously present in the depletion region, can radiatively recombine to generate photons through either spontaneous or stimulated emission processes. In the spontaneous emission process, which is an incoherent process, the photons are randomly emitted in all directions with no phase relation among them. In contrast, the stimulated emission process is a coherent process, where the emitted photons not only have the same energy as the excitation photon, but also propagate in the same direction. In addition to the radiative recombination processes, electrons and holes can recombine nonradiatively. Nonradiative recombination processes include recombination at trap and defect sites, surface, and Augur recombination [2]. In Augur recombination, which is important in long-wavelength semiconductor lasers, the electron-hole recombination energy is given to another electron or hole as kinetic energy rather than being emitted as photons. The internal quantum efficiency of the p-n junction, which is given by [2]

$$\eta_{int} = \frac{R_{rr}}{R_{rr} + R_{nr}} = \frac{\tau_{nr}}{\tau_{rr} + \tau_{nr}}$$

(3.2)

is the radiative recombination rate (R_{rr}) relative to the total recombination rate. The radiative and nonradiative recombination times are defined as N/R_i, where N is the carrier density, and i = rr or nr. In direct bandgap semiconductors such as GaAs, where the conduction band minimum and the valence band maximum occur at the same electron wave vector ($\vec{k}$), the probability of radiative recombination is large ($\eta_{int} \approx 1$). In contrast, $\eta_{int} \ll 1$ for indirect bandgap semiconductors such as Si or Ge.

---

[1] The Fermi level is the energy at which the probability of occupation by an electron is equal to ½.

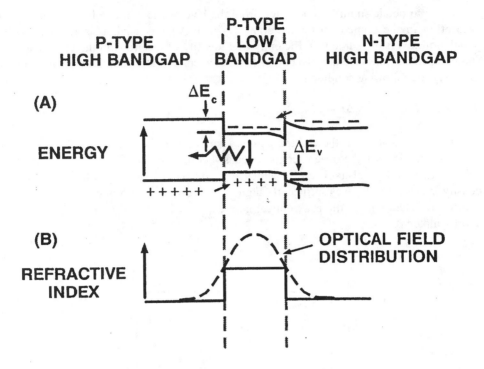

Figure 3.1 Schematic of the simultaneous confinement of the charged carriers and the generated optical mode in a double-heterostructure design. After Ref. [3] (© 1991 IEEE).

So far, we have discussed p-n junctions that are homojunctions, namely, the same semiconductor material is used on both sides of the junction. The problem with homojunctions is that the mobile carriers are not confined to the immediate vicinity of the junction due to diffusion, resulting in lower carrier densities. Inserting a thin semiconductor layer between the p-type and n-type layers with the same lattice constant but with a smaller bandgap and higher index of refraction solves this problem. Such a semiconductor structure is called a heterostructure. The thin semiconductor layer with a typical thickness of about 0.3 μm or less is called the active layer since light is generated inside it. The active layer provides two important functions. First, it confines the electrons and the holes due to the bandgap energy difference between the active layer and the surrounding layers, which are called cladding layers. Second, it also confines the optical modes like a dielectric waveguide since its index of refraction is higher than the cladding layers. The number of optical modes can be con trolled by the active layer thickness. Figure 3.1 illustrates the simultaneous confinement of the charged carriers and the generated optical mode to the active layer in a double-heterostructure (DH) design [3].

In order to obtain stimulated emission from the heterostructure, one has to overcome the absorption in the semiconductor material. To achieve this goal, it is required to have population inversion, namely, sufficiently high density of electrons in the conduction band and holes in the valence band. To obtain stimulated emission, the emitted photon energy has to satisfy the following condition

$$E_g < h\nu < E_{Fc} - E_{Fv} \equiv \Delta E_F \tag{3.3}$$

Equation (3.3) states that the emitted photon energy has to be larger than the bandgap energy, but smaller than the difference in the quasi-Fermi level energies in the conduction and valence bands. Optical gain can be achieved in the active layer when the injected carrier density in the active layer is sufficiently high to overcome the absorption. When the optical gain is sufficiently high, the peak optical gain dependence on the carrier density N can be approximated as [2]

$$g_p = g_0 \left[N - N_0 \right] \tag{3.4}$$

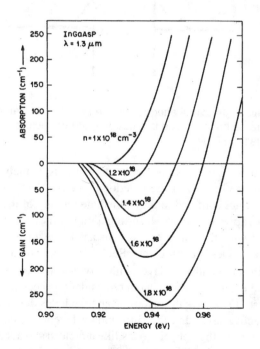

Figure 3.2 Calculated optical gain coefficient versus the photon energy at various carrier densities for 1.3-μm InGaAsP active layer, after Ref. [4] (© 1980 American Institute of Physics).

where N_0 is the carrier density at transparency, which is typically in $1-1.5 \cdot 10^{18}$ cm^{-3} for InGaAsP lasers, and g_0 is the differential gain coefficient. In general, the optical gain is a function of both the carrier density and the photon energy g(N,ν). Figure 3.2 shows the calculated optical gain at various carrier densities as a function of the emitted photon energy for 1.3-μm InGaAsP active layer [4,5]. The differential gain coefficient can be increased by at least a factor of two in multiple quantum well (MQW) laser diodes (see Section 3.2.2).

Thus, we learned that in order to obtain optical gain, the injected carrier density has to be sufficiently high. However, a semiconductor laser diode is not an optical amplifier, but an optical oscillator. Thus, the optical feedback mechanism is required to achieve optical oscillations. The optical feedback is provided by cleaving both facets of the semiconductor laser, and applying high-reflectivity coatings to one or both facets. An optical cavity, which is formed by placing a gain medium between two optical mirrors, is often called a *Fabry-Perot* (FP) cavity. In the case of semiconductor lasers, the cleaved facets act as mirrors with reflectivity of $(n-1/n+1)^2$ for $n = 3.5$, or about 30% reflectivity.

To achieve lasing, the two following conditions must be satisfied:

- The optical gain in the cavity has to be equal to the loss.
- Optical feedback is needed, which is provided by the cleaved facets in FP lasers.

The laser threshold condition is the minimum optical gain that is necessary to overcome the FP cavity losses and operate the laser. To obtain the laser threshold condition, let us assume an electric field with amplitude E_0, angular frequency ω, and propagation constant β = $n\omega/c$, inside a cavity with length L, and mirror reflectivities of R_1 and R_2. During one cavity round trip, the electric field amplitude changes by $(R_1 \cdot R_2)^{1/2} \exp(-\alpha_i L)$, where α_i is the internal loss of the cavity due to free-carrier absorption and scattering. At the same time, the amplitude of the electric field also increases as $\exp[g/2(2L)]$, where g is the optical gain of the cavity, and its phase changes by $2\beta \cdot L$. Thus, after one round trip and in steady state, the electric field should remain unchanged, namely

$$E_0 = E_0 \exp(g_{th}L)\sqrt{R_1 R_2} \, \exp(-\alpha_i L) \cdot \exp(2j\beta L) \tag{3.5}$$

From the real part of Equation (3.5), we obtain the first lasing condition at threshold:

$$g_{th} = \alpha_i - \frac{1}{2L}\ln[R_1 R_2] = \alpha_{cavity} \tag{3.6}$$

Equation (3.6) states that the cavity losses must equal the laser gain at threshold. For GaAs lasers with cavity length $L = 300$-μm, $\alpha_i = 10$ cm^{-1}, $R = 0.36$, the threshold gain is 44 cm^{-1}. The second lasing condition is obtained from the imaginary part of Equation (3.5):

$$2\beta_m L = 2m\pi \tag{3.7}$$

where m is an integer. Thus, the laser wavelength must satisfy the following condition:

$$\lambda_m = 2nL / m \tag{3.8}$$

Equation (3.8) defines the longitudinal modes or the FP modes of the semiconductor laser. The longitudinal mode spacing is given by

$$\Delta\lambda = \frac{\lambda^2}{2[n - \lambda(dn / d\lambda)]L} = \frac{\lambda^2}{2n_g L} \cong \frac{\lambda^2}{2nL} \tag{3.9}$$

where n_g is the group index, which can be approximated to the index of refraction of the active layer if its wavelength dependence can be ignored. Using the above example for GaAs lasers with λ = 850-nm and n = 3.5, the mode spacing is 0.385-nm. Since the gain bandwidth of a FP semiconductor laser is very broad ($\cong$ 30-nm), it can support many longitudinal modes, where each mode can carry a significant amount of the optical power relative to the main mode. Such a laser is often called a *multimode* semiconductor laser. When the light from a multimode laser is coupled into a SMF, each of the longitudinal modes propagates at slightly different speed due to the fiber dispersion, resulting in a limited propagation distance. Consequently, single-wavelength semiconductor lasers are needed, which are discussed in Section 3.2.

3.1.2 Semiconductor Laser Structures

Semiconductor laser diodes can be divided into two main categories: gain-guided and index-guided lasers. Let us first discuss the structures of gain-guided lasers. Figure 3.3 shows, for example, gain-guided semiconductor laser in a stripe geometry [4]. In a simple DH semiconductor laser, there is no mechanism for lateral confinement of the injected carriers in the active layer. Consequently, the threshold current, which is proportional to threshold current density (J_{th}) times the active layer area (LxW), can be as high as 300-mA. One of the approaches to solve this problem was to deposit an oxide stripe (SiO_2) on top of the p-type layer where carriers can be injected [7]. Another approach, for example, was to deposit an n-type layer on top of the p-layer and diffuse Zn over the central region, which converts the n-type to a p-type layer [8]. Thus, the narrow strip ($\leq$ 5 μm) along the cavity laterally confines the injected carriers and the optical modes. These types of semiconductor laser designs have several problems. The stripe geometry leads to a spatially varying carrier density in the lateral direction, which is determined by the carrier diffusion. In addition, the optical gain, which peaks at the center of the stripe, may be significantly reduced beyond the stripe to large carrier absorption. Since the confinement of the optical modes is governed by the

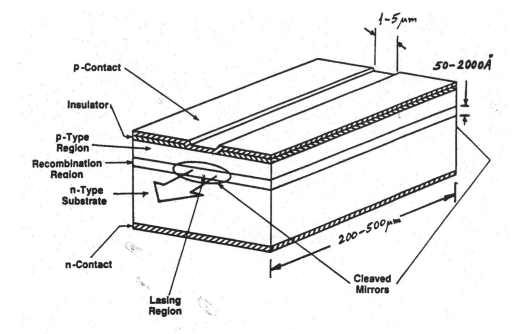

Figure 3.3 A Gain-guided semiconductor laser in a stripe geometry showing its main
features, including the lasing region, n-type and p-type regions and contacts, and the cleaved mirrors. After Ref. [6].

optical gain, such lasers are called gain-guided semiconductor lasers. One of the main disadvantages of gain-guided lasers is their light-versus-current nonlinearity characteristic as
the current is increased. This nonlinearity, which has been attributed to a transition to higher
order modes, limits the maximum amount of optical power that can be coupled into
a SMF [2].

In index-guided lasers, lateral confinement of the injected carriers and the optical modes is
achieved by forming a waveguide in the lateral direction with refractive index distribution.
Index-guided lasers can be subdivided into two classes: weakly index-guided and strongly
index-guided semiconductor lasers. An example of a weakly index-guided design is a *ridge-
waveguide* laser, where the ridge is formed by etching parts of the p-type layer and depositing an SiO_2 layer to block the current flow and create the index guiding. The lateral index
guiding is because of the higher index of refraction of the ridge-waveguide (p-type layer)
compared with the surrounding SiO_2 layer. Although such designs produce a refractive index change of only about 0.01, they are attractive due to their relatively simple design. An
example of strongly index-guided design is a *buried-heterostructure* (BH) laser, where the
active layer is buried from all sides by layers with a lower index of refraction [2,9]. BH
semiconductor lasers typically exhibit a very linear light-versus-current characteristic with

relatively low threshold ($\approx$ 10-mA), modulation bandwidth in excess of 2 GHz, and can operate in the fundamental transverse mode even at high power.

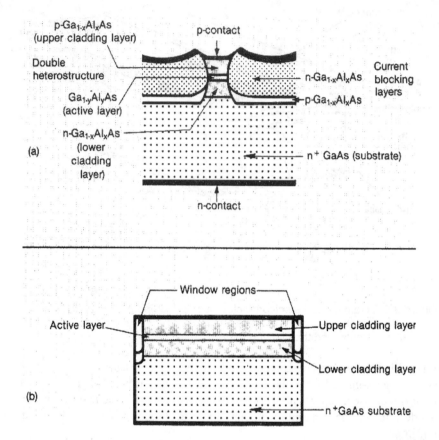

Figure 3.4 (a) Cross section, and (b) side view of window buried-heterostructure semiconductor laser. After Ref. [9].

3.1.3 Light-Current Characteristic

Generally speaking, the most fundamental characteristic of a laser diode is its output optical power versus the injected current or the L–I characteristic. Figure 3.5 shows the measured L–I curves (CW) at operating temperatures ranging from 10°C up to 130°C for a 1.3 µm double-channel planar BH (DCPBH) InGaAsP laser [2]. Two important observations can be made. First, the laser threshold current at room temperature degrades significantly from about 20-mA to more than 100-mA as the operating temperature is increased. In fact, the threshold current was found experimentally to increase exponentially with temperature as:

$$I_{th}(T) = I_0 \exp[T / T_0] \tag{3.10}$$

where I_0 is a constant, and T_0 is a characteristic temperature often used to describe the temperature sensitivity of the threshold current. For example, T_0 is typically in the range of 50 to 70°K for InGaAsP lasers compared with 100–150°K for GaAs lasers. The exponential increase of the threshold current with temperature is primarily attributed to the reduction of the optical gain with temperature (gain $\propto 1/T$). Because of the high temperature sensitivity of InGaAsP lasers, their temperature is often controlled in a laser transmitter system using a thermoelectric cooler (see Section 3.5.3).

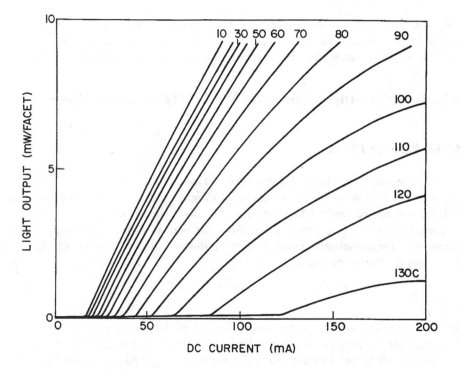

Figure 3.5 Measured L–I curve at various operating temperatures for 1.3-μm InGaAsP DCPBH semiconductor laser. After Ref. [2] (with kind permission from Kluwer Academic Publishers).

The second important observation from Figure 3.5 is the slope (dP/dI) of the $L–I$ curve, which is being considerably reduced as the operating temperature is increased. Above threshold ($I > I_{th}$), the slope efficiency is given by [2]

$$\frac{dP}{dI} = \frac{hv}{q}\eta_D = \frac{hv}{q}\left[\frac{\alpha_i\alpha_m}{\alpha_i + \alpha_m}\right] \tag{3.11}$$

where η_D is the so-called differential quantum efficiency, which measures the efficiency of the laser to convert current to output light, and αm is the mirror loss. Using Equation (3.6), the differential quantum efficiency can be written as

$$\eta_D = \eta_{int}\frac{\ln[1/R_1R_2]}{2\alpha_iL + \ln[1/R_1R_2]} \tag{3.12}$$

As the operating temperature is increased, the internal quantum efficiency as defined in Equation (3.2) is reduced due to the exponential temperature dependence of the nonradiative recombination processes such as Auger recombination and leakage currents in BH lasers

3.2 DFB and Multiple-Quantum-Well (MQW) Laser Diodes

3.2.1 DFB Laser Diodes

As we have learned in Section 3.1.1, the cleaved facets of the semiconductor cavity provide the optical feedback in FP lasers. Another type of optical feedback is a *distributed feedback* (DFB) throughout the semiconductor cavity. The distributed feedback is achieved by periodic perturbation of the refractive index from the built-in diffraction grating. The DFB mechanism is a mode-selective mechanism that allows oscillation to occur only for the mode that satisfies the Bragg condition:

$$\lambda_B = \frac{2n\cdot\Lambda}{m} \tag{3.13}$$

where n is the refractive index of the mode, Λ is the grating spatial period, and m is an integer representing the Bragg diffraction order. Typically $m = 1$ is used because of the strongest coupling between the forward and backward oscillating modes, but sometimes $m = 2$ is used. For a 1.55-μm DFB laser, the gating period is only 235 nm for $m = 1$ and $n = 3.3$.

The grating in DFB lasers is typically etched onto either side of the active layer, creating periodic variations in the waveguide layer next to the active layer, which translate to refractive index modulation of the propagating modes along the cavity. It should be pointed out that the cleaved facets of the DFB laser act as mirrors whose reflectivity is maximized for the Bragg wavelength that satisfies Equation (3.13).

In order to obtain a single longitudinal operation, the cavity losses have to be significantly higher for the side modes than the main mode. The longitudinal mode with the smallest cavity losses can reach threshold and becomes the dominant mode. However, depending on the coupling coefficient between the forward and backward modes over the grating length (κL), two dominant modes may coexist in the DFB laser cavity. DFB lasers with two competing longitudinal modes with a typical power difference of about 10 dB are undesirable from an optical transmission point of view. To discriminate between the two modes, the grating is shifted by $\lambda_B/4$ in the middle of the laser cavity to produce a $\pi/2$ phase shift to one of the modes. The $\pi/2$ phase shift detunes the Bragg wavelength from the gain peak, thus allowing only one mode to exist. Furthermore, the mode detuning is adjusted on the shorter wavelength side of the gain peak in order to obtain a higher differential gain. These types of DFB lasers are called $\lambda/4$-shifted DFB lasers [2].

3.2.2 Multiple-Quantum-Well (MQW) Lasers

In conventional DH semiconductor lasers, the active layer thickness (0.1–0.3 µm) is sufficiently thin to confine both the carriers and the optical mode, but its electronic and optical properties remain the same as the bulk semiconductor material. If the active layer thickness is further reduced to about 10 nm or less, its electronic and optical properties are drastically changed since the three-dimensional free-electron motion is reduced to two-dimensional motion. Consequently, the energy levels of the confined carriers are quantized in the z-direction, and given by

$$E_n = \frac{h}{4\pi m^*}\left[\frac{n\pi}{L_z}\right]^2 \tag{3.14}$$

where h is Planck's constant, m^* is the effective mass of the carrier, L_z is the quantum-well thickness, and n is an integer. The carriers are free to move in the x and y directions. The quantization of the energy levels changes the density of states from a parabolic dependence to a step-like behavior as shown in Figure 3.6 [3]. Notice that the density of states is constant at each energy sub-band, rather than gradually increasing from zero. This means that there is a group of electrons with nearly the same energy that is available to recombine with a group of holes of nearly the same energy. Therefore, a much higher optical gain can be achieved in these MQW structures than in conventional DH designs. Consequently, the required carrier density to reach threshold (N_0) is lower, resulting in very low threshold currents.

Strained-layer (SL) QW lasers are an important class of QW lasers with an active layer whose lattice constant is slightly different from the cladding layers and the substrate. These lasers, which have been extensively investigated over the last decade, exhibit many desirable characteristics, including very low threshold currents, narrower CW and modulated linewidth than conventional QW lasers, high relaxation oscillation frequency, and low nonlinear distortions [3]. In fact, threshold currents as low as 1-mA or less have been demonstrated for 1.3-μm SL-MQW InGaAsP semiconductor lasers operating at room temperature [10, 11]. The improved laser characteristics have been explained by the changes in the band structure caused by the lattice mismatch-induced strain. The induced strain splits the heavy-hole and the light-hole valence bands at the Γ point of the Brillouin zone, which is the minimum bandgap energy in the direct bandgap semiconductors [3]. One of the widely used material systems for SL-MQW lasers is $In_xGa_{1-x}As$ QW layers with InGaAsP barriers whose composition is lattice match to InP. For x < 0.53, the QW active layer is under a tensile stress, while for x > 0.53, the QW active layer is under a compressive stress.

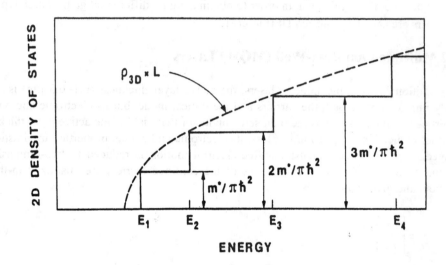

Figure 3.6 Density of states of a two-dimensional confined carrier in a quantum well compared to that of a bulk semiconductor. After Ref. [3] (© 1991 IEEE).

3.3 Laser Dynamic Characteristics

Until now, only the static characteristics of semiconductor lasers have been discussed. In the following subsections, we will examine the dynamic characteristics of semiconductor lasers and define key parameters that play an important role in the laser transmitter design for cable TV transmission. First, the temporal behavior of a semiconductor laser is analyzed using the modal rate equations. The aim of this section is to summarize some of the important results and concepts rather than to present detailed derivations of the various equations.

3.3.1 Small-Signal Response

The temporal behavior of semiconductor lasers is governed by the laser rate equations, which describe the interaction between the injected carriers and the photons inside the active region. Let us assume a single-longitudinal laser diode such as DFB laser with uniform distribution of electrons with density N over the volume $V (= dxLxW)$ of the active layer, and photon density S for the lasing mode. The single-mode rate equations can be written as [2]:

$$\frac{dN}{dt} = \frac{I}{qV} - \frac{N}{\tau_n} - \Gamma v_g g_0 (N - N_0)(1 - \varepsilon S)S \tag{3.15}$$

$$\frac{dS}{dt} = \Gamma v_g g_0 (N - N_0)(1 - \varepsilon S)S - \frac{S}{\tau_p} + \Gamma \gamma_{sp} \frac{N}{\tau_n} \tag{3.16}$$

where τ_n and τ_p are the electron and photon lifetimes, Γ is the carrier confinement factor in the active layer, v_g is the group velocity, γ_{sp} is the fraction of the spontaneous emission coupled into the cavity mode. In Equation (3.15), the electron density rate of change is proportional to (A) the injected electron rate, (B) the electron-density rate reduction due to spontaneous emission, and (C) the rate reduction due to stimulated emission of the lasing mode. Notice that the mode gain has been modified from Equation (3.4) to include the effect of gain compression. Without gain compression, the net rate of peak gain due to stimulated emission is given by

$$G = v_g g_0 (N - N_0) \tag{3.17}$$

However, due to an important phenomenon called spectral-hole burning (SHB), the optical gain is compressed, requiring modifications to Equation (3.17). SHB is a small-localized reduction of the optical gain curve at the lasing mode wavelength [12]. It occurs when the rate of stimulated emission is very high and the large number of available electrons in the conduction band cannot immediately fill the localized reduction in the number of conduction band states. This is because the intraband scattering processes such as electron-electron and electron-phonon scattering have a relaxation time on the order of 1-ps [13]. The gain rate reduction caused by SHB is proportional to the stimulated emission rate, and can be written as

$$\Delta G = v_g g_0 (N - N_0) S \cdot \varepsilon \tag{3.18}$$

where ε is the gain compression coefficient, which is also called the nonlinear gain parameter. Thus, the total net gain rate is given by

$$G(n, S) = v_g g_0 (N - N_0) - \Delta G = v_g g_0 (N - N_0)(1 - \varepsilon S) \tag{3.19}$$

The rate of change for the photon density in Equation (3.16) is proportional to (A) the fraction Γ of photons confined in the active layer due to stimulated emission, (B) the reduction of the photon density due to internal absorption or transmission through the facets, and (C) the rate increase of the photon density caused by the coupling of the spontaneous emission into the lasing mode. Table 3.1 summarizes the typical values of DFB laser diodes that are used for cable TV transmission.

Table 3.1 Typical values for 1.3-μm InGaAsP DFB laser diodes.

Parameter	Typical Values
Active Layer Volume	2-μmx0.2-μmx300-μm
Electron Lifetime (τ_n)	1–3-ns
Photon Lifetime (τ_p)	1–3-ps
Optical Mode Confinement Factor (Γ)	0.3–0.5
Electron Density at Transparency (N_0)	$1-1.5 \times 10^{18} \ cm^{-3}$
Gain Compression Coefficient (ε)	$1-5 \times 10^{-17} \ cm^{-3}$
Differential Gain Coefficient (g_0)	$2-3 \times 10^{-16} \ cm^2$

Due to the nonlinear behavior of the rate equations, it is necessary to solve them numerically. However, analytical solution can be obtained for the small-signal modulation response. The small-signal response solution is important for cable TV applications since it reveals the design parameters that need to be controlled in order to obtain high modulation bandwidth of DFB laser diodes. To obtain an analytical solution, let us assume that the general solution to the rate equations is in the form

$$N = n_0 + n e^{j\omega t}$$
$$S = S_0 + s e^{j\omega t} \tag{3.20}$$
$$I = I_0 + i e^{j\omega t}$$

where ω is the angular frequency of the modulating signal, and n_0, S_0, and I_0 are the steady-state values for the electron density, photon density, and current modulation, respectively. Substituting Equation (3.20) into the rate equations, equating terms on both sides of the rate equations with exp($j\omega t$), and neglecting second-order small-signal products, the intensity modulation response is given by [14]

$$H(\omega) = \frac{s(\omega)/S_0}{i(\omega)/I_0} = \frac{B\omega_R^2}{B\omega_R^2 - \omega^2 + j\omega\left[\dfrac{\Gamma\gamma_{sp}I_{th}}{qVS_0} + \dfrac{1}{\tau_n} + S_0\left(v_g g_0 + \dfrac{\varepsilon}{\tau_p}\right)\right] + \dfrac{\Gamma\gamma_{sp}I_{th}}{qVS_0 \cdot \tau_n} + \dfrac{\gamma_{sp} + \varepsilon S_0}{\tau_n\tau_p}}$$

$$(3.21)$$

where I_{th} is the threshold current, $B = 1 - \varepsilon S_0$, and ω_R is the so-called laser resonance frequency or relaxation oscillation frequency, which is given by

$$\omega_R = \left[\frac{v_g g_0 S_0}{\tau_p}\right]^{1/2} = \left[\frac{\Gamma v_g g_0}{qV}(I - I_{th})\right]^{1/2}$$

$$(3.22)$$

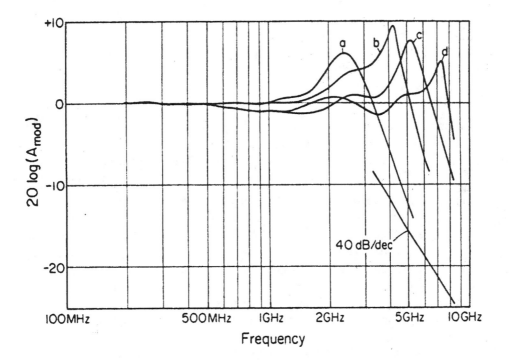

Figure 3.7 Measured modulation characteristics of 120-μm cavity of BH laser on semi-insulating substrate at optical output powers of (a) 1-mW, (b) 2-mW, (c) 2.7-mW, and (d) 5-mW. After Ref. [15] (© 1985 IEEE).

According to Equation (3.21), the small-signal modulation response of a semiconductor laser behaves as a standard second-order low-pass transfer function with a resonance frequency ω_R and with a damping factor, whose magnitude is given by the coefficient of the $j\omega$ term. For laser operation above threshold, $\gamma_{sp} = 0$, the $1/\tau_n$ and $S_0 g_0$ terms in the damping coefficient are small and can be neglected, and Equation (3.21) simplifies to

$$H(\omega) = \frac{\omega_R^2}{\omega_R^2 - \omega^2 + j\omega\left[S_0\varepsilon / \tau_p\right]} \qquad (3.23)$$

The electrical power spectrum of the small-signal frequency response is given by

$$|H(\omega)|^2 = \frac{B^2\omega_R^4}{\left[B\omega_R^2 - \omega^2\right]^2 + \omega^2\left[S_0\varepsilon / \tau_p\right]^2} \qquad (3.24)$$

In most lasers the damping coefficient is much smaller than the resonance frequency. Consequently, the 3 dB modulation frequency can be obtained from Equation (3.24) as [15]

$$\omega_{3dB} = \omega_R\sqrt{1+\sqrt{2}} \cong 1.55\omega_R \qquad (3.25)$$

where the approximation $B^{1/2}\omega_R \approx \omega_R$ was used in Equation (3.24). If, on the other hand, the damping coefficient is large, the height of the resonance peak will be reduced, and its frequency will be different than ω_R. Notice that the 3 dB modulation frequency is proportional to $(I-I_{th})^{1/2}$, which has been verified by many experimental results [16]. In addition, from Equation (3.24), the power spectrum at the resonance frequency is inversely proportional to S_0, which means that the resonance peak is reduced due to gain saturation as the output power is increased. Figure 3.7 shows, for example, the measured modulation characteristics of 120-μm cavity of BH semiconductor laser on semi-insulating substrate at optical output powers of (a) 1-mW, (b) 2-mW, (c) 2.7-mW, and (d) 5-mW [15]. Notice that the low frequency region (up to 1 GHz) of the frequency response of the BH laser is essentially flat. This region can be used for SCM lightwave transmission. In addition, above the resonance frequency, the magnitude of the frequency response approaches asymptotically a slope of −40-dB/decade.

It is well known that the laser behavior is highly nonlinear near its resonance frequency. For SCM lightwave transmission, the RF carriers must be sufficiently far away from the resonance frequency in order to minimize any dynamically induced nonlinear distortions. Therefore, the semiconductor laser designer must find ways to expand the laser modulation bandwidth, increasing the frequency separation between the operating frequency range and the resonance frequency. According to Equation (3.22), the laser modulation bandwidth can be increased by (A) increasing the output power by operating at higher bias current, (B) increasing the differential gain coefficient g_0, and (C) reducing the active layer volume. In fact, recent results have shown that SL-MQW DFB lasers have 30% and 90% higher resonance frequency compared with MQW DFB lasers and conventional DFB lasers, respectively.

3.3.2 Large-Signal Response Circuit Model

Generally speaking, it is convenient to model the high-speed characteristics of a packaged semiconductor laser as a two-port system as shown in Figure 3.8 [14]. It consists of three subsections: (A) the package or mount parasitics, (B) the parasitics associated with the semiconductor chip, and (C) the intrinsic laser namely, the active region and the cavity. The package parasitics include, for example, bond-wire inductance and some capacitance between the input terminals, which can be entirely eliminated by monolithically integrating the laser with the drive circuit elements, but can also behave as nonlinear elements for some types of drive circuitry. The laser chip parasitics are the stray capacitance and resistance of the various semiconductor layers surrounding the active region, which typically behave as linear current in cases such as input step current waveform. The laser chip parasitics, which greatly vary among different laser designs, are diverting the high-frequency components of the drive current away from the intrinsic laser. Figure 3.9 illustrates, for example, the equivalent circuit model of the package and chip parasitics for a 1.3-μm InGaAsP etched mesa BH laser [14]. The chip parasitics take the form of a resistance in series combined with a shunt capacitance. L_p and R_p represent the bond-wire inductance and resistance, while C_s and R_s are the parasitic capacitance and resistance of the intrinsic laser. The current source I_L models the DC leakage current around the active region, and R_{IN} is the source resistance.

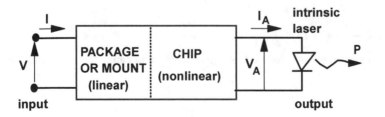

Figure 3.8 Two-port model of semiconductor laser with the package and chip parasitics. After Ref. [14] (© 1985 IEEE).

The small-signal analysis, which was introduced in the previous section, does not apply to a directly modulated DFB laser transmitter for cable TV applications. For a large-signal

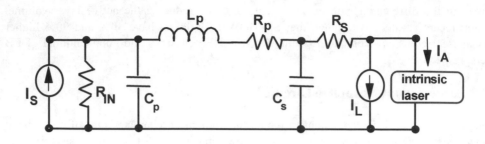

Figure 3.9 Equivalent circuit model of the laser's package and chip parasitics. After Ref. [14] (© 1985 IEEE).

response, the rate Equations (3.15) and (3.16) must be solved numerically. Another convenient method is to transform the rate equations into an equivalent circuit model, which can be solved using standard circuit-analysis techniques. Various authors have developed a number of large-signal models [14, 17]. The following example demonstrates the circuit-modeling methodology for a large-signal response. Figure 3.10 shows the equivalent circuit model of an intrinsic semiconductor laser [14]. To transform the rate Equations (3.15) and (3.16) to the circuit model, the equivalent spontaneous recombination current is defined as:

$$I_{spon} = \frac{qV \cdot N}{\tau_n} = I_s \exp(qV_A / \theta k_B T) \qquad (3.26)$$

where $\theta \cong 2$, and I_s the saturation current. The equivalent stimulated emission current is defined as

$$I_{stim} = qV \cdot v_g g_0 (N - N_0)(1 - \varepsilon S) \qquad (3.27)$$

The photon loss and storage are modeled as a resistor R_{ph} and capacitance C_{ph} defined as

$$R_{ph} = \frac{\tau_p}{qV} \qquad (3.28)$$

$$C_{ph} = qV \qquad (3.29)$$

Using the above definitions, the rate Equations (3.15) and (3.16) become

$$\tau_n \frac{dI_{spon}}{dt} = I - I_{spon} - I_{stim} \qquad (3.30)$$

$$C_{ph} \frac{dS}{dt} = I_{stim} - \frac{S}{R_{ph}} + \gamma_{sp} I_{spon} \tag{3.31}$$

Notice that the diode shown in Figure 3.10 models the spontaneous recombination current. The stimulated emission current as well the other current sources in Figure 3.10 model dN/dt in Equation (3.15). The stimulated emission current as well as the spontaneous emission coupled into the lasing mode is modeled by two different current sources.

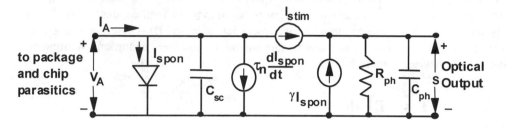

Figure 3.10 Large-signal circuit model of intrinsic laser. After Ref. [14] (© IEEE 1985).

Direct amplitude modulation of a semiconductor laser is always accompanied by a phase modulation. The variation of the electron density (N) in the active region during the current modulation also produces a change in the refractive index in the gain medium. The parameter α, which is called the linewidth enhancement factor, is defined as [18, 19]

$$\alpha = -\frac{4\pi}{\lambda} \left[\frac{\left(\frac{dn}{dN}\right)}{\left(\frac{dg}{dN}\right)} \right] \tag{3.32}$$

which is the ratio of the changes in the real part to the imaginary part of the refractive index. Thus, as the optical power varies with the current modulation, it produces changes in both the real and imaginary parts of the refractive index, dynamically shifting the optical frequency of the longitudinal mode. The so-called frequency chirp is approximately related to the time-dependent optical power as [20]

$$\Delta v = -\frac{\alpha}{4\pi} \left[\frac{d}{dt} \ln P(t) + \kappa P(t) \right] \tag{3.33}$$

where $\kappa = 2\Gamma\varepsilon/V\eta_d h\nu$. According to Equation (3.33), the frequency chirp consists of a "transient chirp" (first term) that arises from the frequency relaxation oscillation during the onset of the laser pulse, and an "adiabatic chirp" (second term) that arises from the damping of the relaxation oscillation. The result is a wavelength shift between the high and low power points in the optical waveform. At high modulation frequencies, lasers with strong damping are likely to have good chirp characteristics due to the reduction in the rate of change of $ln[P(t)]$. However, excessive damping may lead to a reduced modulation band-width. Consequently, the laser designer must compromise between the amount of damping of the relaxation oscillation and the "transient chip." As we will see in Chapter 9, the frequency chirp of directly modulated DFB lasers has important implications for cable TV transmission. The interaction between the frequency chirp of DFB lasers and the fiber chromatic dispersion produces an unacceptable amount of nonlinear distortions, limiting the transmission distance that can be used.

3.4 Noise in Laser Diodes

The subject of noise in semiconductor lasers has an important implication for cable TV transmission. The following sections will discuss the two fundamental noise mechanisms in the laser output, namely, its intensity and phase noise. The laser intensity and phase noise can be analyzed by adding the so-called *Langevin* noise sources to the single-mode rate equations [21, 22]. The purpose of the following subsections is to summarize some of the important results and their implications for laser transmitters in cable TV networks.

3.4.1 Relative Intensity Noise (RIN)

The output power of semiconductor laser fluctuates around its steady-state value due to quantum fluctuations in the electron density as well as spontaneous emission events that are converted to intensity noise. The statistical properties of the laser noise can be studied by adding to each of the rate equations the corresponding Langevin noise source. Assuming a spatially uniform laser, which is driven by Langevin forces F_N and F_S for the carrier density and photon density, respectively, the stochastic rate equations can be written as follows [21, 22]:

$$\dot{N} = I/qV - G(N)S - \gamma_e N + F_N(t)$$
$$\dot{S} = G(N)S - \gamma_p S + R_{sp} + F_S(t)$$

(3.34)

where $\gamma_e = 1/\tau_N$, $\gamma_p = 1/\tau_p$, and $R_{sp} = \gamma_{sp} n_{sp} N/\tau_N$ is the spontaneous emission rate. The solution to the stochastic rate equations is considerably simplified if the Langevin forces are assumed to be Gaussian random processes with zero mean, and under the Markovian approximation satisfies the following relation [21]:

$$\left\langle F_i(t)F_j(t')\right\rangle = 2D_{ij}\delta(t-t')$$ (3.35)

where the angle brackets denote ensemble average, and D_{ij} are the corresponding diffusion coefficients. It can be shown using a rigorous analysis that the nonzero diffusion coefficents are $D_{SS} = R_{sp}S$, and $D_{NN} = R_{sp}S + \gamma_N N$ [2], where N and S are the steady-state values. With these assumptions, the stochastic rate equations can be solved in the frequency domain using Fourier analysis. The laser relative intensity noise (RIN) is defined as [2]

$$RIN = \frac{\left\langle |\delta S(\omega)|^2\right\rangle}{S^2}$$ (3.36)

where δS is the output power fluctuation from the average power value S. The laser RIN can be written as [2]

$$RIN(\omega) = \frac{2R_{sp}\left[\Gamma_N^2 + \omega^2 + G_N^2 S^2\left(1 + \frac{\gamma_N N}{R_{sp}S}\right)\right]}{S\left[\left(\omega_R^2 - \omega^2\right)^2 + 2\Gamma_R^2\left(\omega_R^2 + \omega^2\right) + \Gamma_R^4\right]}$$ (3.37)

where $\Gamma_R \equiv (\Gamma_N + \Gamma_S)/2$ is the relaxation oscillation decay rate, $\Gamma_N = \gamma_N + N(\partial\gamma_N/\partial N) + G_N S$ is the small-signal decay rate, $\Gamma_S = R_{sp}/S - G_S S$ is the photon decay rate, $G_N = \Gamma v_g[\partial g/\partial N]$, $G_S = \Gamma v_g[\partial g/\partial S]$. Figure 3.11 illustrates, for example, the RIN spectra of a self-pulsating ridge-waveguide QW laser with a 15-μm absorber at bias currents of 48% and 90% above threshold [23]. At low frequencies ($\omega \ll \omega_R$), the laser RIN is almost frequency independent, but it is significantly enhanced in the vicinity of $\omega = \omega_R$. At a given frequency, the RIN decreases with the bias current as $(I-I_{th})^{-3}$. As the bias current is increased, the RIN decreases more slowly as $1/(I-I_{th})$. As we will discuss in Chapter 5, the laser RIN imposes an upper limit on the maximum achievable CNR at the fiber node receiver. Consequently, the RIN of DFB laser transmitters, which are used for analog video transmission, are typically equal to −155 dB/Hz or better. It should be pointed out that the laser RIN can significantly degraded by multiple optical reflections [24]. This issue is discussed in detail in Section 9.3.

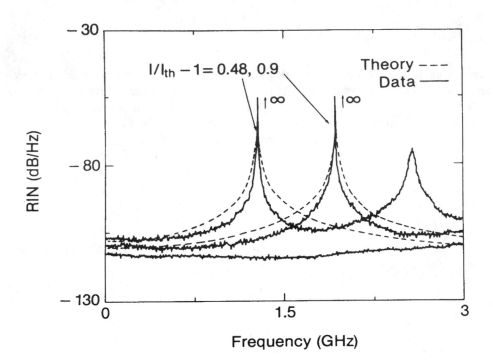

Figure 3.11 Measured and calculated RIN spectra of a self-pulsating ridge-waveguide QW laser with a 15-μm absorber at bias currents of 48% and 90% above threshold. The bottom trace is the instrument noise floor spectra. After Ref. [23] (© 1992 IEEE).

3.4.2 Laser Phase Noise

It is well known that spontaneous emission events in the laser cavity change both the phase and amplitude of the optical field. Coupling of the spontaneous emission into the lasing modes as well as fluctuation in the electron density induce changes in both the real and imaginary parts of the refractive index, and produce phase noise. Following the theoretical work by C. Henry, the spectral linewidth of a semiconductor laser due to spontaneous emission can be written as [18, 19, 25]

$$\Delta v = \frac{v_g \cdot hv \cdot n_{sp}(\alpha_i + \alpha_m)\alpha_m(1+\alpha^2)}{8\pi S} \tag{3.38}$$

where α_i and α_m are internal and mirror losses, respectively; α is the linewidth enhancement factor given by Equation (3.32), and n_{sp} is the spontaneous emission factor. According to Equation (3.38), the laser linewidth is inversely proportional to the output power and it is enhanced by the factor $(1 + \alpha^2)$. The factor α is about 2 for QW lasers and about 4–7 for InGaAsP lasers. The smaller α factor is attributed to the quantum size effect in MQW lasers [26]. The unmodulated or CW linewidth of semiconductor DFB lasers is typically in the range of 0.5 MHz to 10 MHz [19]. When a DFB laser is directly modulated, the modulated linewidth, which is typically greater than 1 GHz, is considerably larger than the CW linewidth. The larger DFB laser linewidth is undesirable, particularly for relatively long-distance transmission (> 30-km).

3.5 DFB Laser Transmitter

3.5.1 DFB Laser Transmitter System Design

In order to design a DFB laser transmitter for cable TV applications, the laser diode must be properly packaged, coupled into an SMF, and mounted with additional components and circuitry. Figure 3.12 shows the three primary subsystems of a typical DFB laser transmitter.

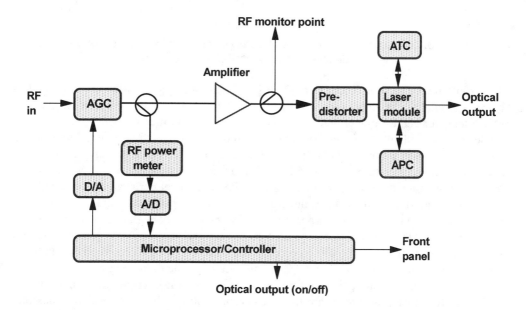

Figure 3.12 Basic block diagram of a directly modulated DFB laser transmitter.

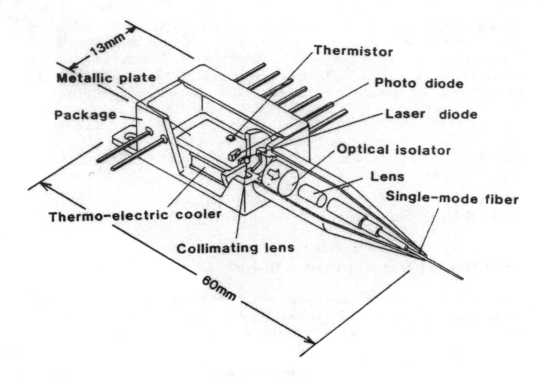

Figure 3.13 Three dimensional view of a DFB laser diode module configuration with a single-mode fiber pigtail. After Ref. [27] (© 1990 IEEE).

The front-end RF section functions to amplify the RF input signal to the proper amplitude before modulating the laser diode. It consists of one or more RF preamplifiers with a variable attenuator and RF power detector. The laser diode section contains the packaged laser diode module with all the necessary circuitry to control its operating point. This includes the automatic gain control (AGC), linearization circuitry, automatic power control (APC), and automatic temperature control (ATC). The role of the AGC circuitry, which is controlled by the microprocessor, is to maintain the RF input level per channel to a fixed level for an optimum operation. It turns out that this feature is desired by cable operators since with no AGC, the laser transmitter performance can be severely degraded, depending on the RF input level. The linearization or predistortion circuitry suppresses the generated nonlinear distortions by the laser module in order to meet the transmission requirements at the fiber node (see Section 3.5.1). The ATC circuitry controls the thermoelectric cooler (TEC), which is necessary for stable temperature operation of the DFB laser (see Section 3.5.3).

 Finally, the microprocessor/controller subsystem controls all the critical and auxiliary functions within the laser transmitter, including the status of alarm indicators, front-panel display. To maintain the laser transmitter peak performance, the microprocessor can access the optimum operating point, which is typically stored in a nonvolatile memory, and adjusts the gain of the RF signal driving the laser. In addition, the microprocessor can report the laser transmitter status to an external status monitoring system. The most important and expensive part of the laser transmitter is the packaged DFB laser diode module. Figure 3.13 shows, for example, a three-dimensional schematic view of a DFB laser diode mounted in a hermetically sealed butterfly package with an SMF pigtail [27]. The DFB laser diode, which is mounted on a metallic plate with a back-facet photodiode, is in direct contact with the TEC. Collimating optics is typically used in front of the laser to provide efficient coupling to the SMF.

3.5.2 Optical Isolators

The packaged laser module often includes another important component, namely, optical isolators. Optical isolators are passive, nonreciprocal optical devices based on the Faraday effect. In 1842, Michael Faraday discovered that the plane of polarized light is rotated while transmitting through glass that is contained in a magnetic field. The direction of rotation is independent of the direction of the light propagation, and only dependent on the direction of the magnetic field. Generally speaking, optical isolators can be divided into two categories,

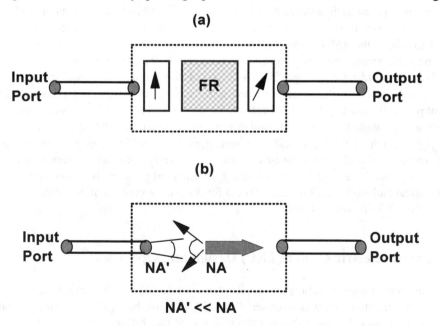

Figure 3.14 Schematic configurations of (a) polarization-sensitive optical isolators, and (b) polarization-insensitive optical isolators. FR = Faraday rotator, NA = numerical aperture. After Ref. [28] (© 1994 IEEE).

polarization-sensitive and polarization-insensitive [28]. Figure 3.14 illustrates the operating principles of the two types of optical isolators, which will be referred to as type "A" and type "B" [29]. In a type A optical isolator, shown in Figure 3.14(a), a Faraday rotator with a cylindrical magnet, which applies the magnetic field in the light direction, is sandwiched between a polarizer and an analyzer. When a light beam is passed through the Faraday rotator, its polarization plane is rotated by 45°. Since the polarization plane of the analyzer is already rotated by 45° relative to the polarizer, the propagating beam passes through the isolator with almost no attenuation loss. If a reflected light with an arbitrary polarization is incident on the analyzer, only the reflected light, which matches the polarization of the analyzer, will pass through it. Then, the Faraday rotator will rotate the polarization plane by 45° in the opposite direction, resulting in a 90° rotation relative to the polarizer. Thus, the polarizer blocks the backward-reflected light. The primary disadvantages of a type A isolator are its polarization sensitivity, which may result in additional insertion losses if the input light is randomly polarized, and its packaging reliability issues.

A type B optical isolator, which is polarization insensitive, consists of a Faraday rotator sandwiched between two wedge-shaped birefringent prisms with a collimating lens on each side. When an input light with an arbitrary polarization in incident on the first birefringent wedge-shaped prism, it is separated into ordinary and extraordinary beams caused by the different refractive indices of the prism. Since the second birefringent prism is already aligned to match the 45° rotation by the Faraday rotator, the ordinary and extraordinary beams become parallel, and the output light is coupled to the fiber with minimal attenuation loss. The backward-reflected light is also separated into ordinary and extraordinary beams, which are rotated 45° by the Faraday rotator in the opposite direction. After the second birefringent prism, the existing backward-reflected ordinary and extraordinary beams are diverging with a large angle between them, and thus are not coupled into the input fiber.

For operation in the 1.3 μm and 1.55 μm, garnet crystals such as *yttrium-iron-garnet* (YIG) and bismuth-substituted rare-earth iron garnet (BIG) are most commonly used for high-performance isolators [28]. Single-stage commercial isolators typically provide about 40-dB optical isolation and as much as 60-dB for dual-stage configurations with an insertion loss less than 0.3-dB (single stage).

3.5.3 Thermoelectric Cooler (TEC) Design and Operation

A TEC module is a small solid-state device that operates as a heat pump, utilizing the so-called Peltier effect, which was discovered in 1834, to move heat [30]. Generally speaking, when electrical current passes through the junction of two different types of conductors, a temperature gradient is produced. The Peltier effect requires the development of semiconductors that are good electrical conductors but poor conductors of heat. For example, heavily doped n-type and p-type bismuth telluride is primarily used as the semiconductor material for the TEC. Figure 3.15 illustrates a schematic configuration of a typical TEC assembly. The TEC consists of a number of p-type and n-type pairs (couples) connected electrically in series and sandwiched between two ceramic plates. The purpose of the ceramic plates on each side of the TEC elements is to prevent shorting of the laser diode by the TEC circuit. When connected to a DC power supply, the current causes the heat to move from one side of the TEC to the other, creating a "hot" side and a "cold" side on the TEC. In a typical application, the cold side of the TEC is exposed to the laser diode to be cooled and the hot side is exposed to the heat sink, which dissipates the heat to the environment. The dissipated heat at the hot side (Q_H) is simply the sum of the removed heat from the cold side (Q_C) and the input power to the TEC, which can be written as

$$Q_H = Q_C + I_{TEC}^2 R \qquad\qquad (3.39)$$

where R is the total resistance of the serially connected TEC elements. To achieve temperature stability over the operation of the laser diode, a closed-loop temperature control is needed. In this approach, the controlled cold-side temperature is compared to some reference temperature, which is typically the ambient or the hot-side temperature.

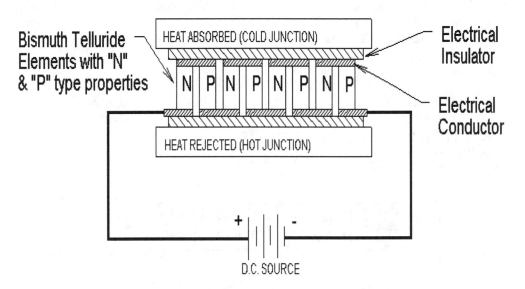

Figure 3.15 A schematic of typical thermoelectric module assembly (courtesy of Melcor Corporation). Note that the bismuth telluride elements are electrically in series and thermally in parallel.

3.5.4 Linearization Methods

As mentioned before, the nonlinear distortions generated by the semiconductor laser are often too large to meet the performance requirements for cable TV networks. Employing a predistortion or linearization circuitry before the laser module as shown in Figure 3.12 typically solves this problem. There are several linearization or predistortion methods to suppress the effect of CSO and CTB distortions in DM-DFB laser transmitters [31,32]. Figure 3.16 shows, for example, a block diagram of an electronic predistortion circuit to minimize the effect of second- and third-order nonlinear distortions [31]. The RF input signal is split into two paths with the primary path applied directly to the DFB laser with some time delay to compensate for the delays in the secondary path. A predistortion circuit in the secondary path generates second- and third-order distortions with equal amplitude and opposite phase to the nonlinear distortions generated by the DFB laser at the desired operating point. The phases of signals in the secondary path are synchronized before they are recombined with the delayed RF signal at the primary path to produce the modulating signal with the nonlinear distortions. A push-pull amplifier with unbalanced bias current can be used, for example, to generate the required even and odd harmonic distortions. There are other types of linearization methods such as a feedforward linearization and an optical linearization, which are sometimes used for an EM-DFB laser transmitter (see Sections 4.3 and 4.4).

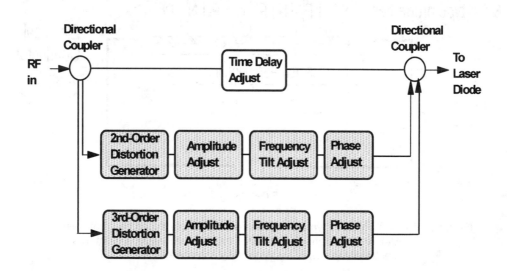

Figure 3.16 Block diagram of a predistortion circuit that minimizes the effect of second- and third-order distortions in DFB laser transmitters. After Ref. [31].

3.5.5 DFB Laser Transmitter Performance Requirements

In order for cable operators to transmit downstream multiple AM/QAM video channels over their networks, the DFB laser transmitter must be able to comply with cable TV transport requirements as well as environmental conditions [33]. In particular, the DFB laser transmitter is required to provide the following:

(A) Single longitudinal mode with high side-mode suppression ratio (> 30-dB)
(B) High-output optical power with high differential quantum efficiency
(C) Low RIN (< −155 dBc/Hz)
(D) Low nonlinear distortions over the cable TV spectrum (50–860 MHz)
(E) Stable operating characteristics over the required temperature and humidity conditions
(F) Small frequency chirp for 1550-nm operation (< 100 MHz/mA)

Generally speaking, the laser transmitter is designed to transport multiple AM/QAM video signals with the maximum possible CNR for a given optical received power at the fiber node. To achieve the maximum CNR for a given optical link budget, the output optical power as well as the optical modulation index per channel (m) must be as large as possible to overcome the laser RIN, fiber losses, and receiver noise. On the other hand, the maximum optical modulation index is limited to avoid unacceptably large nonlinear distortions and a clipping effect [34,35]. This is because when transporting multiple AM-VSB channels, the composite RF input signal is clipped by temporarily driving the laser current below threshold. The static clipping behavior, which will be discussed in detail in Chapter 9, increases the laser's nonlinear distortions and imposes an upper limit on the channel capacity. Consider a SCM system with N channels, each with modulation index m and a total photocurrent at the receiver $I(t)$. For large N, the photocurrent $I(t)$ can be modeled as a Gaussian random process with mean I_p and standard deviation $\sigma_p = mI_p(N/2)^{1/2}$. The normalized RMS modulation index can be defined as

$$\mu = \frac{\sigma_p}{I_p} = m \cdot \sqrt{\frac{N}{2}} \qquad (3.40)$$

It can be shown that the carrier-to-nonlinear distortions caused by clipping for small μ is approximately given as [36]

$$C / NLD = \frac{1}{\Gamma} \sqrt{\frac{\pi}{2}} \mu^{-3} \left(1 + 6\mu^2\right) \exp(1 / 2\mu^2) \qquad (3.41)$$

where Γ ($\approx 1/2$) is the fraction of the distortion power within the cable TV band. The resultant CNR is the sum of the CNR and the C/NLD according to Equation (2.10). Thus, for a given CNR at the fiber node receiver, the AM-VSB channel capacity (N) can be determined using Equation (3.41). This upper limit of the number of transmitted channels due to non-linear clipping distortions is sometimes called the Saleh limit [36].

Table 3.2 summarizes the required performance characteristics of a directly modulated DFB laser transmitter for multiple AM/QAM channel downstream transmission over cable TV networks [37]. For example, SL-MQW InGaAsP DFB laser diodes operating at 1310 nm have been shown to meet these performance requirements.

Table 3.2 Typical characteristics of DFB laser transmitter for multichannel AM/QAM video transmission over cable TV network.

Parameter	Specification
Optical Wavelength	1310 nm or 1550 nm ± 10 nm
Optical Power	3 mW to 14 mW
Operating Bandwidth	50 to 860 MHz
RF Input Level Range (video signals)	+10 to +25 dBmV per channel
Optical Modulation Index	3 to 4% per channel
CNR (for 80 channel loading)	52 dB (with 0-dBm at the receiver)
CSO Distortion	−62 dBc
CTB Distortion	−65 dBc
Frequency Response Variations	± 1 dB (50 to 860-MHz)
Return Loss	> 16 dB (at RF input)
Operating Temperature range	0° to +50°C
Storage Temperature Range	−40° to +70°C
Relative Humidity	Maximum 85% noncondensing

3.6 Return-Path Laser Transmitters

So far, we have discussed directly modulated DFB laser transmitters for multiple AM/QAM downstream transmission over cable TV networks. To enable a mass deployment of high-speed multimedia return-path transmitters, low-cost, preferably uncooled, DFB laser transmitters or even FP laser transmitters operating at either 1310-nm or 1550-nm wavelengths are needed [38]. For example, high-performance uncooled 1.3-μm $Al_xGa_yIn_{1-x-y}As/InP$ SL-MQW lasers suited for fiber-in-the-loop and return-path applications have been developed recently. The use of the AlGaInAs/InP material system instead of the conventional $Ga_xIn_{1-x}As_yP_{1-y}/InP$ material system provides a better carrier confinement under high temperature operation. Preliminary test data of these lasers show a mean-time-to-failure of 9.4 years at operating temperature of 85°C with an output power of more than 5-mW. One of the main problems in these quasi-single-mode or multimode lasers is the observation of mode-partition noise and mode-hopping noise. In the next section, both noise mechanisms are reviewed, and their impact on return-path laser transmitter design for digital transmission is discussed.

3.6.1 Mode-Partition Noise and Mode-Hopping Noise

A mode-partition noise (MPN) arises when the optical power is distributed among many different laser modes such that the instantaneous power in each individual mode can fluctuate, but the total power in all the modes remains relatively unchanged. Figure 3.17 shows the calculated and measured intensity fluctuations of a dominant mode of a multimode AlGaAs laser [39, 40]. The intensity noise spectral density of the dominant mode is essentially unchanged at low frequencies (< 50-MHz), and decreases at 6-dB per octave above the low frequency region. Generally speaking, at the low-frequency region, the laser intensity noise can be very larger and tends to mask the MPN. However, at higher frequencies, say above 1-GHz, the MPN is the dominant noise, masking the laser intensity noise. The reason semiconductor lasers are subjected to MPN is that the different optical modes are competing with each other for the same optical gain. The MPN introduces several problems

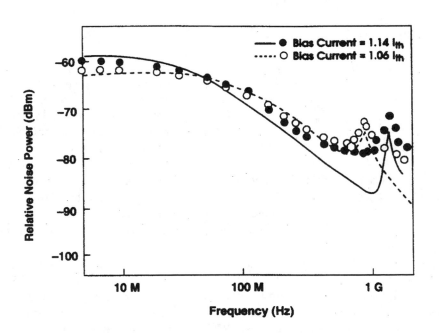

Figure 3.17 Calculated (lines) and Measured (symbols) intensity fluctuations of a dominant mode of a multimode AlGaAs laser. After Ref. [39] (© 1994 IEEE).

that limit the performance of multiple-longitudinal-mode FP lasers. First, the laser's mode distribution was found to vary cyclically with the bias current and temperature, which can dramatically alter the laser transmitter performance [41]. Second, when a multiple longitudinal-mode FP laser is used with a dispersive fiber, the instantaneous power fluctuations of the individual modes become separated in time because each mode experiences a different propagation delay [42]. Consequently, the power fluctuations of the various modes at any given time do not cancel each other. Thus, using multimode FP lasers limits the transmission of video signals in an SCM lightwave system. The MPN problem is usually eliminated with the use of single-longitudinal mode DFB laser diodes.

Another type of noise, which has been observed in multiple modes or quasi-single-mode FP lasers, is called mode-hopping noise (MHN). MHN is attributed to random hopping of among the various oscillating modes of the laser when the laser bias current or temperature is varied. The mode-hopping mechanism can cause a significant intensity noise enhancement (< -130 dB/Hz), particularly at low frequencies (< 30-MHz). Controlling the bias current and/or temperature of the semiconductor lasers can eliminate the mode-hopping mechanism. Another way to suppress the MHN can be achieved by introducing a saturable absorber in index-guided semiconductor lasers, for example, by using Te-doping of the n-type cladding layer of AlGaAs/GaAs lasers [43].

3.6.2 Performance Requirements

Directly modulated semiconductor uncooled FP lasers can be used to transport digital channels such as QPSK and/or 16-QAM in the return-path portion of the cable TV networks [37]. The return-path laser transmitters are not required to provide the same CNR or linearity as the downstream laser transmitters since they are not intended to transport AM-VSB video channels. The optical signals can be transmitted at either 1310-nm or 1550-nm. For cost-effective laser transmitters, the TEC or the optical isolator in the laser diode module can be eliminated [44, 45]. Figure 3.18 shows, for example, the measured uncoded BER versus the optical modulation index per channel (m) for a differentially encoded QPSK channel transmitted over 24-km of a standard SMF using an uncooled FP laser with a RIN of -127-dBc/Hz. Notice that at room temperature, an error-free QPSK transmission can be achieved over a wide range of optical modulation indices, from about 1% to 11%. The operating range of optical modulation indices is reduced by 5-dB at a higher temperature ($\approx 80°C$) indicating an upper limit on an upstream channel capacity. Table 3.3 summaries the performance specifications for return-path laser transmitters. For example, assuming a 256-kb/s QPSK channel, the required upstream C/(N+I) must be equal or greater than 20-dB, which includes a 4-dB margin caused by the simultaneous presence of multiple impairments.

The upstream impairments include a laser RIN, thermal noise, and optical reflections, as well as cumulative ingress noise from all the homes connected to a given fiber node. Since the return-path transmitters are typically installed inside the downstream optical receiver at the fiber node, they are required to operate over a much larger temperature range than the downstream DFB laser transmitters. The performance requirements of the return-path laser

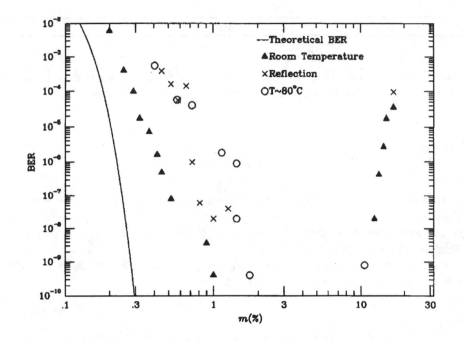

Figure 3.18 BER versus the optical modulation index per channel (m) for differentially encoded QPSK for an uncooled FP laser diode. After Ref. [44] (© IEEE 1996).

Table 3.3 Typical characteristics for return-path laser transmitters.

Parameter	Specification
Optical Wavelength	1310 nm or 1550 nm ± 10-nm
Optical Power	3 to 5 mW
Operating Bandwidth	5 to 42-MHz
RF Input Level Range (video signals)	+10 to +25 dBmV per channel
Dynamic Range	15-dB
CNR (QPSK channels)	≥ 16 dB at BER $\leq 10^{-7}$
C/(N+I) QPSK channels	≥ 20 dB^2 at BER $\leq 10^{-7}$
Laser Transmitter Spurious Output Level	≤ -25 dBc
Return Loss	> 16 dB (at RF input)
Operating Temperature range	−20° to +65°C
Storage Temperature Range	−40° to +70°C

---

[2] Includes the simultaneous presence of all the impairments in the upstream portion of the cable TV network.

Relative Humidity	Maximum 85% noncondensing

transmitters change when the return-path portion of the cable TV access network is digitized, particularly in a DWDM network. For more discussion, see Section 10.4.2.

References

1. S. M. Sze, *Semiconductor Devices Physics and Technology*, John Wiley & Sons (New York, 1985).
2. G. P. Agrawal and N. K. Dutta, *Long-Wavelength Semiconductor Lasers*, Van Nostrand Reinhold Company (New York, 1986).
3. T. P. Lee, "Recent Advances in Long Wavelength Semiconductor Lasers for Optical Fiber Communications," *Proceedings of IEEE* (1991).
4. N. K. Dutta, "Calculated Absorption, Emission, and Gain in $In_{0.72}Ga_{0.28}As_{0.6}P_{0.4}$," *Journal of Applied Physics* 51, 6095–6100 (1980).
5. N. K. Dutta and R. J. Nelson, *Journal of Applied Physics* 53, 74 (1982).
6. E. Kapon, Semiconductor Lasers: Physics and Technology, lecture notes (1988).
7. K. Oe and K. Sugiyama, *Japanese Journal of Applied Physics* 47, 3533 (1976).
8. K. Oe, S. Ando, and K. Sugiyama, "GaInAsP/InP Planar Stripe Lasers Prepared by Using Sputtered SiO2 Film as a Zn Diffusion Mask," *Journal of Applied Physics* 51, 43–49 (1980).
9. J. Ungar, N. Bar-Chaim, and I. Ury, "High-Power GaAlAs Window Lasers," *Laser and Applications* 9, 111–114 (1985).
10. K. Uomi, T. Suchiya, M. Komori, A. Oka, K. Shinoda, and A. Oishi, "Extremely Low Threshold (0.56 mA) Operation in 1.3-μm InGaAsP/InP Compressive-Strained-MQW Lasers," *IEE Electronic Letters* 30, 2037–2038 (1994).
11. H. Nobuhara, T. Inoue, T. Watanabe, K. Takana, T. Odagawa, T. Abe, and K. Wakao, "1.3-μm Wavelength, Low-Threshold Strained Quantum Well Laser on p-Type Substrate," *IEE Electronic Letters* 30, 1292–1293 (1994).
12. G. P. Agrawal, "Effects of Gain and Index Nonlinearities on Single-Mode Dynamics in Semiconductor Lasers," *IEEE Journal of Quantum Electronics* 26, 1901 (1990).
13. M. Asada and Y. Suematsu, "Density Matrix Theory of Semiconductor Lasers with Relaxation Broadening Modal-Gain and Gain-Suppression in Semiconductor Lasers," *IEEE Journal of Quantum Electronics* QE-21, 434–442 (1985).
14. R. S. Tucker, "High-Speed Modulation of Semiconductor Lasers," *IEEE Journal of Lightwave Technology* LT-3, 1180–1192 (1985).

15. K. Y. Lau and A. Yariv, "Ultra-High Speed Semiconductor Lasers," *IEEE Journal of Quantum Electronics* **QE-21**, 121–138 (1985).

16. P. L. Derry, T. R. Chen, Y. H. Zhuang, J. Paslaski, M. Mittelstein, K. Vahala, and A. Yariv, "Spectral and Dynamic Characteristics of Buried-Heterostructure Single Quantum Well (Al,Ga) As Lasers," *Applied Physics Letters* **53**, 271–273 (1988).

17. I. Habermayer, "Nonlinear Circuit Model for Semiconductor Lasers," *IEEE Journal of Optical and Quantum Electronics* **13**, 461–468 (1984).

18. C. H. Henry, "Theory of the Linewidth of Semiconductor Lasers," *IEEE Journal of Quantum Electronics* **QE-18**, 259–264 (1982).

19. M. Osinski and J. Buus, "Linewidth Broadening Factor in Semiconductor Lasers - An Overview," *IEEE Journal of Quantum Electronics* **QE-23**, 9–29 (1987).

20. T. L. Koch and R. A. Linke, "Effect of Nonlinear Gain reduction on Semiconductor Laser Wavelength Chirping," *Applied Physics Letters* **48**, 613–615 (1986).

21. K. Vahala and A. Yariv, "Semiclassical Theory of Noise in Semiconductor Lasers-Part I," *IEEE Journal of Quantum Electronics* **QE-19**, 1096–1101 (1983).

22. K. Vahala and A. Yariv, "Semiclassical Theory of Noise in Semiconductor Lasers-Part II," *IEEE Journal of Quantum Electronics* **QE-19**, 1102–1109 (1983).

23. S. Ovadia and K. Y. Lau, "Low-frequency Relative Intensity Noise in Self-Pulsating Ridge-Waveguide Quantum Well Lasers," *IEEE Photonics Technology Letters* **4**, 336–338 (1992).

24. W. I. Way et al., "Multiple-Reflection-Induced Intensity Noise Studies in a Lightwave System for Multichannel AM-VSB Television Signal Distribution," *IEEE Photonics Technology Letters* **2**, 360–362 (1990).

25. C. H. Henry, "Theory of the Phase Noise and Power Spectrum of a Single Mode Injection Laser," *IEEE Journal of Quantum Electronics* **QE-19**, 1391–1397 (1983).

26. K. Uomi, S. Sasaki, T. Tsuchiya, M. Okai, M. Aoki, and N. Chinone, "Spectral Linewidth Reduction by Low Spatial Hole Burning in 1.5 μm Multi-Quantum Well λ/4 Shifted DFB Lasers," *IEE Electronics Letters* **26**, 52–53 (1990).

27. A. Takemoto, H. Watanabe, Y. Nakajima, Y. Sakakibara, S. Kakimoto, J. Yamashita, T. Hatta, and Y. Miyake, "Distributed Feedback Laser Diode and Module for CATV Systems," *IEEE Journal of Selected Areas in Communications* **8**, 1359–1364 (1990).

28. K. Kikushima, K.-I. Suto, H. Yoshinaga, and E. Yoneda, "Polarization Dependent Distortion in AM-SCM Video Transmission Systems," *IEEE Journal of Lightwave Technology* **12**, 650–657 (1994).

29. SR-NWT-002855, *Optical Isolators: Reliability Issues and Proposed Tests*, Issue 1 (Bellcore, December 1993).

30. P. W. Shumate, "Semiconductor Laser Transmitters," *Optoelectronic Technology and Lightwave Communication Systems*, Ed. Chinlon Lin, Van Nostrand-Reinhold (New Jersey, 1990).

31. H. A. Blauvelt and H. L. Loboda, "Predistorter for Linearization of Electronic and Optical Signals," U.S. Patent Number 4,992,754, February (1991).

32. M. Nazarathy, C. H. Gall, and C. Y. Kuo, "Predistorter for High frequency Optical Communications Devices," U.S. Patent Number 5,424,680, June (1995).

33. H. Yonetani, I. Ushijima, T. Takada, and K. Shima, "Transmission Characteristics of DFB Laser Modules for Analog Applications," *IEEE Journal of Lightwave Technology* **11**, 147–153 (1993).

34. C. Y. Kou, "Fundamental Second-order Nonlinear Distortions in Analog AM CATV transport Systems Based on Single Frequency Semiconductor Lasers," *IEEE Journal of Lightwave Technology* **10**, 235–243 (1992).

35. K. Maeda, H. Nakata, and K. Fujito, "Analysis of BER of 16QAM Signal in AM/16QAM Optical Transmission System," *IEE Electronics Letters* **29**, 640–641 (1993).

36. A. A. M. Saleh, "Fundamental Limit on Number of Channels in Subcarrier Multiplexed Lightwave CATV Systems," *IEE Electronics Letters* **25**, 776–777 (1989).

37. GR-2853-CORE, *Generic Requirements for AM/Digital Video Laser Transmitters and Receivers*, Issue 2 (Bellcore, December 1995).

38. C. E. Zah, R. Bhat, B. N. Pathak, F. Favire, W. Lin, M. C. Wang, N. C. Andreadakis, D. M. Hwang, M. A. Koza, T. P. Lee, Z. Wang, D. Darby, D. Flanders, and J. J. Hsieh, "High Performance Uncooled 1.3-μm $Al_xGa_yIn_{1-x-y}As/InP$ Strained-Layer Quantum-Well Lasers for Subscriber Loop Applications," *IEEE Journal of Quantum Electronics* **30**, 511–523 (1994).

39. G. J. Meslener, "Mode-Partition Noise in Microwave Subcarrier transmission Systems," *IEEE Journal of Lightwave Technology* **12**, 118-126 (1994).

40. T. Ito, S. Machida, K. Nawata, and T. Ikegami, "Intensity Fluctuations in Each Longitudinal Mode of a Multimode AlGaAs Laser," *IEEE Journal of Quantum Electronics* **13**, 574–579 (1977).

41. G. J. Meslener, "Temperature Dependence of Distribution, Intensity Noise, and Mode-Partition Noise in Subcarrier Multiplexed Transmission System," *IEEE Photonics Technology Letters* **4**, 939–941 (1992).

42. R. H. Wentworth, G. E. Bodeep, and T. E. Darcie, "Laser Mode Partition Noise in Lightwave Systems Using Dispersive Optical Fiber," *IEEE Journal of Lightwave Technology* **10**, 84–89 (1992).

43. N. Chinone, T. Kuroda, T. Ohtoshi, T. Takahashi, and T. Kajimura, "Mode Hopping Noise in Index-Guided Semiconductor and its Reduction by Saturable Absorbers," *IEEE Journal of Quantum Electronics* **21**, 1264–1270 (1985).

44. S. L. Woodward and G. E. Bodeep, "Uncooled Fabry-Perot Lasers for QPSK Transmission," *IEEE Photonics Technology Letters* **7**, 558–560 (1995).

45. S. L. Woodward, V. Swaminathan, G. E. Bodeep, and A. K. Singh, "Transmission of QPSK Signals Using Unisolated DFB Lasers," *IEEE Photonics Technology Letters* **8**, 127–129 (1996).

CHAPTER 4

EXTERNALLY MODULATED CABLE TV
LIGHTWAVE LASER TRANSMITTERS

Direct modulation of a laser diode has several limitations, depending on the operating wavelength. One of the key limitations is the induced nonlinear distortion due to the interaction between the frequency chirp of directly modulated DFB laser transmitters operating at 1550-nm and the dispersion of standard SMF (see Chapter 9). Transmission at 1550-nm wavelength band is desirable in order to take advantage of the low fiber losses and the use of Erbium-doped fiber amplifiers (EDFAs). The higher required optical power from DFB laser transmitters operating at 1310-nm compared with the 1550-nm due to higher SMF losses imposes another limitation. In addition, the performance of directly modulated DFB laser transmitters can also be severely degraded in the presence of multiple optical reflections (see Chapter 9). These problems, which significantly limit the transmission distance over standard SMF, can be overcome by using an externally modulated laser transmitter with an optical modulator.

We shall start Chapter 4 with a review of the basic operation and distortion characteristics of LiNbO$_3$ optical modulators, which are the most widely used in the cable TV industry. In Section 4.2, the different electronic linearization methods and circuits are reviewed. There are also optical linearization methods, including optical dual parallel linearization and optical dual cascade linearization, which are discussed in Section 4.3. Finally, we will discuss the design and performance characteristics of the two different types of externally modulated laser transmitters in Section 4.4.

There are two different types of commercially available externally modulated laser transmitters operating at 1.3-μm and 1.55-μm. The first type is externally modulated diode-pumped Nd:YAG laser operating at 1319-nm, which has many desirable characteristics such as extremely narrow linewidth (< 200-kHz) and very low laser RIN (< −165-dBc/Hz) [3]. The second type is an externally modulated DFB laser transmitter operating at 1550-nm. Since the mid-1990s, high-power (> 20 mW) with low RIN (<−155-dBc/Hz) SL-MQW DFB laser diodes operating 1550-nm have been developed to meet the cable TV performance requirements. The early generation of DFB laser diodes had limited output power levels, which were reduced due to the insertion loss of the LiNbO$_3$ modulator. Consequently, EDFAs were incorporated in the laser transmitter design. Chapter 6 is devoted to the design, operation, and characteristics of EDFAs, since they are one of the key building blocks of cable TV networks.

4.1 LiNbO$_3$ Optical Modulators

Optical modulators such as Lithium Niobate (LiNbO$_3$) modulators are one of the key components in the externally modulated laser transmitter. Consequently, their basic operation, design, and distortion characteristics are discussed in this section.

4.1.1 Basic Operation of LiNbO$_3$ Intensity Modulators

Lithium Niobate (LiNbO$_3$) is an important dielectric material for many applications in optical communications, optical sensors, and signal processing due to its unique physical properties [1]. One of the most important properties of LiNbO$_3$ is its large electro-optic coefficient. The linear electro-optic effect, which is also called the Pockels effect, is a change in the material refractive index when an electric field is applied [2]. Figure 4.1 shows schematic cross-sections of a ridge-type LiNbO$_3$ waveguide modulator with the relative orientation of the electrodes to the waveguide. When a voltage V is applied to the electrodes, which are placed alongside the waveguide with a gap D, an internal electric field is created whose magnitude is approximately given by: $|E| = V/D$. If the electric field is applied in the z-direction (E$_z$), one can obtain the largest electro-optic induced refractive index change, which is given by

$$\Delta n \cong -n^3 r_{33} \cdot \Gamma \left[\frac{V}{2D} \right] \tag{4.1}$$

where n is either the ordinary or the extraordinary refractive index of the waveguide, r_{33} is the largest electro-optic coefficient, and Γ is the overlap integral between the RF and optical mode fields. To generate E$_z$, the electrode orientation relative to the waveguide depends on the LiNbO$_3$ crystal orientation. The crystal orientation is often specified as a "cut," which is perpendicular to the flat surface on which the waveguide is fabricated. For example, in X-cut LiNbO$_3$, the electric field is propagating in the y direction along the waveguide as shown in Figure 4.1(a). The resulting phase shift from the refractive index change over the effective waveguide length is given by

$$\Delta \phi = \frac{2\pi \cdot \Delta n \cdot L}{\lambda} = -\pi n^3 r_{33} \cdot \Gamma \left[\frac{V \cdot L}{\lambda D} \right] \tag{4.2}$$

In order to obtain the largest phase shift, one needs to optimize the waveguide geometry, namely, minimizing the lateral waveguide parameter D/Γ and maximizing the waveguide length. However, if the electrode gap, which is typically in the range of 10–20-μm, is too small, the electrode capacitance may be too large, resulting in a reduced modulation bandwidth and an increased amount of DC drift [3]. It should be pointed out that both the driving

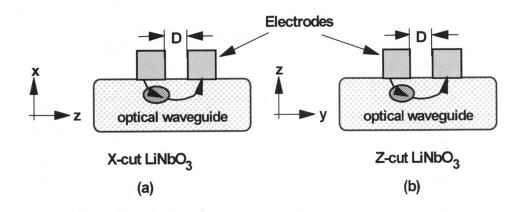

Figure 4.1 Schematic cross sections of ridge-type LiNbO3 waveguide with the corresponding electrodes with their orientation for (a) X-cut, and (b) Z-cut
LiNbO3 waveguides.

voltage and the modulation bandwidth are proportional to $(1/L)$. Since cable TV networks typically operate below 1-GHz, reducing the driving voltage is very desirable. The required voltage to switch the optical output from on to off, which is also called *half-wave voltage* (V_π), is typically in the range of 3–5 V [3, 4].

Generally speaking, integrated optical modulators are constructed by patterning optical waveguides onto the LiN_bO_3 substrate using standard photolithography techniques, and depositing the metal electrodes on top of the waveguide. The two primary methods to fabricate the optical waveguides on the $LiNbO_3$ substrate are annealed proton exchange (APE) and titanium indiffusion [4, 5]. Although both methods produce waveguides with somewhat similar characteristics, the recently developed APE has been more widely adopted for the following reasons. First, the APE waveguide supports only a single polarization light because the APE process increases the extraordinary refractive index (n_e), while the ordinary refractive index (n_o) decreases slightly. Typical polarization extinction ratios are greater than 50 dB. Thus, one can avoid additional waveguide loss due to the different polarization modes, but it also means that one must properly align the input polarization to the waveguide for efficient coupling. Second, the APE method can produce a much large refractive index change, which has advantages for more compact waveguide designs.

Optical modulators can be divided into two main configurations, which are called a lumped circuit-element modulator and a traveling-wave modulator. In the lumped circuit-element modulator configuration, the RF input signal terminates at a pair of electrodes in parallel with a matched load. The problem with this modulator configuration is that its modulation bandwidth is primarily limited by the product of the load resistance and the electrode capacitance. However, reducing the electrodes' length or increasing their gap also increases the driving voltage, which is undesirable. For operation below 1-GHz, a possible solution can be obtained by using a matching circuitry between the RF signal source and the electrodes. Another approach is to use the traveling-wave modulator, where the electrodes are constructed along the optical waveguide and acting as transmission lines whose imped-ance is designed to match the RF source. The electrodes are terminated with a matched load. The propagating RF input signal along the transmission line interacts with the co-propagating optical wave inside the waveguide, resulting in efficient phase modulation of the optical wave along the waveguide. Consequently, the modulation bandwidth is not lim-ited by the RC time constant, but by the inherent velocity mismatch between the propagating optical wave and RF signal.

For transmission over cable TV networks, externally modulated laser transmitters use pri-marily optical amplitude or intensity modulators. The most widely used intensity modula-tors, which are illustrated in Figure 4.2, are Mach-Zehnder interferometer (MZI) modulators and balanced-bridge interferometer (BBI) modulators. In the MZI modulator, a single input waveguide is split into two waveguides by a 3 dB Y junction, and then recombined by

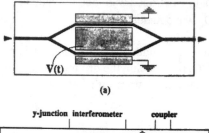

(a)

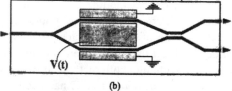

(b)

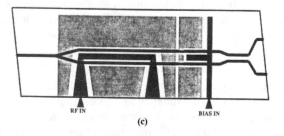

(c)

Figure 4.2 Schematic of different optical intensity modulators: (a) Mach-Zehnder inter-ferometer, (b) balanced-bridge interferometer modulator, and (c) state-of-the-art balanced-bridge interferometer modulator, including traveling-wave electrodes, a separate DC bias electrode, and slanted facets. After Ref. [3] (© 1993 IEEE).

a second Y junction into a single output waveguide. In the traveling-wave MZI modulator, the RF modulation signal is traveling along the transmission lines causing a phase shift between the two arms. The phase-modulated optical wave at each arm of the waveguide can recombine either constructively or destructively, resulting in amplitude modulation of the output optical wave. The modulated optical power output from a MZI modulator can be written as:

$$P(V) = \langle P \rangle \left[1 + \sin\left(\frac{\pi V}{V_\pi} + \phi_b \right) \right]$$

(4.3)

where $\langle P \rangle = P_{in}/2L_m$ is the average optical output power, P_{in} is the input CW optical power, L_m is the excess loss of the modulator, V_π is the half-wave voltage, and ϕ_b is the bias phase.

One of the main problems with external modulated transmitter was the relatively high insertion loss of LiNbO$_3$ optical modulators. The MZI modulator insertion loss factors include a coupling loss at the input facet of the waveguide (≈ 0.5-dB), an inherent 3 dB waveguide loss at the Y-junction combiner, and an optical propagation loss, which is typically negligible. There is an additional 3-dB loss since the modulator has to be biased at a half-power point (see explanation later), and an electrode loading loss due to impedance mismatch with the RF signal source. Thus, the typical insertion loss of the MZI modulator is about 7 dB.

To understand the inherent 3-dB loss at the Y-junction combiner, let us assume that the optical field distribution in the two arms is represented as a superposition of symmetric mode and antisymmetric mode. The symmetric mode is gradually coupled into the single output waveguide, while the antisymmetric mode, which cannot be supported by the output waveguide, radiates into the substrate. Thus, 50% (3 dB) of optical power is being wasted in the substrate. This limitation of the MZI modulator is overcome by incorporating two optical output ports by replacing the Y-junction combiner with a 2x2 directional coupler as shown in Figure 4.2(b). In this type of intensity modulator, which is called the balanced-bridge interferometer (BBI) modulator, the 2x2 directional coupler mixes the two optical waves and routes them to the two optical outputs. Consequently, both optical outputs can be utilized, and the 3-dB inherent loss of the MZI modulator is being eliminated. The modulated optical power of the BBI modulator at each output port ($P_\pm$) is given by

$$P_\pm = \langle P \rangle \left[1 \pm \sin(\theta + \phi_b) \right]$$

(4.4)

where $\theta = \pi V/V_\pi$ is the normalized modulating voltage. The two transfer functions of the BBI modulator are plotted in Figure 4.3. Notice that the sum of the two optical outputs from the MZI modulator is constant and equal to $2\langle P \rangle$. Thus, compared with the MZI modulator, the BBI modulator has twice the optical power utilization efficiency. Figure 4.2 (c) shows a schematic of state-of-the art BBI modulator, which includes a traveling-wave electrode design, a separate DC bias electrode to separate the RF and DC paths, and slanted input and output facets to minimize the effect of interferometric noise. The advantage of the traveling-wave electrode design is a flatter frequency response over the operating bandwidth that is critical for the modulator linearization (see Section 4.3).

The ideal transfer characteristics of the BBI modulator may be modified due to wavelength and polarization effects. If the laser source is not a single-mode laser, the resulting transfer function will be a superposition of Equation (4.4) with a different V_π and ϕ_b for each wavelength. This is undesirable for a cable TV transmission system. In addition, if there is a polarization mismatch between the launched optical beam and the BBI modulator, significant optical coupling losses may occur. Consequently, a polarization-maintaining fiber is typically used to couple the light into the single-mode optical modulator waveguide.

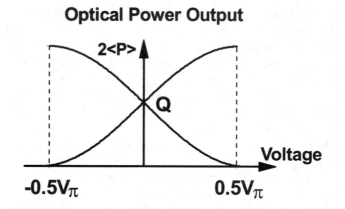

Figure 4.3 The optical power versus the modulating voltage for the two output ports of the balanced-bridge interferometer modulator. After Ref. [3] (© 1993 IEEE).

4.1.2 Distortion Characteristics of MZI and BBI Modulators

The linearity characteristics of MZI and BBI modulators are key for the operation of these modulators in the transmission of analog video signals. Following Reference [3], let us define the AC component of the modulator output as

$$p = \frac{P - \langle P \rangle}{\langle P \rangle} = \sin(\theta + \phi_b) \qquad (4.5)$$

Let us assume a multichannel cable TV modulation signal consisting of N unmodulated carriers plus a DC bias term V_B, which can be written as:

$$V(t) = \sum_{i=1}^{N} V_0 \cdot \cos[\omega_i t + \psi_i] + V_B \tag{4.6}$$

and it is applied to Equation (4.5). Expanding $\sin(\theta)$ and $\cos(\theta)$ in a power series and retaining terms up to third order, Equation (4.5) becomes

$$p = \cos(\phi_T)\left[\beta \sum_{i=1}^{N} \cos(\omega_i t + \psi_i) - \frac{\beta^3}{6}\left(\sum_{i=1}^{N} \cos(\omega_i t + \psi_i)\right)^3\right] + \sin(\phi_T)\left[1 - \frac{\beta^2}{2}\left(\sum_{i=1}^{N} \cos(\omega_i t + \psi_i)\right)^2\right] \tag{4.7}$$

where $\beta = \pi V/V_\pi$, and $\phi_T = \phi_b + \pi V_B/V_\pi$ is the total retardation bias, which is the sum of the intrinsic and applied DC phase biases. In Equation (4.7), the linear term in β is the desirable term, the term in β^2 is the CSO component, while the term in β^3 is the CTB component. Notice that if $\phi_T = 0$, the CSO term is canceled, the CTB term is maximized, and the ratio of the linear term and the CTB term is independent on the bias point ϕ_T. Consequently, the ideal bias point of the BBI modulator is the inflection point $\phi_T = 0$, which is also known as the quadrature point (Q), as shown in Figure 4.3. Furthermore, when the BBI modulator is biased at the Q point, all the even orders of the intermodulation products are canceled. In practice, temperature variations and stress may cause the intrinsic bias to slowly drift with time constants of minutes and even hours [6]. This problem is typically solved using a tracking DC voltage V_B, which is applied to the bias electrode, whose magnitude is given by

$$V_B(t) = -\frac{\phi_b(t) \cdot V_\pi}{\pi} \tag{4.8}$$

This condition is satisfied using a parametric feedback control system for CSO suppression, and it is discussed in Section 4.4. If the BBI modulator is biased at the Q point, its normalized output from Equation (4.7) is reduced to:

$$p = p^{(1)} + p^{(3)} = \beta \sum_{i=1}^{N} \cos(\omega_i t + \psi_i) - \frac{\beta^3}{6}\left[\sum_{i_1=1}^{N}\sum_{i_2=1}^{N}\sum_{i_3=1}^{N} \cos(\omega_{i_1} t + \psi_{i_1}) \cdot \cos(\omega_{i_2} t + \psi_{i_2}) \cdot \cos(\omega_{i_3} t + \psi_{i_3})\right] \tag{4.9}$$

where the first term $p^{(1)}$ is the linear term, and the second term $p^{(3)}$ is the CTB contribution, which can be further expanded to yield

$$p^{(3)} = -\frac{\beta^3}{24}\sum_{\pm}\sum_{i_1=1}^{N}\sum_{i_2=1}^{N}\sum_{i_3=1}^{N} \cos\left[\left(\omega_{i_1} \pm \omega_{i_2} \pm \omega_{i_3}\right)t + \psi_{i_1} \pm \psi_{i_2} \pm \psi_{i_3}\right] \tag{4.10}$$

where the $\pm$ implies summation over all four possible pairs of signs. To evaluate the different combinations of summation in Equation (4.10), one has to remember that the cable TV channels are spaced 6 MHz apart from channel 2, and many of the CTB products could potentially fall on the same beat frequency ω_b whenever the following condition is satisfied:

$$\omega_b = \omega_{i_1} + \omega_{i_2} + \omega_{i_3}$$

(4.11)

where the following convention is used for negative indices: $\omega_{-i} = -\omega_i$. In addition, the beat contribution depends on whether the various beats are added coherently (i.e., equal phases) or incoherently. For a given triple $\{i_1, i_2, i_3\}$ with different indices, the beats associated with all of the other five possible permutations of these three indices are mutually coherent. Thus, the total beat contribution associated with a given combination of three different indices with an arbitrary order is six times larger than any individual beat associated with any particular order. Similarly, for a given triple $\{i_1, i_1, i_2\}$ with only two distinct indices, the beats associated with the other two possible permutations of these indices are mutually coherent. Thus, the total beat contribution associated with a triplet combination with only two distinct indices is three times larger than any other individual beat with a particular index permutation. Using these results, the CTB component of the modulator output is given by

$$p^{(3)} = \sum_{\pm} \left[6 \sum_{i_1=1}^{N-2} \sum_{i_2=i_1+1}^{N-1} \sum_{i_3=i_2+1}^{N} B(i_1, \pm i_2, \pm i_3) + 3 \sum_{i_1=1}^{N-1} \sum_{i_2=i_1+1}^{N} \left[B(i_1, \pm i_2, \pm i_2) + B(i_2, \pm i_1, \pm i_1) \right] + \sum_{i=1}^{N} B(i, \pm i, \pm i) \right]$$

(4.12)

where the triple-beat mixing product B is defined as

$$B(i_1, i_2, i_3) = -\frac{\beta^3}{24} \cos\left[\left(\omega_{i_1} + \omega_{i_2} + \omega_{i_3} \right) t + \psi_{i_1} + \psi_{i_2} + \psi_{i_3} \right]$$

(4.13)

The CTB component of the normalized modulator output can easily be evaluated using computer simulations.

4.2 Linearization Methods of Optical Modulators

Many linearization methods of optical modulators have been proposed and implemented in externally modulated laser transmitters [7–11]. Generally speaking, the linearization methods can be divided into feedforward linearization and predistortion linearization categories. The operation principle of each of the linearization methods is explained in the following subsection with a practical implementation example. The intention here is not to cover all the possible implementations, but to explain the main advantages and disadvantages of the different methods. Optical linearization methods of external modulators, including optical feedforward linearization, dual parallel linearization, and dual cascade linearization methods are discussed separately in Section 4.3.

4.2.1 Feedforward Linearization Method

Figure 4.4 shows a practical implementation block diagram of the feedforward linearization method for the BBI modulator [3]. The RF input signal is split before the modulator and it is delayed and amplified relative to the applied RF signal. A small fraction of the output power from each arm of the optical modulator is also split and detected using a simple photodetector. The RF error signal, which is generated by combining the delayed RF signal with the photodetector output signal using a broadband 180° RF hybrid comparator, is applied to an auxiliary DFB laser. The DFB laser must be tuned to a wavelength close to the source laser, but it is not allowed to have the same wavelength in order to prevent coherent interference. The optical output of each of the auxiliary DFB lasers is then combined with

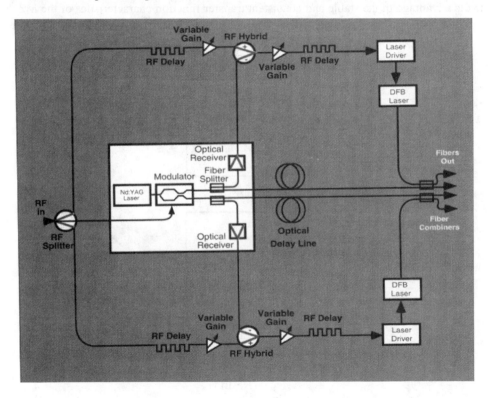

Figure 4.4 Block diagram of the feedforward linearization method for the balanced bridge interferometer modulator. After Ref. [3] (© 1993 IEEE).

the corresponding arm of the BBI modulator output. It should be pointed out that the auxiliary DFB lasers have to be sufficiently powerful to provide a signal swing four times larger than the correction signal swing in the output fiber, assuming a 20/80% optical coupler. For example, the required DFB laser peak power is of 2.72-mW, assuming 60 NTSC channels at an optical modulation index of 3.6% per channel.

To compensate for the modulator nonlinearity, the feedforward linearization method requires frequency flattening and gain matching at the auxiliary photodetector, DFB laser, and its drive circuitry. When combining the optical signals from two different lasers operating at two different wavelengths, a differential time delay is generated due to the fiber dispersion, which also has to be compensated. The differential delay is proportional to the fiber length.

4.2.2 Predistortion Linearization Method

The predistortion linearization method, which is simpler than the feedforward linearization, takes advantage of the stable and consistent transfer function characteristics of the MZI/BBI modulators. The basic concept behind the predistortion linearization method is illustrated in

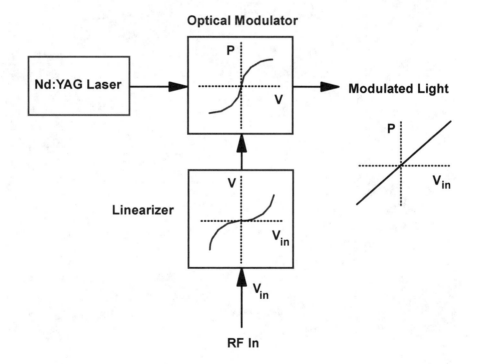

Figure 4.5 Principle of electronic predistortion linearization method. After Ref. [3] (© 1993 IEEE).

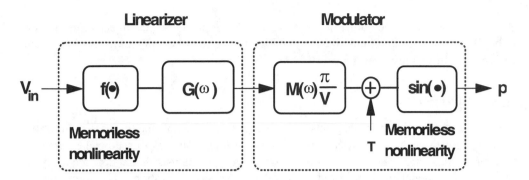

Figure 4.6 General model of the predistortion linearization method used in externally modulated laser transmitters. After Ref. [3] (© 1993 IEEE).

Figure 4.5. The linearized light-versus-voltage of the optical modulator is obtained by inserting a linearizer between the transmitter RF input and the modulator with an arcsine-shaped transfer characteristic. The general model of the predistortion linearization method employed in externally modulated laser transmitters in shown in Figure 4.6 [3]. Assuming that the modulator is biased at the Q point ($\phi_T = 0$), and both the modulator and linearizer can be described by memoryless nonlinearity, then the linearizer transfer characteristic can be written as a power series with odd terms:

$$V = f(V_{in}) = \alpha_1 V_{in} + \alpha_3 V_{in}^3 + \alpha_5 V_{in}^5 + ... \qquad (4.14)$$

where α_i (i = 1,2,3,...) represents the small signal gain coefficients. Furthermore, assuming $M(\omega) = 1$, the normalized optical power output is given by

$$p = \sin(K \cdot V) = \sin\left[K \cdot \alpha_1 \cdot V_{in} + K \cdot \alpha_3 \cdot V_{in}^3 + K \cdot \alpha_5 \cdot V_{in}^5 + ... \right] \qquad (4.15)$$

where $K = \pi G/V\pi$ is the effective transfer gain of the linearizer, assuming $G(\omega)$ is a constant G. Using the power series expansion of sin(•), and neglecting fifth-order or higher terms for simplicity, we obtain

$$p = K \cdot \alpha_1 \cdot V_{in} + \left[K \cdot \alpha_3 - \frac{\alpha_1 \cdot K^3}{6} \right] V_{in}^3 + O(V_{in}^5) \qquad (4.16)$$

To cancel the CTB contribution of the modulator, G has to satisfy the following condition:

$$G = \frac{V_\pi}{\pi} \sqrt{\frac{6\alpha_3}{\alpha_1}} \qquad (4.17)$$

Since fifth- and higher odd order distortions are still present, a practical linearizer design has to be further optimized to minimize their contribution at the fiber node receiver. In the general model, the effective transfer function $K(\omega)$ becomes

$$K(\omega) = G(\omega)\frac{\pi}{V_\pi}M(\omega) \qquad\qquad (4.18)$$

If the power series expansion neglects fifth- or higher order terms, the linearization problem is equivalent to analyzing a system consisting of a linear filter $K(\omega)$ sandwiched between two memoriless nonlinearities. It follows from such an analysis that a flat frequency response of the modulator is required for good frequency linearization. The frequency response flatness can be affected by various factors such as parasitics of the linearizer circuits, RF amplifier ripples, electrode reactance and losses, and a large velocity mismatch between the RF and optical fields in the modulator waveguide.

4.2.3 Linearizer Circuits

So far, we have learned about the predistortion linearization theory and operation principles.

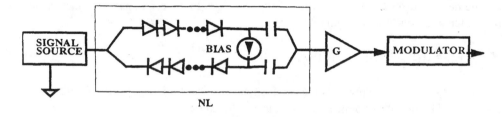

(a)

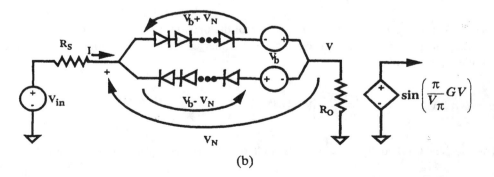

(b)

Figure 4.7 Predistortion circuit (a) and the equivalent circuit model (b). After Ref. [5] (© 1990 IEEE).

Let us examine several examples of linearizer circuits for external optical modulators. Figure 4.7 shows an example of a linearizer circuit and the equivalent circuit model [9]. The circuit takes advantage of a parallel combination but in opposite directions of two or more balanced diodes to suppress the magnitude of third-order distortion. The DC current source biases the diodes to a given bias, and the capacitors are charged to first approximation to a fixed large voltage and are modeled as a voltage source as shown in Figure 4.7(b). Using the well-known diode characteristic, $I = I_s \exp(V/V_T)$, where I_s is the leakage current, $V_T = k_B T/e$, and the output current through the resistor R_0 is given by:

$$I = I_s\left[e^{(V_B+V_N)/NV_T} - e^{(V_B-V_N)/NV_T} \right] = 2I_s \cdot e^{V_B/NV_T} \sinh\left[\frac{V_N}{NV_T} \right] = 2I_B \sinh\left[\frac{V_N}{NV_T} \right] \qquad (4.19)$$

where N is the number of diodes in each branch, V_B and V_N are defined in Figure 4.7(b), and I_B is the current flowing through the diode with no applied signal. The voltage-current characteristic can easily be obtained by inverting Equation (4.19):

$$V_N = NV_T \cdot a\sinh\left[\frac{I}{2I_B} \right] \qquad (4.20)$$

Using Equation (4.20), we can expand V_N in power series as

$$V_N = r_1 \cdot I + r_3 \cdot I^3 + ... = \left[\frac{NV_T}{2I_B} \right] I - \left[\frac{NV_T}{48I_B^3} \right] I^3 + ... \qquad (4.21)$$

From the equivalent circuit model, the input voltage is related to the diode current by

$$V_{in} = R_s \cdot I + (r_1 \cdot I + r_3 \cdot I^3) + R_0 \cdot I = (R_0 + R_s + r_1)I + r_3 \cdot I^3 \qquad (4.22)$$

Inverting Equation (4.22) yields the following power series

$$I = g_1 \cdot V_{in} + g_3 \cdot V_{in}^3 = \frac{V_{in}}{R_s + R_0 + r_1} - \left[\frac{r_3}{(R_s + R_0 + r_1)^4} \right] V_{in}^3 \qquad (4.23)$$

The linearizer circuit output voltage is given by

$$V_{out} = R_0 \cdot I = \alpha_1 \cdot V_{in} + \alpha_3 \cdot V_{in}^3 = R_0 \cdot g_1 \cdot V_{in} + R_0 \cdot g_3 \cdot V_{in}^3 \qquad (4.24)$$

where the coefficients g_1 and g_3 are given by Equation (4.23). Equation (4.24) provides the dependence of the output voltage on the bias current and the number of diodes, allowing one to complete the linearizer circuit design by applying Equation (4.17) to determine the linearization gain.

4.3 Optical Linearization Methods

Optical linearization can also be used instead of electronic linearization circuit. Optical linearization methods can be classified into three categories as follows: (A) an optical dual parallel linearization, (B) an optical dual cascade linearization, and (C) an optical feedforward linearization. The operation principles as well as the practicality of each method will be briefly reviewed in the following sections.

4.3.1 Optical Dual Parallel Linearization Method

The optical dual parallel linearization method has been proposed by Korotky et al. [11] and implemented by Brooks et al. [12]. A schematic configuration of an MZI modulator using the optical dual parallel linearization method is shown in Figure 4.8. In this configuration, two MZI modulators, which are called "primary" (top) and "secondary" (bottom), are biased at two 180° out-of-phase Q points such that their optical output power is given by Equation (4.4). In practice, the static phase of each MZI modulator may be different. Consequently, a

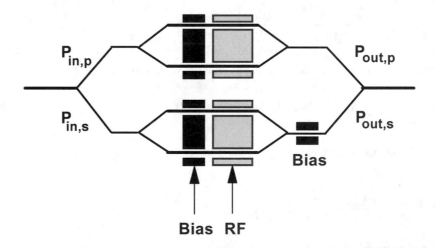

Figure 4.8 External modulator configuration employing the dual parallel linearization method. After Ref. [10] (© 1993 IEEE).

phase modulator at the output of the secondary MZI modulator is used to maintain the required π phase shift between this modulator and the primary MZI modulator. In order to understand the operation principal of the dual parallel linearization method, let us assume that the RF drive signal is asymmetrically split between the two modulators with ratio 1:α^2 ($\alpha > 1$) such that the secondary MZI modulator operates at higher modulation depth and generates greater distortion than the primary MZI modulator. Assuming both modulators are biased at their Q point with π relative phase shift, the output optical power of the primary and secondary MZI modulators is given by

$$P_{primary}(t) = \langle P_p \rangle \left[1 - \sin\left(\frac{\pi \cdot V(t)}{V_\pi}\right) \right] \approx \langle P_p \rangle \left[1 - \left(\frac{\pi \cdot V(t)}{V_\pi}\right) + \frac{1}{6}\left(\frac{\pi \cdot V(t)}{V_\pi}\right)^3 \right] \qquad (4.25)$$

$$P_{sec\,ondary}(t) = \langle P_s \rangle \left[1 + \sin\left(\frac{\alpha \cdot \pi \cdot V(t)}{V_\pi}\right) \right] \approx \langle P_s \rangle \left[1 + \left(\frac{\alpha \cdot \pi \cdot V(t)}{V_\pi}\right) - \frac{1}{6}\left(\frac{\alpha \cdot \pi \cdot V(t)}{V_\pi}\right)^3 \right] \qquad (4.26)$$

where $\langle P_p \rangle$ and $\langle P_s \rangle$ are the average optical output power of the primary and secondary MZI modulators. The combined output power from both modulators is given by

$$P_{total} = \langle P_p \rangle + \langle P_s \rangle - \left[\frac{\pi \cdot V(t)}{V_\pi}\right]\left(\langle P_p \rangle - \alpha\langle P_s \rangle\right) + \frac{1}{6}\left[\frac{\pi \cdot V(t)}{V_\pi}\right]^3\left(\langle P_p \rangle - \alpha^3\langle P_s \rangle\right) \qquad (4.27)$$

The condition to cancel the third-order distortion of the MZI modulator is easily obtained from Equation (4.27) as

$$\langle P_p \rangle = \alpha^3 \langle P_s \rangle \qquad (4.28)$$

Substituting Equation (4.28) into Equation (4.27), the magnitude of the fundamental component of the signal $P^{(1)}$ is given by

$$P^{(1)} = \frac{\pi |V(t)|}{V_\pi}\langle P_p \rangle\left[1 - \frac{1}{\alpha^2}\right] \qquad (4.29)$$

and the magnitude of the DC component $P^{(0)}$ is given by

$$P^{(0)} = \langle P_p \rangle\left[1 + \frac{1}{\alpha^3}\right] \qquad (4.30)$$

Using Equations (4.29) and (4.30), we find that the optical modulation index after the linearization is reduced by

$$\frac{1-\alpha^{-2}}{1+\alpha^{-3}} = \frac{\alpha(\alpha-1)}{1-\alpha+\alpha^2} \tag{4.31}$$

compared with the unlinearized MZI modulator. For example, for $\alpha = 3$, the optical modulation index is reduced to about 86% of that of a single MZI modulator.

    This linearization method is not as practical as it may seems since the accuracy of the relative phase shift between the two MZI has to be controlled extremely tightly by the phase modulator. For example, to achieve a CSO distortion of –65-dBc with $\alpha = 3.3$, the relative phase shift must be $|\Delta\phi| < 0.06°$, and to achieve a –70 dBc CTB distortion, the relative phase shift must be $|\Delta\phi| < 0.54°$. Another possible solution is to replace the phase modulator with two independent optical sources that must be accurately maintained at a constant optical power ratio. However, the cost of two optical sources as well as the use of a secondary MZI modulator to linearize the primary MZI modulator makes this solution prohibitively expensive and unpractical.

4.3.2 Optical Dual Cascade Linearization Method

Another optically linearized modulator configuration is shown in Figure 4.9 [17–20]. The dual stage modulator consists of Y-junction splitter, two push-pull phase modulators with their separate RF electrodes and DC bias voltages, and two directional couplers, resulting in two output waveguide ports. For a 2x2 directional coupler modulator, the output optical power from the upper-waveguide port, for example, is modified from Equation (4.4) as:

$$P_1 = \langle P \rangle [1 - \sin(\theta + \phi_b)\sin(2\Gamma)] \tag{4.32}$$

where $\theta = \pi V/V_\pi$ is the normalized modulating voltage, $\Gamma = \kappa L$, κ is the coupling coefficient between the two waveguides, and L is the interaction length. In the dual cascade optically linearized modulator, the bias voltages are required to maintain the modulator operation at the Q point. To analyze the performance of this optical linearization method, let us assume that the RF input signal is split into a normalized voltage signal θ for RF_1 and $\rho\theta$ for RF_2. In addition, let us assume that the couplings between the two branches of the first and second directional couplers are Γ_1 and Γ_2, respectively. It can be shown that by multiplying the transfer matrix for each section of this modulator its output power transfer function is given by

$$P(V) = \langle P \rangle \begin{Bmatrix} 1 - \sin(2\Gamma_2)\cos(\theta + \phi_1)\sin(\rho\theta + \phi_2) \\ -[\sin(2\Gamma_1)\cos(2\Gamma_2) + 2\Gamma_1 \cdot \sin(2\Gamma_2)\cos(\rho\theta + \phi_2)]\sin(\theta + \phi_1) \end{Bmatrix} \tag{4.33}$$

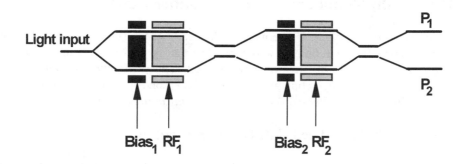

Figure 4.9 Basic configuration of a dual cascade optically linearized modulator. After Ref. [19] (© 1995 IEEE).

where ϕ_1 and ϕ_2 are the bias offset angles. The transfer function $P(V)$ can be linearized by first biasing the modulator at the Q point, namely, setting $\phi_1 = \phi_2 = 0$, to ensure that even higher order distortion contributions are eliminated. The third-order distortions can be canceled by requiring that $d^3P/dV^3 = 0$. This can be achieved by either fixing both Γ_1 and Γ_2 with an adjustable ρ or vice versa. From a practical perspective, it is easier to fix $\Gamma_1 = \Gamma_2 = \Gamma$ when the waveguide is fabricated, allowing ρ to vary. Consequently, the modulator transfer function is reduced to

$$P(V) = \langle P \rangle \left\{ 1 - \frac{1}{2} \sin(4\Gamma)\sin(\theta)\left[1 + \cos(\rho\theta)\right] - \sin(2\Gamma)\cos(\theta)\sin(\rho\theta) \right\} \tag{4.34}$$

The optimum coupling values for this modulator are in the range of 58°–65°. For $\rho = 0$, there is no linearization since the transfer function is pure sinusoidal. The optimum linearization of Equation (4.33) is obtained for $\rho = 0.5$. However, in practice, a slightly larger value of ρ is optimal due to the contribution of the odd higher order distortions. The optimal value of ρ is determined by the number of cable TV channels to be modulated and optical modulation index per channel.

This optical linearization method has two main disadvantages. First, due to asymmetric losses between the upper and lower branches of each of the directional couplers, it is difficult to obtain the same linearization for both optical outputs simultaneously compared with a single optimized output. Consequently, only one linearized optical output is typically being implemented. Second, this method has double the insertion losses compared with a single linearized stage modulator, which may limit the maximum optical output power.

4.4 Externally Modulated Laser Transmitter Design

There are two primary types of externally modulated laser transmitters that are used in the cable TV industry and will be reviewed here. The first type, which became commercially available in the early 1990s, is an externally modulated (EM) *yttrium aluminum garnet* (YAG) laser transmitter operating at 1319-nm. The second type, which became commercially available later, is an EM-DFB laser transmitter operating at 1550-nm.

4.4.1 Externally Modulated YAG Laser Transmitter

The first type of an EM laser transmitter, which was developed in the early 1990s, is the high-power Nd:YAG laser transmitter, which operates at 1319-nm [3]. Figure 4.10 shows a simplified energy level diagram of a YAG crystal doped with Nd atoms. When pumped with 808-nm light, the Nd atoms are excited to pump bands, which decay nonradiatively to a metastable state. The stimulated emission at 1319-nm is emitted from the metastable state as shown in Figure 4.10. The Nd:YAG laser cavity, which is shown in Figure 4.11, consists of a high-power (0.5–1 W) AlGaAs diode array with a broad active area mounted on a TEC, a Nd:YAG crystal, and micro-optic collimating and coupling lenses to shape and focus the 1319-nm light. Using selective coatings on the Nd:YAG crystal and output mirror, the 1319-nm light is confined to a resonant cavity overlapping the pump volume. Only a small fraction (1–2%) of the intracavity light is coupled through the output mirror. The output po-

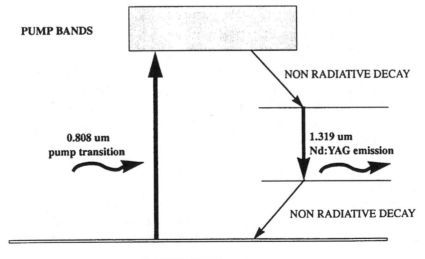

Figure 4.10 Energy-level diagram of Nd:YAG. After Ref. [3] (© 1993 IEEE).

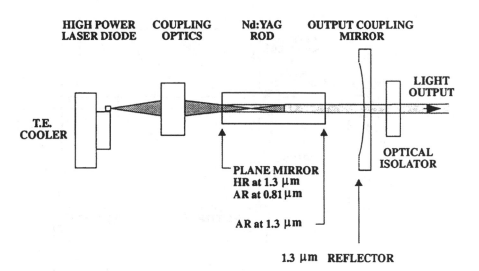

Figure 4.11 Schematic of laser diode array-pumped Nd:YAG laser cavity. After Ref. [3] (© 1993 IEEE).

wer from Nd:YAG laser typically ranges from 50 to more than 200 mW. The laser output beam is a almost perfectly shaped Gaussian beam, which allows efficient coupling to single-mode waveguides. In addition, the output beam is linearly polarized with an extinction ratio typically better than 23 dB. Due to the very low cutoff frequency ($\approx$ 200-kHz) of the Nd:YAG crystal, the high-frequency noise is filtered, resulting in a very low RIN (< −165 dBc/Hz). The external optical modulator typically has a total loss of about 8 dB, including a 3-dB loss due to the DC bias at the quadrature point, a 3-dB loss due to the directional coupler, waveguide-fiber coupling and waveguide propagation losses. A block diagram of the EM Nd:YAG laser transmitter with predistortion linearization and parametric CSO and CTB control is illustrated in Figure 4.12. To achieve the long-term CSO and CTB performance requirements of the EM linearized laser transmitter, parametric feedback loops are implemented. In order to maintain the modulator bias at the Q point ($\phi_T = 0$), an error signal, which is proportional to ϕ_T, is detected and processed to generate a corrected bias voltage to the modulator. A separate closed loop is used to control the CTB performance by controlling the relationship between the gain and the nonlinearity of the linearizer. For an optical budget link of 13 dB, assuming 0 dBm at the optical receiver with 77 AM-VSB video channels, the CSO and CTB distortions are typically less than −70-dBc and −65-dBc using the EM Nd:YAG laser transmitter. ·

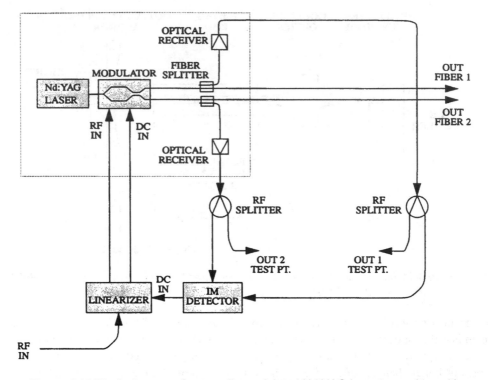

Figure 4.12 Block diagram of externally modulated Nd:YAG laser transmitter with pre-distortion linearization and parametric CSO and CTB control. After Ref. [3] (© 1993 IEEE).

Cascading predistorted externally modulated laser transmitters has an additional advantage in terms of CTB performance compared with directly modulated DFB laser transmitters [21]. When cascading two directly modulated DFB laser transmitter based links, the CTB distortions behave according to Equation (2.30), which is also called the "20 log" rule. This rule means that if the CTB distortion of each link is –65 dBc, the total CTB distortion of the cascaded link is degraded by 6-dB to –59-dBc. It turns out that the CTB distortions of cascaded predistorted EM YAG laser transmitters follow instead the "10 log" rule. This is because the distortion beats are randomized and uncorrelated with one another between the two externally modulated links. Therefore, the resultant CTB distortion of the cascaded link is degraded by only 3-dB to –62-dBc. In general, the CTB distortions are added according to the "X log" rule, where X can vary from 0 to 20, depending on the type of the type of linearization scheme and the cable TV channel frequency. The CTB performance advantage of the EM YAG laser transmitters was a very popular solution for cable TV supertrunk applications with links exceeding 30-km before low-cost high-power EDFAs became readily available.

4.4.2 Externally Modulated DFB Laser Transmitter

An externally modulated DFB laser transmitter operating at 1550-nm is the most widely used transmitter for broadcasting and narrowcasting applications over cable TV networks. Figure 4.13 shows a simplified block diagram of an externally modulated DFB laser transmitter operating at 1550-nm with a built-in boost optical amplifier. The optimal RF input signal range to the external modulator is controlled by the AGC circuitry, while the modulator bias point is controlled by the microprocessor. As was discussed previously in Chapter 1, to meet the CNR performance requirements for many cable TV network configurations, it is necessary to significantly increase the output optical power of these transmitters by incorporating a booster fiber amplifier. However, the maximum optical output power of these laser transmitters that can be launched into the fiber may be limited by a stimulated Brillouin scattering (SBS) effect [22]. The SBS optical power limit, which is discussed in detail in Chapter 9 can be increased by applying a single frequency tone whose frequency is more than twice the maximum transmitted signal frequency ($\approx$ 2-GHz), or by dithering the optical frequency of the DFB laser transmitter [23].

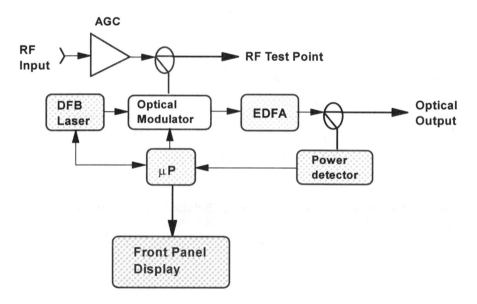

Figure 4.13 A simplified block diagram of a 1550-nm externally modulated DFB laser transmitter with a built-in EDFA.

Figure 4.14 shows, for example, the measured worst-case CSO and CTB distortions as a function of the optical modulation index per channel for externally modulated optically linearized 1550-nm DFB laser transmitter [24]. Seventy-nine simulated AM video channels were transmitted 50-km over a standard SMF. The optimal modulation index per channel of this transmitter is 3.1% to meet the CTB distortion requirement of –65 dBc. Notice that the CTB distortion rapidly degrades as the optical modulation index increases beyond its optimal value. Consequently, the maximum possible CNR becomes limited at the fiber node receiver.

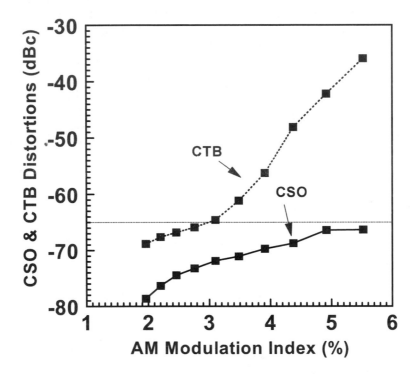

Figure 4.14 Measured worst-case CSO and CTB distortions versus the optical modulation index per channel using an 1550-nm externally modulated optically
linearized DFB laser transmitter. After Ref. [24] (© 1998 IEEE).

References

1.   R. S. Weis and T. K. Gaylord, "Lithium Niobate: Summary of Physical Properties and Crystal Structure," *Applied Physics A* **37**, 191–203 (1985).

2. M. Nazarathy, J. Berger, A. J. Ley, I. M. Levi, and Y. Kagan, "Progress in Externally Modulated AM CATV Transmission Systems," *IEEE Journal of Lightwave Technology* **11**, 82–105 (1983).

3. F. W. Willems, W. Muys, and J. S. Leong, "Simultaneous Suppression of Stimulated Brillouin and Interferometric Noise in Externally Modulated Lightwave AM-SCM Systems," *IEEE Photonics Technology Letters* **12**, 1476–1478 (1994).

4. H. Skeie, "An Optically Linearized Modulator for CATV Applications," *Proceedings of SPIE*, **2291**, 227–238 (1994).

5. R. B. Childs and V. A. O'Byrne, "Multichannel AM Video Transmission Using a High-Power Nd:YAG Laser and Linearized External Modulator," *IEEE Journal on Selected Areas in Communications* **8**, 1369–1376 (1990).

6. R. S. Cheng, W. L. Chen, and W. S. Wang, "Mach-Zehnder Modulators with Lithium Niobate Ridge Waveguides Fabricated by Proton-Exchange Wet-Etch and Nickel Indiffusion," *IEEE Photonics Technology Letters* **7**, 1282–1284 (1995).

7. H. Nagata, K. Kiuchi, and T. Saito, "Studies of Thermal Drift as a Source of Output Instabilities in Ti:LiNbO$_3$ Optical Modulators," *Journal of Applied Physics* **75**, 4762–4764 (1994).

8. P. G. Suchoski Jr. and G. R. Boivin, "Reliability and Accelerated Aging of LiNbO3 Integrated Optic Fiber Gyro Circuits," OE Fibers, Components, Networks, and Sensors conference, Boston, September 8-11 (1992).

9. H. Skeie, An Optically Linearized Modulator for CATV Applications," *Proceedings of SPI,* **2291**, 227–238 (1994).

10. K. D. LaViolette, "CTB Performance of Cascaded Externally Modulated and Directly Modulated CATV Transmitters," *IEEE Photonics Technology Letters* **8**, 281–283 (1996).

11. W. Wang, R. Tavlykaev, and R. V. Ramaswamy, "Bandpass Traveling-Wave Mach-Zehnder Modulator in LiNbO$_3$ with Domain Reversal," *IEEE Photonics Technology Letters* **9**, 610–612 (1997).

12. M. Seino, T. Nakazawa, Y. Kubota, M. Doi, T. Yamane, and H. Hakogi, "A low DC-drift Ti:LiNbO$_3$ Modulator Assured over 15 Years," *Technical Digest of Optical Fiber Communications Conference*, Post-deadline paper PD-3 (1992).

13. D. A. Atlas, "On the Overmodulation Limit in Externally Modulated Lightwave AM-VSB CATV Systems," *IEEE Photonics Technology Letters* **8**, 697–699 (1996).

14. G. C. Wilson, "Optimized Predistortion of Overmodulated Mach-Zehnder Modulators with Multicarrier Input," *IEEE Photonics Technology Letters* **9**, 1535–1537 (1997).

15. J. L. Brooks, G. S. Maurer, and R. A. Becker, "Implementation and Evaluation of a Dual Parallel Linearization System for AM-SCM Video Transmission," *IEEE Journal of Lightwave Technology* **11**, 34–41 (1993).

16. S. K. Korotky and R. M. deRidder, "Dual Parallel Modulation Schemes for Low-Distortion Analog Optical Transmission," *IEEE Journal on Selected Areas in Communications* **8**, 1377–1381 (1990).

17. H. Skeie and R. V. Johnson, "Linearization of Electro-Optic Modulators by a Cascade Coupling of Phase Modulating Electrodes," Integrated Optical Circuits, *SPIE proceedings*, **1583**, 153–164 (1991).

18. W. K. Burns, "Linearized Optical Modulator with Fifth Order Correction," *IEEE Journal of Lightwave Technology* **13**, 1724–1727 (1995).

19. D. J. M. Sabido IX, M. Tabara, T. K. Fong, C.-Li Lu, and L. Kazovsky, "Improving the Dynamic Range of a Coherent AM Analog Optical Link Using a Cascaded Linearized Modulator," *IEEE Photonics Technology Letters* **7**, 813–815 (1995).

20. J. D. Farina, B. R. Higgins, and J. P. Farina, "New Linearization Technique for Analog Fiber-Optic Links," Technical Digest, Optical Fiber Communications conference, paper ThR6, San Jose, CA (1996).

21. M. Nazarathy and Y. Simler, "Integrated Networks for CATV Transmission and Distribution," *Cable Telecommunication Engineering* **16**, Numbers 3 and 4 (1994).

22. X. P. Mao, G. E. Bodeep, R. W. Tkach, A. R. Chraplyvy, T. E. Darcie, and R. M. Derosier, "Brillouin Scattering in Externally Modulated Lightwave AM-VSB CATV Transmission Systems," *IEEE Photonics Technology Letters* **4**, 287–289 (1992).

23. F. W. Willems, W. Muys, and J. S. Leong, "Simultaneous Suppression of Stimulated Brillouin Scattering and Interferometric Noise in Externally Modulated Lightwave AM-SCM Systems," *IEEE Photonics Technology Letters* **12**, 1476–1478 (1994).

24. S. Ovadia and C. Lin, "Performance Characteristics and Applications of Hybrid Multichannel AM/M-QAM Video Lightwave Transmission Systems," *IEEE Journal of Lightwave Technology* **16**, 1187–1207 (1998).

CHAPTER 5

LIGHTWAVE RECEIVERS FOR CABLE TV NETWORKS

The primary role of optical receivers in cable TV networks, which are typically located at both the fiber nodes and at the cable TV headends, is to convert the transmitted optical signals to RF signals. To understand the design and performance requirements of lightwave receivers, let us first discuss the physics of a p-i-n photodetector and define its basic parameters. In Section 5.2, we will review the three primary noise sources in optical receivers, namely, shot noise, thermal noise, and laser relative-intensity noise (RIN). Using this knowledge, the calculation of the SNR or CNR at the optical receiver for an SCM fiber-optics system is explained in Section 5.3. Section 5.4 will explain the nonlinear behavior of the p-i-n photodetector in order to be effectively used in cable TV networks. Finally, the basic cable TV receiver designs, including low- and high-impedance front-end designs, transimpedance front-end design, and various circuit configurations for high-performance operation are discussed in Section 5.5.

5.1 p-i-n Photodiode

Photodetectors are semiconductor devices that convert optical signals to electrical signals. The p-i-n photodiode, which is the most commonly used photodetector, consists of p-type and n-type semiconductors separated by lightly doped intrinsic semiconductor material. In n-type semiconductors, the majority carriers are electrons, while in p-type semiconductors, the majority carriers are holes. In an intrinsic semiconductor, which contains only small amounts of impurities compared with the thermally generated electron and holes, the density of electrons (per unit volume) in the conduction band is equal to the density of the holes in the valence band ($n = p = n_i$).

Before we examine the characteristics of p-i-n photodiodes, let us first explain the basic concepts of p-n photodiodes. A p-n junction is formed by physically joining uniformly doped p-type and n-type semiconductors. The importance of the p-n junction will be clear later. When an electric field is applied to an n-type semiconductor in thermal equilibrium (assuming uniform doping), the electrons will be accelerated between collisions in the opposite direction of the field. The composite displacement of an electron is due to its random thermal motion and the drift velocity is due to the electric field. The drift velocity can be written as $-\mu_n E$, where the factor μ_n is called the electron mobility. A similar expression can be written for the holes in the valence band. In general, the equations for the drift current densities of the holes and electrons can be written as [1]

$$J_n^{drift} = -q \cdot n(x,t) \cdot v_n(E)$$
$$J_p^{drift} = q \cdot p(x,t) \cdot v_p(E)$$

(5.1)

where q is the electronic charge, $n(x,t)$ and $p(x,t)$ are the densities of the electrons and holes, respectively, and $v_n(E)$ and $v_p(E)$ are the electric-field dependent hole and electron drift velocities, respectively, which will be discussed in Section 5.4.

Another important process in semiconductors is called the diffusion process, which describes the spatial variation of the carrier concentration in the semiconductor material. When an electric field is applied to a semiconductor at uniform temperature, assuming only that the electron and hole densities vary along the x-axis, then the total current density of the electrons and holes is simply the sum of the drift and diffusion components, and can be written as [2]:

$$J_n = J_n^{drift} + J_n^{diff} = qn \cdot \mu_n \cdot E + qD_n \cdot \frac{dn}{dx}$$
$$J_p = J_p^{drift} + J_p^{diff} = q\mu_p \cdot pE - qD_p \cdot \frac{dp}{dx}$$

(5.2)

where, D_n and D_p are the electron and hole diffusion constants, respectively, which are related to the electron and hole mobilities through the *Einstein relation* as follows [2]:

$$D_i = \left[\frac{kT}{q} \right] \mu_i$$

(5.3)

where i = n or p. Notice that the diffusion current is proportional to the spatial derivative of the electron density. Now, consider a direct band-gap semiconductor in thermal equilibrium, where the relation $np = n_i^2$ is valid. When the p-n junction is forward-biased and/or a light is incident on the p-type semiconductor, the thermal equilibrium is disturbed (i.e., $np > n_i^2$) by the injection of additional carriers, which are called excess carriers. The process of recombination of the injected minority carriers with the majority carriers restores the p-n junction to thermal equilibrium. The recombination process can be radiative with the emission of photons or nonradiative. The rate of direct recombination is proportional to the number of available electrons in the conduction band and holes in the valence band. To maintain thermal equilibrium, the carrier generation rate $G(x,t)$ must be equal to the carrier recombination rate $R(x,t)$. Combining the effects of carrier drift, diffusion, recombination, and generation, the operation of the p-n junction in one dimension is governed by the continuity equations as follows [1]:

$$\frac{\partial n(x,t)}{\partial t} = G(x,t) - R(x,t) + v_n \frac{\partial n}{\partial x} + n \frac{\partial v_n}{\partial x} + \frac{1}{q} \frac{\partial J_n^{diff}}{\partial x}$$

(5.4)

$$\frac{\partial p(x,t)}{\partial t} = G(x,t) - R(x,t) - v_p \frac{\partial p}{\partial x} - p \frac{\partial v_p}{\partial x} - \frac{1}{q} \frac{\partial J_p^{diff}}{\partial x}$$

where v_n and v_p are the electron and hole velocities, respectively. The continuity equations are linear only if the carrier velocities (i.e., v_n and v_p) are independent of the carrier densities.

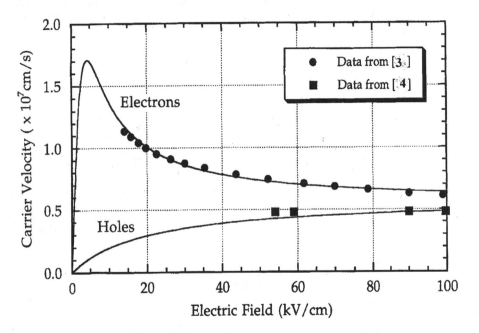

Figure 5.1 Measured and fitted electron and hole velocities versus the electric field in InGaAs. After Ref. [1] (© IEEE 1996).

Figure 5.1 shows the measured and fitted electron and hole velocities versus the electric fields in n-type InGaAs at room temperature. The electron velocity measurements have been fitted using the following empirical expression [3]:

$$v_n(E) = \frac{E\left[\mu_n + v_n^{hf} \cdot \beta|E|\right]}{1 + \beta \cdot E^2} \tag{5.5}$$

where v_n^{hf} is the high-field electron velocity, and β is a fitting parameter. Similarly, hole velocity measurements have also been fitted using the following expression [4]:

$$v_p(E) = \frac{\mu_p \cdot v_p^{hf} \cdot E}{\left[(v_p^{hf})^\gamma + (\mu_p \cdot E)^\gamma\right]^{1/\gamma}} \tag{5.6}$$

where v_p^{hf} is the high-field hole velocity, and γ is a temperature-dependent factor. The measured low-field electron and hole mobilities are 8000 cm^2/Vs and 300 cm^2/Vs, respectively [3, 4]. Notice that for $\beta = 0$, the carrier drift velocities become the low-field carrier mobility times the applied electric field.

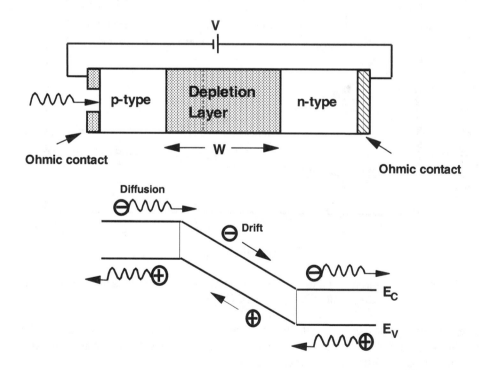

Figure 5.2 (a) Operation of p-i-n photodiode under reverse bias, and (b) energy-band diagram under reverse bias.

The large carrier concentration gradient across the p-n junction, holes from the p-side diffuse into the n-side, and electrons from the n-side diffuse into the p-side. Consequently, a negative space charge is formed on the p-side and a positive space charge is formed at the n-side, creating an electric field pointing from the n-side to the p-side. The electrostatic potential ψ distribution across the p-n junction is given by the Poisson's equation [1, 2]

$$\frac{d^2\psi}{dx^2} = -\frac{dE}{dx} = -\frac{q}{\varepsilon_s}\left[N_D - N_A + p - n\right] \tag{5.7}$$

where N_D and N_A are the densities of ionized donor and acceptor dopants present in the crystal, respectively, and ε_s is the dielectric constant. The total electrostatic potential difference between the p-side and n-type at thermal equilibrium is given by

$$V_{bi} = \psi_n - \psi_p = \frac{k_B T}{q}\ln\left[\frac{N_A N_D}{n_i^2}\right] \tag{5.8}$$

where V_{bi} is called the built-in potential. The space-charged region across the p-n junction, where the number of the mobile carriers is essentially zero, is called the depletion region, and its width W is given by [2]:

$$W = \left[\frac{2\varepsilon_s}{q}\left(V_{bi} - V\right)\left(N_A^{-1} + N_D^{-1}\right)\right]^{1/2} \tag{5.9}$$

where V is the applied voltage across the p-n junction. The built-in potential value, which depends on the specific semiconductor material, is, for example, 1.1 volts for GaAs.

According to Equation (5.9) the depletion-region width is reduced when the p-n junction is forward-biased, namely, the positive voltage is applied to the p-side relative to the n-side. To increase the depletion-region width, a layer of lightly doped intrinsic semiconductor is inserted between the p-n junction. Figure 5.2(a) shows the operation of such a structure, which is called p-i-n photodiode under reverse bias. The corresponding energy-band diagram, which is illustrated in Figure 5.2(b), shows the electron and hole movement through the diffusion and drift processes. Increasing the thickness of the intrinsic layer allows one to optimize the width of the depletion layer based on the photodiode requirements. It should be pointed out that increasing the intrinsic layer thickness also increases the photodetector response time, since it takes longer for the carriers to drift across the depletion layer. When light is incident on the p-side of the p-i-n junction, electron-hole pairs are created through absorption in the depletion region, where the large electric field separates the electrons and holes to the n-side and p-side, respectively. The generated photocurrent is proportional to the incident light power, where the proportionality factor R is called the *responsivity* (A/W) of the photodetector. The quantum efficiency (η) of the photodetector is defined as the ratio

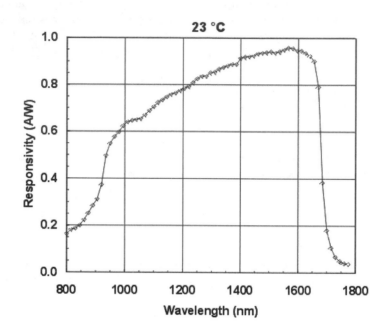

Figure 5.3 Typical spectral response of InGaAs p-i-n photodetector at 23°C. Courtesy
of JDS Uniphase, Inc., Epitaxx division.

of the generated electron rate to the incident photon rate. The responsivity is related to the
quantum efficiency through the following relation [5]:

$$R(\lambda) = \frac{q\eta(\lambda)}{h\nu} = \frac{q\lambda\eta(\lambda)}{hc} \tag{5.10}$$

where the wavelength dependence of the quantum efficiency is through the absorption coefficient α. The measured spectral responsivity at room temperature of a typical InGaAs p-i-n
photodetector is shown in Figure 5.3.   Notice that the spectral responsivity of these
photodetectors peaks around 1550-nm at 0.9-A/W, making them particularly useful for cable
TV networks.   Assuming the incident light has a normal incidence with reflectivity r, and
each absorbed photon generates an electron-hole pair, the quantum efficiency is given by

$$\eta(\lambda) = (1 - r)[1 - \exp(-\alpha(\lambda) \cdot W] \tag{5.11}$$

According to Equation (5.11), almost 100% quantum efficiency can be achieved by increasing the depletion-region width. Notice that, as expected, $\eta \rightarrow 1$ as $\alpha \rightarrow 0$, assuming $r = 0$. To be useful in lightwave communication systems, photodetectors have to operate below the so-called cutoff wavelength ($\lambda < \lambda_c$), where α approaches zero. Both InGaAs and Ge semiconductors have high quantum efficiency (> 85%) in the 1.3-µm to 1.55-µm region. However, compared with InGaAs photodetectors, Ge photodetectors require high reverse bias voltages around 30V, and their quantum efficiency drops rapidly above 1.55-µm. Consequently, InGaAs photodetector are typically preferred.

It should be pointed out that the performance of p-i-n photodetectors could be improved by using a double-heterostructure design, where the intrinsic layer is sandwiched between n-type and p-type layers of a different semiconductor material such that the light is absorbed only in the intrinsic layer. For example, n-type and p-type layers of InP are transparent to the light in the 1.3–1.55-µm wavelength region, but are strongly absorbed by InGaAs intrinsic layer of the photodetector. By applying antireflection coating to the front facet of the photodetector, its quantum efficiency can be increased to almost 100%. Practical p-i-n photodetectors need ohmic contacts not only to apply the necessary voltage bias, but also to connect to other electronic devices. Ohmic contact is a metal-semiconductor contact that has a negligible contact resistance relative to the bulk resistance of the semiconductor.

5.2 Noise Sources in Lightwave Receivers

There are three fundamental noise mechanisms in p-i-n photodetectors when the incident optical signal is converted into electrical current. The noise mechanisms are (A) shot noise, (B) thermal noise, and (C) laser RIN noise. In the following subsections, we will review each of these noise mechanisms, which will better allow us to discuss the signal-to-noise ratio (SNR) or carrier-to-noise-ratio (CNR) of the received signal.

5.2.1 Shot Noise

Shot noise in photodetectors is a quantum noise, which is due to the random generation of electron-hole pairs when the photodetector is illuminated by photons. The shot-noise phenomenon was first investigated by Campbell in 1909 [6], Schottky in 1918 [7], and by many other researchers since then [8–9]. To derive the noise variance of the photocurrent generated in response to an optical signal with constant amplitude, let us make the following assumptions:

- The probability of generating a single electron-hole pair in a very small time interval Δt is proportional to Δt.
- The probability of generating more than a single electron-hole pair in Δt is negligible.
- The electron-hole pair generation events are statistically independent.

Based on these assumptions, the probability of generating exactly n electron-hole pairs per unit time is described by Poisson statistics, and is given by

$$p(n) = \frac{N_0^n \cdot e^{-N_0}}{n!} \tag{5.12}$$

where N_0 is the average number of received photons in the time interval Δt, which is equal to $P_{in}\Delta t/h\nu$, where P_{in} is the incident optical power and $h\nu$ is the photon energy. Let us also assume that every photon generates an electron-hole pair at the receiver (100% quantum efficiency), then, the average photocurrent is simply qN_0. The noise variance of the photo-current at the receiver per unit frequency bandwidth is given by [5]

$$\sigma_{shot}^2 = \left\langle \delta i_R^2 \right\rangle - \left\langle \delta i_R \right\rangle^2 = q^2 \left[\left\langle n^2 \right\rangle - \left\langle n \right\rangle^2 \right] = q^2 \cdot N_0 = q \cdot I_R \tag{5.13}$$

Mathematically speaking, δi_R, which is the photocurrent fluctuation, was assumed to be a stationary random process with Poisson statistics. Equation (5.13) is also the spectral density of the shot noise, which is frequency independent. If only positive frequencies are considered, the single-sided spectral density becomes $2qI_R$. Under reverse-biased operation, the dark current (I_d), which is the residual photocurrent with no light due, also adds to the photodetector shot noise. Thus, the total photocurrent shot noise variance per unit frequency bandwidth is given by:

$$\sigma_{shot}^2 = 2q\left(I_R + I_d \right) \tag{5.14}$$

5.2.2 Thermal Noise

The electrons move randomly in any conductor due to a finite temperature, which manifests itself as random fluctuations in the current even when no electrical voltage is applied. The random photocurrent fluctuations cause random voltage noise over a load-resistor terminal. The thermal noise is also called Johnson noise or Nyquist noise after the two scientists who first analyzed its behavior experimentally and theoretically [10, 11]. To calculate the effect of thermal noise, one has to recall Planck's relationship for the average energy per mode per unit frequency interval of a blackbody radiation, which is given by the following equation:

$$S_p(\nu) = \frac{2h|\nu|}{\exp\left[\dfrac{h|\nu|}{k_B T}\right] - 1} \tag{5.15}$$

where the factor of 2 accounts for the double-sided spectral density, k_B (= 1.38×10^{-23} J/°K) is the Boltzmann constant, and T is the absolute temperature [12]. Assuming low frequencies

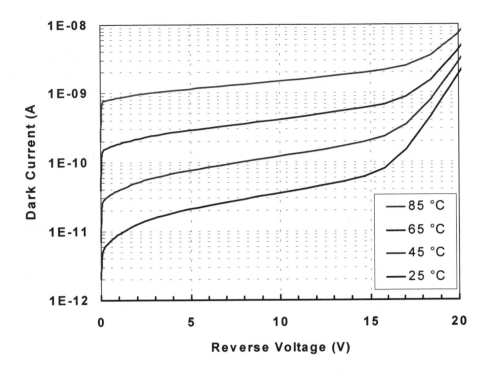

Figure 5.4 Measured dark current versus the reverse voltage at 25° C, 45° C, 65° C, and 85° C for InGaAs p-i-n photodetector. Courtesy of JDS Uniphase, Inc., Epitaxx division.

$(h\nu \ll k_B T)$ operation, and converting the voltage fluctuations to current fluctuations, the double-sided spectral density expression simplifies to

$$S_p(v) = \frac{2k_B T}{R_L} \tag{5.16}$$

where R_L is the load resistor. The open-circuit single-sided spectral density of the photocurrent is given by

$$S_p(v) = \frac{4k_B T}{R_L} = \left\langle \delta i_{th}^2 \right\rangle \tag{5.17}$$

If an RF amplifier with a noise figure F is connected directly to the p-i-n photodetector, then the photocurrent variance per unit frequency interval due to thermal noise is given by

$$\sigma_{th}^2 = \left\langle \delta i_{th}^2 \right\rangle = \frac{4k_B T \cdot F}{R_L} \tag{5.18}$$

At room temperature with a 50-ohm load and a preamplifier with a noise figure of 3, σ_{th} = 31.5–pA/$\sqrt{Hz}$. The effect of thermal noise is demonstrated in Figure 5.4, which shows the measured dark current versus the reverse-bias voltage of a typical InGaAs photodetector at various temperatures. In particular, at low reverse-bias voltages, the dark current increases by almost three orders of magnitude for a 60° C temperature increase.

The thermal noise is sometimes expressed in terms of another useful parameter called noise-equivalent power (NEP), which is defined as the minimum optical power per unit of bandwidth that is required to produce SNR = 1. Therefore, the NEP can be written as

$$NEP = \frac{h\nu \left[\left\langle \delta i_{th}^2 \right\rangle \right]^{1/2}}{q\eta} = \frac{h\nu}{q\eta} \left[\frac{4k_B T \cdot F}{R_L} \right]^{1/2} \tag{5.19}$$

The NEP is useful to estimate the required optical power for a given SNR if the noise bandwidth B is known. Using the NEP definition, the noise-equivalent photocurrent N_R can also be defined as NEPR, which has typical values in the range 1–10-pA/$\sqrt{Hz}$.

5.2.3 Laser RIN Noise

The laser relative intensity noise (RIN) is due to the laser spontaneous emission and fluctuations in the electron density. The laser RIN is defined as

$$RIN \equiv \frac{\left\langle \delta i_{RIN}^2 \right\rangle}{I_R^2} \tag{5.20}$$

where $\langle \delta i_{RIN}^2 \rangle$ is the photocurrent spectral density due to the laser RIN. Thus, the photocurrent variance due to the laser RIN is given by

$$\sigma_{RIN}^2 = \left\langle \delta i_{RIN}^2 \right\rangle = I_R^2 \cdot RIN \tag{5.21}$$

It should be pointed out that the laser RIN has frequency dependence similar to the small-signal modulation response of a DFB laser as was discussed in Chapter 3. In a lightwave communication system, the laser RIN is replaced by the system RIN, which includes the contribution of the various system elements such as the fiber RIN, laser RIN, and EDFA RIN.

5.3 Carrier-to-Noise Ratio at the Receiver

The performance of photodetectors in a lightwave communication system is typically expressed using the signal-to-noise ratio (SNR) or carrier-to-noise ratio (CNR). Let us assume an SCM communication system with m modulation index per RF channel with a DC photocurrent of I_R and an effective noise bandwidth B at the receiver. Then, using the CNR (SNR) definition, the CNR can be written as

$$CNR = \frac{\langle i_R^2 \rangle}{\left[\sigma_{shot}^2 + \sigma_{th}^2 + \sigma_{RIN}^2 \right] \cdot B} \tag{5.22}$$

where $\langle i_R^2 \rangle = (mI_R)^2/2$ is the mean-square signal photocurrent. Substituting Equations (5.13), (5.18), and (5.21) for the shot noise, thermal noise, and RIN noise, respectively, in Equation (5.22) for the CNR at the photodetector, one obtains [13]

$$CNR = \frac{(mI_R)^2}{2B \left[I_R^2 RIN + 2q(I_R + I_d) + \dfrac{4k_B TF}{R_L} \right]} \tag{5.23}$$

To gain further insight into Equation (5.23), let us analyze the CNR in the following three cases. Figure 5.5 illustrates the CNR behavior versus the received photocurrent according to Equation (5.23) as well as the thermal noise, shot noise, and laser RIN contributions to the CNR. We assumed that the photodetector was operating at room temperature with a preamplifier with a noise figure of 3, a 10-kΩ load resistor, laser RIN = −155 dBc/Hz, a 4% modulation index for the transmitted RF channel, a 4-MHz noise bandwidth, and the photodetector dark current was neglected. In most practical cases in which the incident optical power is very small (< -10-dBm), the thermal noise dominates over both the shot noise and the laser RIN in a p-i-n photodetector as shown in Figure 5.5. Therefore, the CNR becomes:

$$CNR = \left[\frac{R_L (mR)^2}{8k_B TBF} \right] P_{in}^2 \tag{5.24}$$

Equation (5.24) shows that the CNR increases as the square of the input optical power in the thermal noise limit. Furthermore, increasing the load resistor and reducing the noise figure of the amplifier can improve the CNR.

Another interesting limit is the shot noise limit, which is where the shot noise dominates over both the thermal noise and the laser RIN. In this limit, the CNR becomes

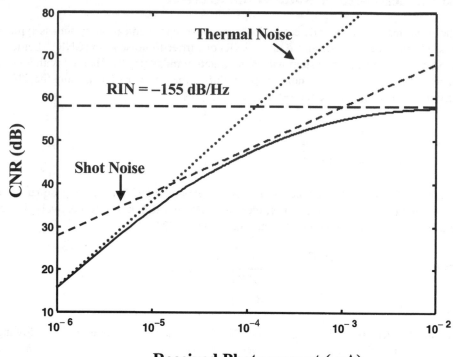

Figure 5.5 Calculated CNR and its components due to shot noise, thermal noise, and laser RIN noise versus the received photocurrent.

$$CNR = \left[\frac{\eta m^2}{4h\nu B}\right]P_{in} \tag{5.25}$$

Notice that the CNR increases linearly with the optical input power. Figure 5.5 shows that the shot noise competes with the thermal noise when a high-impedance load resistor is used (see Section 5.5.1).

The third limit is the laser RIN limit, which is where the laser RIN dominates over both the thermal and the shot noise. In this limit, the CNR does not depend on the photocurrent in the receiver and can be improved by reducing the laser RIN. This limit plays an important role at high optical-input power levels (> 0-dBm), where the CNR at the receiver is upper limited by the laser RIN. Consequently, to maximize the CNR at the fiber-node receiver, directly modulated DFB laser transmitters with RIN of -155 dBc/Hz or less are typically required.

5.4 Nonlinear Behavior of p-i-n Photodetectors

As we have learned in Chapter 2, the transmission of AM-VSB video channels over cable TV networks imposes strict requirements on the magnitudes of the CSO and CTB distortions. This is because visual picture degradation of the transmitted analog channels can be observed if the magnitude of the nonlinear distortions at the subscriber's home is not sufficiently small. To achieve this requirement, the nonlinear distortions not only from the laser transmitter but also from the optical receiver must be sufficiently small to meet the CSO and CTB distortion requirements at the fiber nodes.

The need to develop high-fidelity analog and communications systems has produced many reports on the nonlinearities of p-i-n photodetectors [1, 14–16]. In particular, Williams, et al. analyzed the nonlinear behavior of a single-heterostructure InGaAs p-i-n photodetector [1]. Two primary mechanisms were identified in this device with intrinsic-region length of 0.95-μm, and they are (A) space-charge-induced nonlinearities, and (B) nonlinearities associated with absorption in undepleted semiconductor regions. When the photodetector is illuminated with a high-power optical beam, the carrier velocities and diffusion constant become a function of the carrier densities. The space-charge fields change the carrier velocities according to Equations (5.6) and (5.7) as well as the diffusion constants. Thus, for electric fields below 50-kV/cm, the dominant mechanism for the photodetector nonlinearity is the space-charge effect. As the electric field increases above 10-kV/cm (see Figure 5.1), the dependence of the carrier velocities decreases. Consequently, the contribution of the space-charge field to the photodetector second-order harmonic power is monotonically reduced as shown in Figure 5.6. Contrary to previous work, it was found that the photodetector nonlinearities are insensitive to the load resistor, except for very short devices (0.1–0.2-μm) or large incident beam spot size.

The dominant mechanism for the photodetector nonlinearity at high electric fields is carrier absorption in the undepleted highly doped InGaAs p-contact region, which is adjacent to the edge of the depletion. In contrast to the space-charge field-induced nonlinearity, the p-region carrier absorption mechanism is almost not affected by changing the electric field or the incident beam spot size. This is because the applied bias mainly affects the electric field in the photodetector intrinsic region. Also, the number of available free holes for transport in the p-region is significantly larger than the number of carriers generated by the incident light beam. Figure 5.6 summarizes the dominant regions for the measured second-order harmonic power as a function of the reverse-bias voltage. To minimize the effects of nonlinearities, p-i-n photodiodes should be designed with short intrinsic regions, not long ones. A shorter intrinsic region of the photodiode lends itself to a higher, more uniform, electric field. In addition, the photodiode should be illuminated with uniformly filled light beam spot size and packaged with proper thermal heat sinking.

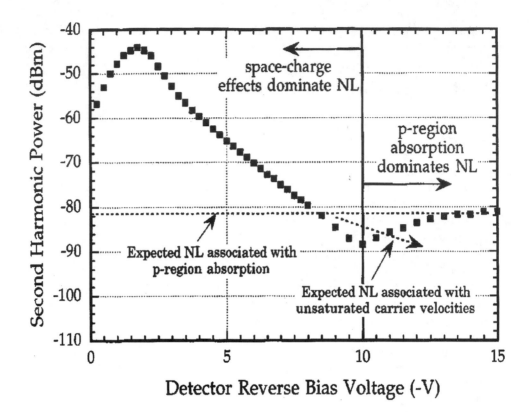

Figure 5.6 Measured second harmonic power versus the photodetector reverse-bias voltage, showing the regimes for the two dominating nonlinear mechanisms. After Ref. [1] (© 1996 IEEE).

The optical cable TV receiver at the fiber node is required not only to maintain its low second- and third-order nonlinear distortions over the reverse-bias voltage range, but also over the optical input power range as well as over the operating temperature range. For example, Figure 5.7(a) shows the measured second-order intermodulation distortion (IM_2) versus the input optical power for an InGaAs p-i-n photodetector operating at reverse bias voltage of $-$12V with 50-ohm load. The IM_2 was measured using two frequency tones at 400-MHz and 450.25-MHz, each with an optical modulation index of 0.7. Notice that the second-order intermodulation distortion of this photodetector is $-$75-dBc or less, even at a high optical input power level. This means that the photodetector nonlinear distortion contributions can essentially be neglected if the transmitted distortions are $-$65-dBc or higher. Figure 5.7(b) shows the measured second-order intermodulation distortions versus the reverse-bias voltage for the same InGaAs p-i-n photodetector for an incident optical power of 0-dBm using the

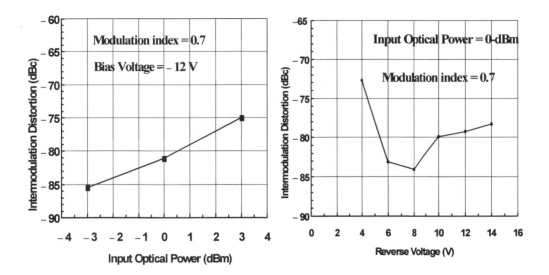

Figure 5.7 Measured second-order intermodulation distortions versus the optical input power and versus the reverse bias voltage for InGaAs p-i-n photodetector operating at modulation index of 0.7. Courtesy of JDS Uniphase, Inc., Epitaxx division.

same frequency tones as in Figure 7.7(a). This result indicates that if the photodetector is reverse-biased below about −8 volts, the nonlinear distortions start to increase rapidly.

Table 5.1 Typical and maximum specifications of an InGaAs p-I-n photodetector used in 860-MHz cable TV networks.

Parameter	Specification
Reverse-Bias Voltage	12 V
Maximum Forward and Reverse Current	10 mA
Maximum Power Dissipation	100 mW
Minimum Responsivity	0.9 A/W at 1550-nm 0.85 A/W at 1310-nm
Maximum Second-Order Intermodulation (IM_2)	−75 dBc
Maximum Dark Current	1 nA
Maximum Capacitance (with grounded case)	0.35-pF
Back Reflection	−45 dB
3-dB Bandwidth (into 50-ohm load)	3-GHz
Operating and Storage Temperature	−40°C to +85°C

Table 5.1 summarizes the typical and maximum performance parameters of an InGaAs   p-i-n photodetector, which is normally used in the optical fiber nodes for 860-MHz cable TV networks.   The photodetector is assembled into a hermetically sealed package with an antireflective-coated lens for efficient coupling to a fiber.   Since the packaged photodetector is required to operate over wide temperature range, its operating bandwidth may be severely degraded due to optical misalignments and other packaging parasitics.

To minimize the impact of the nonlinear distortions of the photodiode in the optical receiver, the interface between the p-i-n photodetector and the preamplifier as well as the design of the preamplifier must be optimized.   The basic front-end receiver design configurations as well as the overall receiver design considerations in cable TV networks are discussed in the next section.

5.5 Basic Cable TV Receiver Design Configurations

There are two primary categories of optical receivers, depending on their applications. Digital receivers are designed to receive digital baseband optical signals, and analog receivers are designed to receive RF or microwave signals, such as SCM analog video channels. Consequently, the performance requirements of digital receivers are different from those of analog receivers. Digital receivers require flat response at very low frequencies as well as high frequencies, depending on the transmitted data rates. For example, SONET receivers operating at OC-48 rate (2.488-Gb/s) must have flat response down to about 300-Hz. Thus, DC-coupled receiver design is needed for these applications. In contrast, analog receivers require flat response only in the frequency range of interest, namely, 50 to 860-MHz for downstream receivers and 5 to 42-MHz for upstream receivers. In addition, the overall analog receiver design, as was discussed in Section 5.3, must have good linearity to meet the strict CSO and CTB distortion requirements generated by the transmitted AM-VSB video signals.

Figure 5.8 illustrates a simplified block diagram of an optical receiver used in cable TV networks. The basic optical receiver configuration can generally be divided into two primary sections: (A) the front-end receiver section, and (B) the signal control and amplification section. The front-end section of the receiver consists of the p-i-n photodiode, a low-noise preamplifier, and some kind of interface circuitry between them. The role of the interface circuitry is to match the impedance between the photodiode and the preamplifier by minimizing the effect of stray capacitance as well as the input capacitance of the preamplifier. The output amplifier provides the necessary low impedance to match the coaxial-cable load. There are various types of interface circuits, depending on whether the front-end design is low impedance, high impedance, or transimpedance. For example, an impedance-matching transformer with some additional circuitry is used in a transimpedance front-end design in a fiber-node receiver. This is because the low-frequency response (i.e., less than 50-MHz) can be essentially ignored, while the high-frequency response is limited to about 860-MHz.

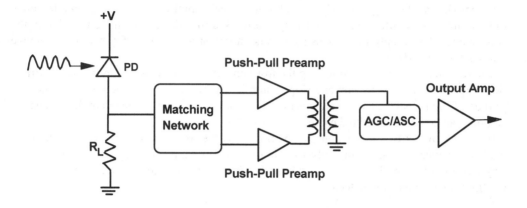

Figure 5.8 Simplified block diagram of a cable TV receiver using push-pull preamplifiers. AGC: automatic gain control. ASC: automatic slope control.

The role of the preamplifier is to convert the photocurrent to voltage with low noise and low nonlinear distortions. The preamplifiers can be, for example, bipolar-junction transistor (BJT) amplifiers or GaAs-junction field-effect transistor (JFET) amplifiers. Without the proper front-end receiver design, the photocurrent amplification may result in enhancement of the nonlinear distortions from the many transmitted analog video signals. The most common solution to this problem involves the use of dual matching preamplifiers in a push-pull configuration as shown in Figure 5.8. The push pull configuration works as follows. Let us assume that the input voltages from the matching network are 180° out of phase, namely, $V_{2,in} = -V_{1,in}$. The output voltage from each amplifier can be written as

$$V_1^{out} = b_1 \cdot V_{1,in} + b_2 \cdot V_{1,in}^2 + O(V_{1,in}^3)$$

$$V_2^{out} = -b_1 \cdot V_{1,in} + b_2 \cdot V_{1,in}^2 - O(V_{1,in}^3)$$

(5.26)

where third-order or higher order distortions are neglected. At the second transformer after the preamplifiers, the output voltages are recombined 180° out of phase, to yield $2b_1 V_{1,in}$. Thus, the result of one amplifier pushing while the second pulling is to cancel the effect of second-order nonlinear distortions. The push pull configuration typically reduces the second-order nonlinear distortions by about 20 dB. The obvious disadvantage of this configuration is that it cannot provide any compensation for the third-order nonlinear distortions.

The push pull configuration offers a gain advantage in bipolar amplifiers, in which emitters are difficult to ground. For heterojunction bipolar transistor (HBT) amplifiers fabricated on an insulating substrate, adjacent push-pull emitters can share a common ballast resistor, thereby raising the gain. The penalty of push pull HBT design is the increased layout and circuit complexity. Another advantage of push pull amplifiers is better isolation between adjacent HBTs, allowing the packaging of a higher gain amplifier. In emitter-coupled logic digital circuitry, a differential output is typically used in order to minimize the kick in the supply line. This supply line desensitization is important in complex ICs that operate above a few hundred MHz.

The receiver configuration following the front-end section consists of additional amplification stages with automatic gain and slope controls. As we learned in Section 2.4.2, the amplitude of the transmitted video signals must be equalized in order to maintain flat frequency response across the transmitted cable TV bandwidth.

In the following two subsections, we will discuss the front-end design considerations of analog receivers that are commonly used in optical communications systems. An improperly designed front-end circuit will often suffer from excessive noise associated with ambient light focused onto the detector.

5.5.1 Low- and High-Impedance Front-End Receiver Design

The front-end design of the photodetector is an essential part of the overall receiver design needed to meet the necessary performance requirements over cable TV networks. The simplest way to convert the photocurrent to voltage is to terminate the reverse-biased photodetector with a load resistor R_L, which is followed by a preamplifier as shown in Figure 5.9(a). The equivalent circuit model is shown in Figure 5.9(b), where the p-i-n photodetector is modeled as a current source in parallel with some shunt capacitance C_p. High-performance InGaAs p-i-n photodetectors are available today with shunt capacitance as low as 0.3-pF. The total capacitance, C_T, is equal to C_p plus the stray capacitance that is associated with the input impedance of the preamplifier. According to Equation (5.23), a large R_L is needed to increase the output CNR at low optical-input power levels. Such a front-end receiver design is called a high-impedance front end. However, the available modulation response bandwidth is proportional to $1/2\pi C_T R_L$. Thus, there is a trade-off consideration between achieving a large bandwidth and high CNR. One common way to extend the high-frequency response of the high-impedance design is to use a voltage equalizer after the preamplifier as shown in Figure 5.9(b). As the operating frequency is increased, the effective voltage drop across R_1 is reduced due to reduced reactance of C_1, while the voltage drop across R_2 is increased, resulting in increased frequency response. The output voltage divided by the input current or the transfer function of the equivalent circuit model with equalization is given by [17]

$$H(f) = \frac{V_R(f)}{I_R} = \frac{G \cdot R_L \cdot R_2}{R_1 + R_2} \tag{5.27}$$

and the extended frequency response is given by [17]

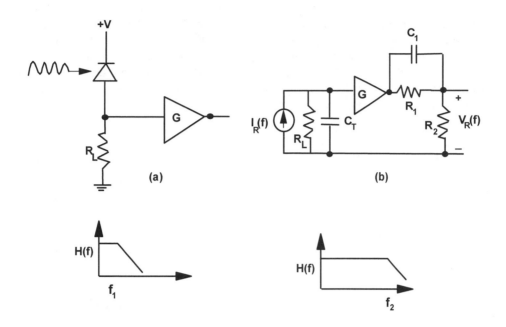

Figure 5.9 (a) A schematic of high-impedance front-end design, and (b) equivalent circuit model with equalization. After Ref. [17].

$$f_2 = \frac{R_1 + R_2}{2\pi R_1 R_2 C_1} = f_1 \left[1 + \frac{R_1}{R_2} \right] \tag{5.28}$$

Even with the frequency equalization, the high-impedance front-end design has some problems. Because of the large load resistor, the high-impedance front-end design does not have a wide dynamic range. In addition, the high-impedance front-end design suffers from nonlinearity, particularly at high optical-input power levels. Reducing the load resistor to 50 or 75-ohms, which is sometimes called a low-impedance front-end design, improves the receiver linearity but also increases the noise level, resulting in a lower CNR. High-impedance front-end design is advantageous for applications in which a sensitive (i.e., low-noise) receiver with narrow dynamic range and frequency response is needed. The problems associated with high-impedance front-end receiver design are solved by transimpedance front-end receiver design, which is discussed in the next section.

5.5.2 Transimpedance Front-End Receiver Design

Figure 5.10 shows a schematic of the transimpedance front-end receiver design with the equivalent circuit model. In this design, the feedback resistor R, which is connecting the output voltage to the input of an inverting amplifier, replaces the load resistor R_L in the high-impedance front-end design. The amplifier acts as a buffer and produces an output voltage proportional to the photodiode current. The feedback resistor should be large to minimize the noise and maximize the output voltage V_R. The equivalent circuit model (Figure 5.10(b)) shows a feedback capacitance C, which can be much lower than the input capacitance, in order to improve the high-frequency response of the photodetector. The transfer function of the equivalent circuit model can be written as [17]

$$H(f) = \frac{V_R(f)}{I_R(f)} = \frac{-R}{1 + j2\pi fR\left[C + C_T G^{-1}\right]}$$
(5.29)

where the 3-dB frequency response is given by

$$f_c = \frac{G}{2\pi R_f\left[C_T + GC\right]}$$
(5.30)

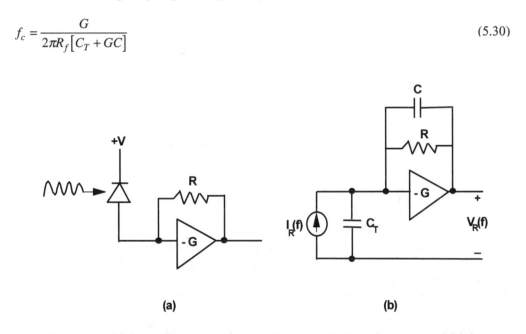

(a)   **(b)**

Figure 5.10 (a) A transimpedance front-end receiver design schematic, and (b) the equivalent circuit model. After Ref. [17].

It is apparent from Equation (5.30) that the reduced effective capacitance of the transimpedance front-end design allows the circuit to work at much higher frequencies. The other advantages of the transimpedance front-end designs are (A) an improved dynamic range over the high-impedance front-end design, (B) no need for external equalization circuits, and (C) an improved sensitivity over the low-impedance front-end design. The primary disadvantage of the transimpedance front-end receiver is a higher noise level than the high-impedance front-end receiver. In addition, similar to the high-impedance front-end design, the circuit still uses a fixed resistor to convert the current to a voltage and is thus prone to saturation and interference from ambient light.

5.5.3 High-Performance Receiver Design for Cable TV

So far, we have shown that the transimpedance front-end design is advantageous for optical receivers operating over cable TV networks. However, the different circuit configurations of the front-end design have not been addressed. In general, two types of circuit configurations are used for transimpedance front-end preamplifiers in optical receivers: an inverter type [18], and a cascode type [19]. Figure 5.11(a) shows a typical circuit diagram of a transimpedance front-end photodetector that has two stages. The inverter-input stage consists of FET Q_2 with an active load Q_1, where the signal is inverted at the input to FET Q_3. The transimpedance preamplifier configuration consists of the input stage together with FETs Q_3 and Q_4, and the feedback resistor R_f. The second stage at the output of the transimpedance preamplifier, which consists of FETs Q_5 and Q_6, acts as a buffer stage to provide 50- or 75-ohms output impedance to the following receiver section. There are two voltage sources, namely, V_{dd} (5 V) and V_{ss} (−2.5 V). Although this configuration is attractive due to its simplicity, it suffers from a large Miller capacitance. It can be shown that the equivalent-input noise current variance for FET preamplifiers is given by [20]

$$\left\langle i_a^2 \right\rangle = \frac{4k_B T \cdot I_2 \cdot B}{R_f} + 2qI_{gate}I_2 \cdot B + \frac{4k_B T \cdot \Gamma}{g_m}\left(2\pi C_T\right)^2 \cdot I_3 \cdot B^3 \tag{5.31}$$

where g_m and Γ are the transconductance and noise factor of the FET at the input stage, respectively, B is the receiver noise bandwidth, and I_2 and I_3 are normalized noise-bandwidth Personick integrals, which depend on the input and output pulse shapes [21]. The first term in Equation (5.31) accounts for the thermal noise of the load resistor, while the last term accounts for the thermal noise associated with the FET channel. The second term in Equation (5.31) arises from the FET gate leakage current. Another noise term, which is associated with FET 1/f noise, was neglected. The equivalent-input noise current using either BJT or JFET shows a cubic dependence on the photodetector bandwidth. However, the noise figures of silicon BJT are much higher, typically in the 7–9-dB range, than that of low-noise GaAs JFETs, which are in the 2–4-dB range. This is the reason why JFET preamplifiers are used, particularly in high-speed optical receivers.

The input capacitance C_T is the sum of the photodiode capacitance and the FET capacitance CFET. The problem is that the FETs at the input stage of the preamplifier are operating in a common-emitter or common-source configuration, and thus suffer from the so-called Miller effect [20]. The input capacitance of the preamplifier consists of the gate-source capacitance and the Miller capacitance, which is the gate-drain capacitance multiplied by the FET gain. Thus, due to the Miller capacitance of the preamplifier, its input capacitance is increased, which reduces the preamplifier high-frequency response.

Minimizing the Miller capacitance by using the cascode preamplifier configuration as shown in Figure 5.11(b) solves this problem.

The cascode circuit configuration consists of a common-source or common-emitter FET Q_3, and a common-gate or common-base FET Q_2. Notice that the common-gate FET Q_2 controls the gain of the common-emitter FET Q_3. By matching the transconductance of Q_2 and Q_3, the current gain of the common-emitter FET can be unity, which minimizes the input capacitance to the preamplifier, producing a large feedback resistance. Further improvements in the transimpedance front-end receiver design can be achieved using monolithically integrated receivers, which is beyond the discussion here.

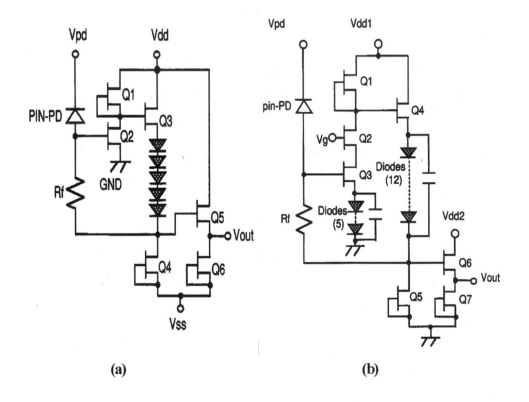

Figure 5.11 Transimpedance circuit configurations for photoreceivers (a) inverter-type preamplifier with multiple power supply, and (b) cascode preamplifier. After Ref. [18] (© 1996 IEEE).

References

1. K. J. Williams, R. D. Esman, and M. Dagenais, "Nonlinearities in p-i-n Microwave Photodetectors," *IEEE Journal of Lightwave Technology* **14**, 84–96 (1996).
2. S. M. Sze, *Semiconductor Devices Physics and Technology*, John Wiley & Sons, New York (1985).
3. T. H. Windhorn et al., "The Electron Velocity-Field Characteristics for n-InGaAs at 300K," *IEEE Electron Device Letters*, **EDL-3**, 18 (1982).
4. P. Hill et al., "Measurement of Hole Velocity in n-type InGaAs," *Applied Physics Letters* **50**, 1260 (1987).
5. G. P. Agrawal, *Fiber-Optic Communication Systems*, John Wiley & Sons, New York (1992).
6. N. Campbell, "Discontinuities in Light Emission," *Proc. Cambr. Phil. Soc.*, **15**, 310–328, 1909.
7. W. Schottky, "Über Spontane Stromschwankungen in Verschiedenen Elektrizitätsleitern," *Annalen der Physik*, **57**, 541–567 (1918).
8. E. N. Gilbert and H. O. Pollak, "Amplitude Distribution of Shot Noise," *Bell System Technical Journal*, **39**, 333–350 (1960).
9. J. E. Mazo and J. Salz, "On Optical Data Communication via Direct Detection of Light Pulses," *Bell System Technical Journal*, **55**, no. 3, 347–369, (1976).
10. J. B. Johnson, Physical Review **32**, 97 (1928).
11. H. Nyquist, Physical Review **32**, 110 (1928).
12. F. Grum and R. J. Becherer, *Optical Radiation Measurements Volume 1: Radiometry*, Academic Press (New York, 1979).
13. T. E. Darcie, "Subcarrier Multiplexing for Lightwave Networks and Video Distribution Systems," *IEEE Journal on Selected Areas in Communications* **8**, 1240–1248 (1990).
14. T. Ozeki and E. H. Hara, "Measurement of Nonlinear Distortion in Photodiodes," *IEE Electronics Letters* **12**, 80 (1976).
15. M. Dentan and B. deCremoux, "Numerical Simulation of the Nonlinear response of a p-I-n Photodiode Under High Illumination," *IEEE Journal of Lightwave Technology* **8**, 1137 (1990).
16. R. R. Hayes and D. L. Persechini, "Nonlinearity of p-i-n Photodetectors," *IEEE Photonics Technology Letters* **5**, 70–72 (1993).
17. T. V. Muoi, "Optical Receivers," *Optoelectronic Technology and Lightwave Communications Systems*, Ed. Chinlon Lin, Chapter 16, Van Nostrand-Reinhold, Princeton (1989).
18. J. Yoshida, Y. Akahori, M. Ikeda, N. Uchida, and A. Kozen, "Sensitivity Limits of Long-Wavelength Monolithically Integrated p-i-n JFET Photoreceivers," *IEEE Journal of Lightwave Technology* **14**, 770–779 (1996).

19. N. Uchida, Y. Akahori, M. Ikeda, A. Kohzen, J. Yoshida, T. Kokubun, and K. Suto, "622 Mb/s High-Sensitivity Monolithic InGaAs-InP pin-FET Receiver OEIC Employing a Cascode Preamplifier," *IEEE Photonics Technology Letters* **3**, 540–542 (1991).

20. R. G. Smith and S. D. Personick, *Semiconductor Devices for Optical Communication*, Springer-Verlag, New York (1982).

21. B. L. Kasper and J. C. Campbell, "Multigigabit-per-second avalanche photodiode lightwave receivers," *IEEE Journal of Lightwave Technology* **5**, 1351–1364 (1987).

CHAPTER 6

OPTICAL FIBER AMPLIFIERS FOR CABLE TV NETWORKS

As we learned in Chapter 1, optical fiber amplifiers, particularly Erbium-doped fiber amplifiers (EDFAs), have been a key enabling technology in the revolution of cable TV access networks. The discussion in this chapter on EDFAs is not intended to be exhaustive, but to explain their basic characteristics and performance requirements for cable TV networks. We shall start this chapter with the review of the technology and operating principles of the key optical-fiber amplifier components, including WDMs, Erbium-doped fibers (EDFs), and pump lasers. In Section 6.2, the basic EDFA system configurations are introduced, including single-stage and dual-stage EDFA system configurations. The noise characteristics of optical amplifiers in different operating regions as well as the SNR and the amplifier's noise figure are derived for SCM lightwave transmission systems in Section 6.3. Finally, we will discuss in Section 6.4 the EDFA selection process for various applications such as long-distance super-trunking and broadband networks based on passive optical networks. In addition, the required EDFA's CNR and gain flatness for a single- and for multiple-wavelength cable TV networks are discussed.

Other types of optical amplifiers operating at 1300-nm, including praseodymium-doped fluoride fiber amplifiers, Raman fiber amplifiers, and semiconductor optical amplifiers, will not be discussed here. From a practical point of view, there are major technical and economical challenges that currently prevent these optical amplifiers from being used in cable TV networks.

6.1 Optical-Fiber Amplifier Components

The key EDFA system components are WDM multiplexers (WDM muxes) and demultiplexers (WDM demuxes), EDFs, and pump lasers. The next subsection discusses the characteristics and technologies for muxes and demuxes.

6.1.1 Wavelength-Division Multiplexers

Optical transmission through conventional SMF is typically characterized by the fiber's wavelength-dependent attenuation (dB/km) and chromatic dispersion (ps/nm/km). There are two low-attenuation "windows" around 1310-nm and 1550-nm with average optical losses of 0.35-dB/km and 0.21-dB/km, respectively, for telecommunication applications [1, 2]. The total corresponding available bandwidths around 1310-nm and 1550-nm are 60-nm and 110-nm. The available electrical bandwidth can easily be calculated as follows:

$$\Delta f = -\left[\frac{c}{\lambda^2}\right] \cdot \Delta \lambda \qquad\qquad\qquad (6.1)$$

For $\Delta \lambda$ = 1-nm, the electrical bandwidth is 172-GHz and 122-GHz around 1310-nm and 1550-nm, respectively. Thus, a standard SMF can provide an enormously large electrical bandwidth, particularly around 1550 nm (172-GHz x110 = 18,920-GHz $\cong$ 19-THz). Since it is impossible for a single-wavelength laser transmitter or a photodetector to utilize this enormous bandwidth, multiple single-wavelength laser transmitters with the corresponding WDM muxes, demuxes, and/or optical filters are used. Generally speaking, there are two categories of WDM filters, namely, passive and active. However, for cable TV access networks, passive WDM filters are preferred due to their lower cost and ease of use. Let us briefly review the different types of passive and active WDM filters.

There are four main technologies for passive WDM filters. They are as follows: (A) fused-fiber couplers, (B) graded-index (GRIN) rods and multilayer filters, (C) planar waveguides, and (D) fiber Bragg gratings. Fused-fiber couplers, which are commercially very common WDM devices, are made when two fibers are fused together; the coupling between the fiber cores determines the device response. The coupled powers in each fiber, namely, P_1 and P_2 can be described using the basic coupled-mode equations [3] as $P_0 \sin^2(\kappa L)$ or $P_0 \cos^2(\kappa L)$, where P_0 is the launch optical power in fiber 1, κ is the coupling coefficient between the two fibers and L is the coupling length. For example, by selecting a specific κL such that $P_2(\lambda_1) = P_0$ and $P_2(\lambda_2) = 0$, only one wavelength is passed through output port 2. The advantage of these devices is that by varying the coupling length, different coupling ratios such as 10/90% or 50/50% can easily be obtained. In addition, these devices can easily be scaled to 1 x N or M x N couplers for DWDM applications. From a practical standpoint, the packaging of these couplers can affect their reliability. In particular, the coupling region of these devices must be protected from environmental changes such as humidity and temperature.

The second type of passive WDM filter technology is a multilayer dielectric stack, which is typically placed between a pair of GRIN glass rods [4, 5]. The GRIN rods are used to collimate and efficiently couple the light beam from one fiber to another. The multilayer filters can be reflective or transmission types, depending on the thickness and refractive index at the designed wavelength. Thus, the reflected or transmitted light from the multilayer dielectric stack can be added constructively at one wavelength, and destructively at another wavelength. Such WDM filters, which are sometimes called interference filters, can be cascaded to produce multiple wavelength channels with desirable characteristics such as low insertion loss, polarization insensitivity, and a high isolation (> 30-dB) between channels. The first major application of these filters was 1550/1480-nm and 1550/980-nm WDM muxes and demuxes for EDFAs [5].   Using cascaded WDM interference filters, eight-channel 200-GHz frequency spacing WDM demux with an insertion loss less than 5.5-dB and channel bandwidth greater than 0.5-nm has been achieved [5].   The advantage of these filters is that they provide flat amplitude across each channel with a low polarization-dependent loss (PDL). With the proper design and packaging, this type of WDM filter can be as reliable as fused-fiber filters.

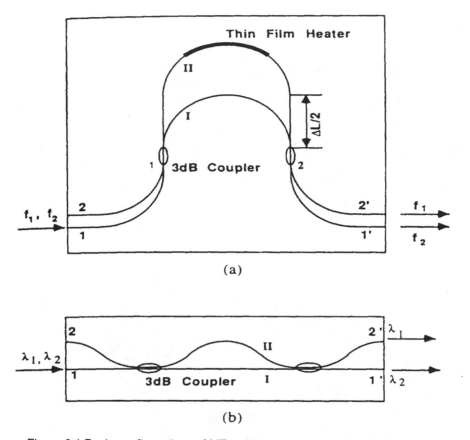

(a)

(b)

Figure 6.1 Basic configurations of MZ multiplexer/demultiplexer with (A) narrow and (B) wide wavelength spacing. After Ref. [6] (© 1990 IEEE).

The third type of passive filter technology is planar waveguides, which are fabricated on silica or silicon substrates using semiconductor-manufacturing processes. Figure 6.1 shows the basic configurations of a 2x2 Mach-Zehnder (MZ) multiplexer/demultiplexer [6].   It consists of two inputs and two output ports, two 3-dB couplers, and two waveguide arms, where there is a fixed length difference of ΔL. The MZ configurations shown in 6.1(a) and 6.2(b) are designed for narrow and wide wavelength spacing (i.e., large or small ΔL), respectively. A thin-film heater is placed on one of the waveguide arms, which acts as a phase shifter, because the light-path difference of the heated waveguide arm changes due to the refractive index change. This method is used for a precise frequency tuning. The output power from ports 1 and 2 is $P_0\cos^2(\pi n\Delta L/\lambda)$ and $P_0\sin^2(\pi n\Delta L/\lambda)$, respectively, where P_0 is the injected power into port 1. In order to obtain λ_1 out of port 2 and λ_2 out of port 1, the injected wavelengths have to satisfy the following condition:

$$\frac{\pi n\cdot\Delta L}{\lambda_1} = \frac{\pi}{2} + \frac{\pi n\cdot\Delta L}{\lambda_2} \tag{6.2}$$

The required wavelength spacing is given by

$$\Delta\lambda = \lambda_2 - \lambda_1 = \frac{\lambda_1\cdot\lambda_2}{2n\cdot\Delta L} \tag{6.3}$$

The corresponding frequency spacing is simply $c/2n\Delta L$. For example, a path-length difference of 20-mm can provide an optical frequency spacing of 5-GHz, or a wavelength spacing of 0.04-nm at 1550-nm. On the other hand, if λ_1 = 1310-nm, and λ_2 = 1550-nm, then the path-length difference is only ΔL = 2.87-µm.

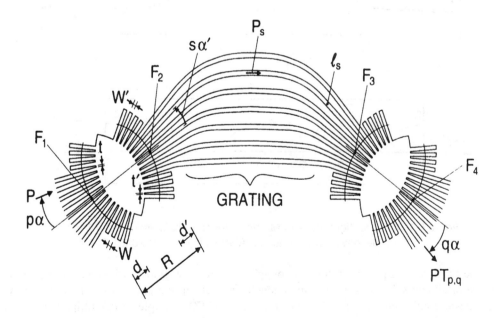

Figure 6.2 NxN multiplexer using two identical couplers and a grating. After Ref. [8] (© 1990 IEEE).

A generalized MZ multiplexer is shown, for example, in Figure 6.2 [7, 8]. An integrated *N*x*N* multiplexer fabricated using an SiO_2/Si waveguide, which consists of two identical star couplers combined with *M* waveguides acting as a grating between the couplers. Each star coupler consists of two confocal arrays of radial waveguides with foci F_1/F_2 and F_3/F_4. The waveguides are made from strips of a constant refractive index n_2 separated by strips of refractive index n_1. The multiplexer operates as follows: The input to the *p*th port is radiated in the free-space region, and most of its power is accepted by the grating. Some of this power is then transferred to the *q*th output port. The grating provides a constant path-length

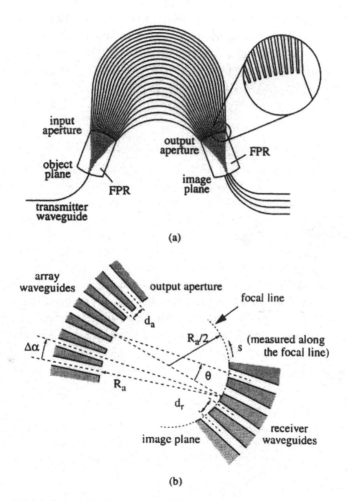

(a)

(b)

Figure 6.3 (a) Schematic layout and (b) receiver side geometry of the AWG demultiplexer. After Ref. [9] (© 1996 IEEE).

difference between adjacent paths such that the phase difference $\phi = \phi_s - \phi_{s-1}$ becomes independent of port s. The transmitted power coefficient varies periodically with $1/\lambda$, producing maximum transmission peaks at wavelengths for which ϕ is an integer multiple of 2π. Such a multiplexer/demultiplexer, which can be scaled up to about 50 input and output wavelength channels, can also be used as a wavelength router. Other types of integrated-optics WDM devices are grating-based or phased array-based devices, which are also called array waveguide gratings (AWGs) [9]. These are imaging devices since they image the field of an input waveguide onto an array of output waveguides in a dispersive way. The schematic layout of the AWG multiplexer is shown in Figure 6.3(a) and its receiver geometry in Figure 6.3(b). Upon entering the free-propagation region (FPR), the laterally confined beam becomes divergent. At the input aperture, the propagating beam is coupled into the waveguide array to the output aperture. The length of the individual waveguides is selected such that the optical path difference between adjacent waveguides is equal to an integer multiple of the central wavelength (λ_c) of the AWG demultiplexer. This condition can be written as [10]

$$\frac{2\pi}{\lambda_C}\left[n_g \cdot \Delta L + n_{FPR} \cdot d_a\left(\theta_i + \theta_a\right)\right] = 2\pi m \qquad (6.4)$$

where n_g and n_{FPR} are the effective refractive indices of the waveguide and the FPR, respectively, d_a is the adjacent waveguide separation, and θ_i and θ_a are the incident and diffracted angles at the input and output aperture of the AWG demux, respectively. For the central wavelength, the fields in the individual waveguides will arrive at the output aperture with equal phase, and the field distribution at the input aperture will be reproduced at the output aperture. Angular dispersion is obtained by linearly increasing the length of the array waveguides. The outgoing beam for each wavelength is tilted, and the focal point is shifted along the image plane as shown in Figure 6.3(b). Consequently, spatial separation is obtained for different wavelengths at the receiver waveguides. It is found that the frequency response of the AWG is a periodic function with a so-called free-spectral range (FSR), where the frequency shift is equal to 2π. The FSR can be obtained from Equation (6.4) as [9]

$$\Delta f_{FSR} = f_1 - f_2 = \frac{c}{n_g \cdot \Delta L} \qquad (6.5)$$

where f_1 and f_2 correspond to adjacent order (i.e., m and $m+1$) frequencies. The AWG demux can be used in other applications such as WDM routers and receivers. For example, multiple-channel WDM receivers can be achieved by integrating the AWG demux with a photodiode array [11].

Another type of passive filter technology is based on Bragg gratings, which are permanently written into the fiber by using either UV light or an excimer laser with a phase mask [12, 13]. The fiber Bragg grating (FBG) filters are formed by a periodic modulation of the refractive index of the fiber core as shown in Figure 6.4(a). When the transmitted light beam is incident on the FBG, the partially reflected beams recombine constructively only if they satisfy the following condition:

$$\frac{\lambda}{n_{av}} = 2d \sin\theta \qquad (6.6)$$

where λ/n_{av} is the wavelength of the propagating beam in the fiber, d is the spatial period of the grating, and θ is the angle of incident of the light beam as shown in Figure 6.4(b). The maximum reflection condition occurs when $\theta = \pi/2$, namely, all the reflected beam recombined constructively, and we have

$$\lambda_B \equiv \lambda / n_{av} = 2d \qquad (6.7)$$

Equation (6.7) is called the Bragg condition, which states that the Bragg wavelength (λ_B) is simply equal to twice the spatial period of the grating. Thus, for a maximum reflection of 1550-nm light, the spatial period of the grating should be 0.53-μm, assuming $n_{av} = 1.47$ for a standard SMF. The FBG concept can be applied to fiber couplers to build optical bandpass filters or routers. For example, an FBG can be fabricated on each output arm of a 1xN fiber coupler in order to transmit or reflect only specific wavelengths. An FBG can also be used as a dispersion-compensation device in both digital and SCM lightwave transmission systems [14, 15].

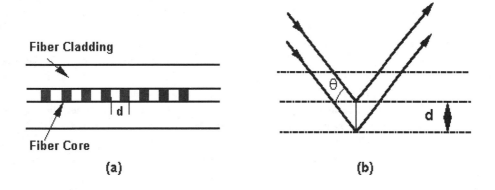

Figure 6.4 (a) Fiber Bragg grating formation by refractive-index modulation, and (b) incident and diffracted light beams at an arbitrary angle θ from the Bragg grating.

In general, active filters can be divided into three categories: (A) Fabry-Perot (FP) filters, (B) liquid-crystal-based filters, and (C) acousto-optic tunable filters (AOTFs). Both the FP and liquid-crystal-based filters, which are similar to the passive ones, are actively tuned by changing the mirror spacing using an RF voltage, and can provide wavelength spacing between 0.5-nm and 10-nm [16, 17]. Another type of active filter is an all-fiber AOTF, whose spectral profile is electronically controllable [18]. The operation of the all-fiber AOTF can be explained as follows. When a flexural acoustic wave, which is generated by a coaxial acoustic transducer, propagates along a standard SMF, it creates antisymmetric microbends that travel along the fiber and introduces periodic refractive-index perturbations. This creates coupling between the input symmetric fundamental mode and the antisymmetric cladding mode when the acoustic wavelength is the same as the beat length between the two modes, which is the phase-matching condition. Thus, the optical filtering is achieved since the coupled light in the cladding mode is significantly attenuated. The all-fiber AOTF acts as an optical notch filter, where the center optical wavelength and the rejection amplitudes are tunable by adjusting the frequency and voltage of the applied RF signals to the acoustic transducer, respectively. As we will see in Section 6.4.2, the all-fiber AOTF can be used for gain flattening of EDFAs.

6.1.2 Erbium-Doped Fibers (EDFs)

Using rare-earth ions such as Erbium (Er^{3+}) to dope glass fiber produces the necessary gain medium in the EDFA system. Figure 6.5 shows the energy-level diagram of Erbium ions in silica fibers, showing both absorption and radiative transitions. The transition wavelengths are in nanometers and indicated only for transitions experimentally observed in silicate and flurozirconate Er-doped fiber. There are many different pump laser wavelengths such as 1480-nm, 980-nm, 800-nm, and 670-nm, which can be used to achieve population inversion. The notation of the possible states of a multi-electron atom such as Er is done according to $^{2S+1}L_J$, where $L = 0,1,2,3,\ldots$, corresponding to the letters S, P, D, F, G,$\ldots$, respectively, is the overall angular momentum, $2S + 1$ is the number of spin configurations, and J is the total angular momentum with $2J + 1$ possible states. For example, the notation for the Er^{3+} ground state is $^4I_{15/2}$, which corresponds to $L = 6$, has a multiplicity $2J + 1 = 8$, and a spin multiplicity $2S + 1 = 4$. Figure 6.5 shows a simplified three-level energy model for the Erbium ions. First, the Erbium ions are excited from the ground state (level 1) to some higher energy state (level 3) by absorbing the pump photons. This process is sometimes called excited state absorption (ESA) [19]. The pump rate is R_{13}, and the stimulated emission to the ground state is R_{31}. The Erbium ions stay at level 3 for an extremely short time ($\approx$1-ps), and quickly relax to a metastable state (level 2), which is characterized by a long lifetime ($\tau \cong$ 10-ms for silica and fluoride-based glasses), predominantly through a nonradiative process. The metastable state (level 2) decays to the ground state through either a stimulated emission at a rate of W_{21} or through a spontaneous emission with $A_{21} = 1/\tau$. The propagating signal is amplified through stimulated emission around 1540-nm, depending on the type of glass medium and the concentration of other doping elements. The absorption process by the ground state from higher energy states is sometimes called ground state absorption (GSA). In addition, there is a stimulated absorption between the ground state and level 2 at a rate of W_{12}. Generally speaking, optical amplifiers are more useful if they have a larger gain bandwidth.

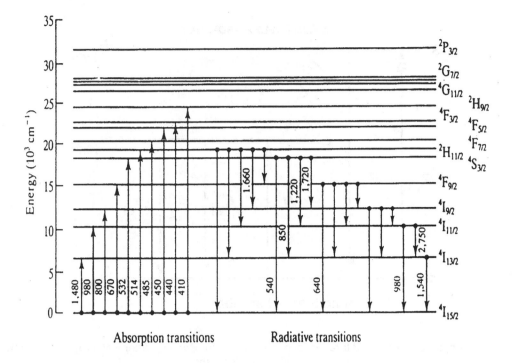

Figure 6.5 Energy level diagram of Er-doped glass, showing absorption and radiative transitions, where the transition wavelengths are in nanometers and indicated only for transitions experimentally observed in silicate and flurozirconate Er-doped fiber. After Ref. [19] (reprinted by permission of John Wiley & Sons, Inc.).

This requirement allows a larger number of optical channels to be multiplexed and transmitted over the communications network. It also relaxes the wavelength tolerance requirement for a single-channel system. It was found that by co-doping the silica-based glass with aluminum (Al) and/or germanium (Ge) together with Erbium ions, one could significantly increase the amplifier 3-dB gain bandwidth to about 40-nm [20]. Recently, a Tellurite-based EDFA with a gain exceeding 20-dB over 80-nm bandwidth from 1530-nm to 1610-nm has been demonstrated [21]. The relatively wide gain bandwidth of these Al/Ge silica-based glasses is attributed to the Stark splitting of the laser transitions induced by the Stark effect. The host's crystalline electric field removes the degeneracy of the energy levels. If, for example, levels 1 and 2 are split into manifolds of m_1 and m_2 nondegenerate sublevels, respectively, then the corresponding laser transition between these two levels is now made of the superposition of $m_1 m_2$ possible transitions. Thus, the laser transition is broadened. Since the Stark sublevels have small energy gaps and are therefore strongly coupled by the effect

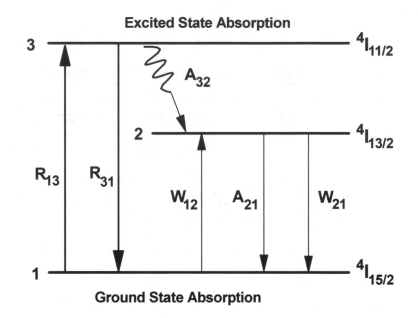

Figure 6.6 Simplified three-level energy model for Erbium. After Ref. [19] (reprinted by permission of John Wiley & Sons, Inc.).

of thermalization, the overall laser line exhibits homogeneous saturation characteristics. However, due to the presence of various defects and dislocations in the host crystals and the amorphous structure of glass hosts, the electric field that each dopant atom experiences varies from site to site, changing the energies of the Stark sublevels [19]. Thus, the spectral lines resulting from the superposition of all of the dopant atom contributions in such a host are inhomogeneously broadened. The inhomogeneous broadening of the emission spectra has been modeled by the Voigt profile, which is a convolution between the Lorentzian line shape and the Gaussian distribution, and has provided a good agreement with the experimental data [22].

The simplified three-level model of the Erbium ions can be analyzed using the rate equations. Let us denote the Erbium ions density by $N = N_1 + N_2 + N_3$, where N_1, N_2, and N_3 are the fractional ion density in energy levels 1, 2, and 3, respectively. The corresponding rate equations for these fractional population densities are given by [19]

$$\frac{dN_1}{dt} = -[R_{13} + W_{12}]N_1 + [W_{21} + A_{21}]N_2 + R_{31}N_3$$

$$\frac{dN_2}{dt} = -[W_{21} + A_{21}]N_2 + W_{12}N_1 + A_{32}N_3 \qquad (6.8)$$

$$\frac{dN_3}{dt} = -[R_{31} + A_{32}]N_3 + R_{13}N_1$$

Let us now consider the steady-state regime, where the population densities are time invariant, namely, $dN_i/dt = 0$ (i = 1,2,3). To simplify the solution to Equations (6.8), let us make the following two assumptions:

1. $A_{32} \gg R_{13}, R_{31}$, which means that the nonradiative decay rate A_{32} dominates over the pumping rates to level 3. Thus, $N_3 \approx 0$, level 3 ions rapidly relax to level 2.
2. $A_{32} \gg A_{21}$, which means that the lifetime of the Erbium ions in level 2 is significantly longer than level 3.

With these two assumptions, the steady-state solution to Equation (6.8) is given by

$$N_1 = N\frac{1 + W_{21}\tau}{1 + R\tau + [W_{12} + W_{21}]\tau} \qquad (6.9)$$

$$N_2 = N\frac{[R + W_{12}]\tau}{1 + R\tau + [W_{12} + W_{21}]\tau} \qquad (6.10)$$

where $R = R_{13}$. The result of the steady-state population densities is important for the calculation of the amplifier's gain coefficient, and will be used later on.

To understand the amplifier gain, one has to consider the pump power, signal power, and amplified spontaneous emission (ASE) power distribution along the EDF length. The length of the EDF typically varies from a few meters to about 30 meters. The spatial evolution of the signal, pump, and ASE power along the fiber, where transverse power variations are neglected, can be described as [23]

$$\frac{dP_s(z,t)}{dz} = P_s\Gamma_s[\sigma_{se}N_2 - \sigma_{sa}N_1] - \alpha_s P_s \qquad (6.11)$$

$$\frac{dP_p^\pm(z,t)}{dz} = \pm P_p^\pm\Gamma_p\left[\sigma_{pe}N_2 - \sigma_{pa}N_1 - \frac{\alpha_p}{\Gamma_p}\right] \qquad (6.12)$$

$$\frac{dP_{ase}^\pm(z,t)}{dz} = \pm P_{ase}\Gamma_s\left[\sigma_{se}N_2 - \sigma_{sa}N_2 - \frac{\alpha_s}{\Gamma_s}\right] \pm 2\sigma_{se}N_2\Gamma_s h\nu_s\Delta\nu_s \qquad (6.13)$$

where σ_a and σ_e are the absorption and emission cross sections for the pump or the signal, and Γ_s and Γ_p are the signal-to-core overlap and pump-to-core overlap factors, respectively. The "+" superscript refers to both the pump and ASE copropagating with the signal, and the "−" superscript refers to when they are counterpropagating with the signal. The second term in Equation (6.13) represents the ASE power produced in the amplifier per unit length with its bandwidth $\Delta\nu$ for both polarization states (factor of 2). The loss terms α_s and α_p represent the internal loss for the signal and pump wavelengths, respectively. Neglecting these terms, Equation (6.13) can be simplified to

$$\frac{dP_{ase}^{\pm}(z,t)}{dz} = \pm\left[P_{ase}^{\pm} + 2n_{sp} \cdot h\nu_s \cdot \Delta\nu_s\right]g(z) \tag{6.14}$$

where the spontaneous emission factor is defined as

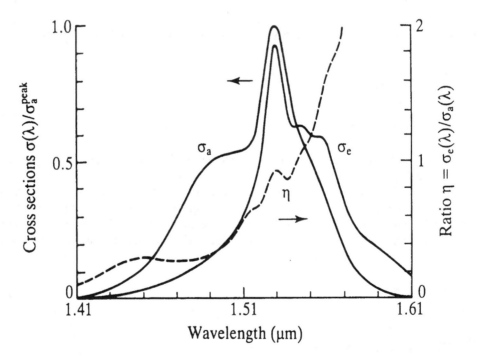

Figure 6.7 Wavelength dependence of absorption and emission cross sections (σ_e and σ_a) as well as their ratio (ηs) for a typical alumino-germanosilicate Er-doped fiber. After Ref. [19] (reprinted by permission of John Wiley & Sons, Inc.).

$$n_{sp} = \frac{\eta_s N_2}{\eta_s N_2 - N_1} \tag{6.15}$$

where $\eta_s = \sigma_{se}/\sigma_{sa}$, and $g(z) = \Gamma_s[\sigma_{se}N_2 - \sigma_{sa}N_1]$ is the gain coefficient independent of z, assuming the amplifier is pumped uniformly by a sufficiently strong pump. The steady-state solution to Equation (6.14) for an amplifier with a length L is given by

$$P_{ase}(z) = e^{gL}P_{ase}(0) + g\int_0^L e^{g(L-z)}\left[2n_{sp}h\nu\Delta\nu\right]dz = 2n_{sp}[G-1]h\nu\Delta\nu \tag{6.16}$$

where $G = \exp(gL)$ is the amplifier gain, and we have assumed the initial condition $P_{ase}(0) = 0$, which is typically the case for a single-stage amplifier. Thus, Equation (6.16) provides the ASE power along a uniformly pumped single-stage amplifier with length L. The steady-state solution for P_p and P_s can be obtained by substituting Equations (6.9) and (6.10) for N_1 and N_2, respectively, into Equations (6.11) and (6.12). By allowing the pumping power to

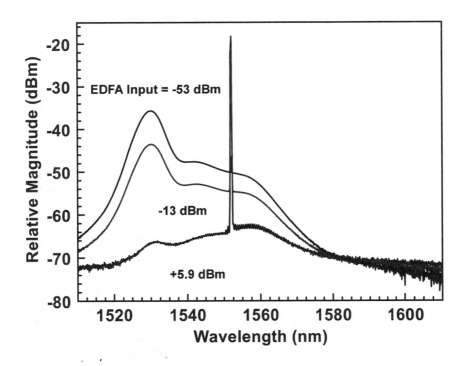

Figure 6.8 Measured optical spectra of a single-stage EDFA at various optical input power levels. Notice the dominant ASE peak at 1530-nm at very low optical input power level.

approach infinity $(R_{13} \rightarrow \infty)$, Equation (6.11) for the signal-power spatial evolution simplifies to

$$\frac{dP_s}{dz} \cong \Gamma_s N \sigma_{se} P_s \tag{6.17}$$

Equation (6.17) implies that when complete population inversion is achieved, the signal gain is proportional to $\Gamma_s N \sigma_{se}$. Assuming the input signal power to the EDFA is $P_s(0)$, the maximum unsaturated gain for an EDFA with length L, which is described by the three-level model, is given by

$$G = \frac{P_s(L)}{P_s(0)} = \exp\left[\Gamma_s \cdot N \cdot \sigma_e \cdot L\right] \tag{6.18}$$

Thus, according to Equation (6.18), there is an optimal EDF length for a given pump power to achieve maximum signal gain. Furthermore, as the EDF length becomes longer than the optimal length, the pump power is not sufficiently strong to deplete the ground-state population, resulting in a partial absorption of the amplified signal. The spatial evolution of P_p and P_{ase} can be obtained numerically by substituting Equations (6.9) and (6.10) into Equations (6.12) and (6.13), respectively.

So far, we have not discussed the spectral shape of the EDFA gain, which is given by

$$g(\lambda) = \Gamma_s \left[\sigma_e(\lambda)N_2 - \sigma_a(\lambda)N_1\right] \tag{6.19}$$

The wavelength dependence of the absorption and emission cross sections as well as their ratio for a typical alumino-germano-silicate EDF is shown in Figure 6.7. Let us make two important points. First, the emission-to-absorption cross-section ratio (η_s) is typically larger at the longer wavelength region (i.e., > 1550-nm) than at the shorter wavelength region. This means that as the pump power is increased, optical amplification will be achieved first at the longer wavelength region. Second, the maximum overlap between the absorption and emission cross sections occurs around 1530-nm, which minimizes the net gain given by Equation (6.19). This means that in the small-signal regime one expects to find the ASE peak around 1530-nm in the EDFA gain spectra as shown in Figure 6.8. As the input signal to the EDFA increases, the gain spectrum near the saturating input signal (λ_s = 1552-nm) compresses unevenly across the EDFA band. In fact, the ASE peak compresses about ten times more than the gain peak at the longer wavelength (> 1552-nm) region. The result is shifting of the gain-spectrum peak to around 1557-nm, while the ASE peak around 1530-nm becomes a secondary peak. The tilted amplifier gain spectrum has a significant impact on the transmission of AM video channels in an SCM lightwave system (see the discussion in Section 10.5). The uneven gain compression can be explained by the following simple argument. For change in the manifold populations induced by the saturating signal $\Delta N_1 = -\Delta N_2$, the local gain coefficient decreases as

$$\Delta g(\lambda) = \Gamma_s \left[\sigma_e(\lambda)\Delta N_2 - \sigma_a(\lambda)\Delta N_1 \right] = \Gamma_s \left[\sigma_e(\lambda) + \sigma_a(\lambda) \right] \Delta N_2 \qquad (6.20)$$

Thus, the maximum gain change occurs where the sum of the emission and absorption cross sections is maximum, which typically occurs at a wavelength of 1531-nm for aluminosilicate EDFAs. It should be pointed out that the amplifier gain is primarily limited by the amplifier self-saturation by the ASE and laser oscillations. As the amplifier gain is increased, the stimulated emission by the ASE is also enhanced, until it competes with the pumping rate, located near the EDF ends. Thus, the amplifier gain becomes saturated. However, increasing the pump power also increases the amplifier gain. Oscillations in the amplifier occur when the optical gain is larger than the return losses of any reflecting element in the path of the ASE signal. Consequently, the reflected back-and-forth ASE in the amplifier operates as a laser, and the laser oscillation saturates to a steady state.

6.1.3 Pump Lasers

Semiconductor laser diodes are considered to be the most practical sources for lightwave communications systems due to their compact size, high gain coefficient (dB/mW), and excellent reliability. From a historical perspective, 1480-nm pump lasers were first employed in EDFA systems in the early 1990s due to their proven reliability [24]. With the development of the necessary packaging technologies with improved reliability, 980-nm pump lasers were used in EDFA systems by 1994 [25]. The performance of a particular EDFA design can be characterized by introducing the so-called EDFA gain coefficient (dB/mW), which is defined as the peak ratio of the small-signal gain to the injected pump power. Figure 6.6 compares the highest recorded gain coefficients (dB/mW) as a function of the input pump power using different pump bands near 1480-nm, 980-nm, 827-nm, 664-nm, and 532-nm.

The best gain efficiency of 11-dB/mW is obtained using 980-nm pump lasers, while the worst gain efficiency of 1.3-dB/mW is obtained using 830-nm pump lasers. Optimization of the EDFA gain coefficient requires (A) confining the Er^{3+} ions doping to the center of the fiber core, and (B) reducing the pump mode size. Er-doping confinement concentrates the ions at the fiber center where the pump has its highest intensity. Reducing the pump mode size increases the pump intensity for a given pump level. Notice that the gain efficiency using 1480-nm pump lasers is only 60% of that obtained using 980-nm pump lasers. This is mainly due to incomplete population inversion (about 70%) with 1480-nm pump lasers compared with a complete inversion with 98-nm pump lasers. It should be pointed out here that because 980-nm pumped EDFA systems achieve complete population inversion, their noise figure is inherently lower than 1480-nm pumped EDFAs, approaching the quantum limit of 3-dB. The dB/mW gain coefficient increases as the length of the EDF is increased. However, it does not increase indefinitely, but levels off due to the amplifier self-saturation effect.   As the maximum gain of the amplifier is increased with longer EDF, more ASE power is generated at both amplifier ends, resulting in gain saturation.   In addition to high output powers, pump lasers are required to produce stimulated emissions in a single trans-verse mode for efficient coupling into standard SMFs, but with multiple longitudinal modes (FP lasers).   The three primary pump laser categories are (A) InGaAsP and multiple-quantum-well (MQW) InGaAs lasers operating at 1480-nm, (B) strained-layer MQW InGaAs lasers operating at 980-nm, and (C) GaAlAs lasers operating at 820-nm.   1480-nm pump lasers based on InGaAs/InGaAsP MQW producing a CW output power of 250-mW have been reported [24].   980-nm pump lasers with improved reliability, which were based on strained-layer InGaAs MQW lasers, produced a maximum output power over 320-mW [25].   Because of the limited output power from conventional FP lasers, a new type of laser device based on a monolithically integrated master oscillator power amplifier (MOPA) has been developed [26, 27].   The MOPA device produced a single-lobed diffraction limited beam with CW output power up to about 2.2-W at 850-nm and 980-nm.   However, the reli-ability of such devices is still under ongoing research.

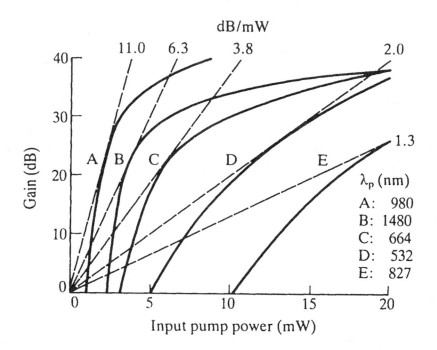

Figure 6.9 Record experimental EDFA gain versus input pump power characteristics using different pump bands near 532-nm, 827-nm, 664-nm, 980-nm, and 1480-nm. After Ref. [19] (reprinted by permission of John Wiley & Sons, Inc.).

6.2 Basic EDFA System Configurations

Three basic EDFA system configurations are illustrated in Figure 6.10 [19]. All of the EDFA system configurations consist of pump laser diodes; one or two WDM filters, which are also called wavelength-selective couplers (WSCs); a section of an EDF; and in-line optical isolators (OI). The first EDFA configuration [Figure 6.10(a)] has unidirectional forward pumping, since the pump wavelength is copropagating forward with the signal wavelength relative to the EDF. EDFs usually have smaller cores and higher numerical apertures (NAs) compared with standard SMFs. Thus, splicing between two dissimilar fibers must be done using nonstandard splicing techniques. In order to couple the light from the pump laser diodes into the SMF, different types of microlenses are used, depending on their characteristics. These microlenses can be, for example, spherical, aspherical, graded-index, or even fabricated directly on the SMF by melting the end of a tapered fiber [28, 29]. There are four basic requirements for good light coupling between pump lasers and SMF, and they are as follows: (A) large NA, (B) matching focal length for both the fiber transverse modes and the laser, (C) low spherical aberrations, and (D) low surface reflections. The last requirement is

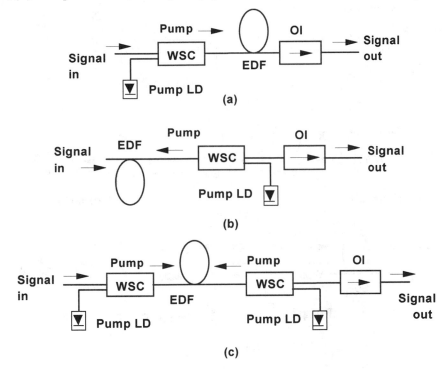

Figure 6.10 Basic EDFA configurations. After Ref. [19] (reprinted by permission of John Wiley & Sons, Inc.).

typically satisfied using antireflection (AR) coating. High LD-SMF coupling efficiency may provide additional power margin for the EDFA system. The proper design of the WDM filters also may be critical to the operation of the EDFA system. There are many requirements for an ideal WDM filter, depending on the particular technology. The ideal WDM filter should have (I) low coupling losses at both the pump and signal wavelengths, (II) polarization insensitivity, (III) low backward reflections, (IV) high-power damage threshold, (V) large optical signal bandwidth, (VI) temperature stability, (VII) compact size, and (VIII) cost-effectiveness.

Another important passive component is the OI, which prevents the backward-scattered light from coupling into the EDFA. In the second EDFA system configuration [Figure 6.10(b)], the propagating signal gain is obtained using unidirectional backward pumping. Thus, the injected signal is counterpropagating to the pump signal. As we will learn later, the lowest EDFA noise figure is always achieved when the pump signal is propagating in the same direction as the signal [19]. In the third EDFA system configuration [Figure 6.10(c)],

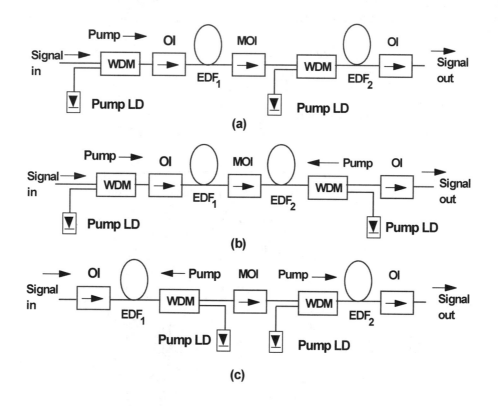

Figure 6.11 Dual-stage EDFA system configurations: (a) unidirectional pumping for both stages, (b) unidirectional pumping for first stage and backward pumping for the second stage, (c) the reverse-configuration (b).

bidirectional forward and backward pumping is employed from both ends of the EDF. Since the output power of today's 980-nm pump lasers is somewhat limited, polarization beam splitters (PBSs) may be used for a polarization-combined bidirectional-pumping scheme. From a practical aspect, it should be pointed out that a generic EDFA system also contains various electronic control subsystems including signal power control, pump power control, and temperature control. For example, the signal power control is achieved by tapping a small portion of the amplifier output power to a photodetector, where its voltage can be used to regulate the signal power using a microprocessor. The pump power can also be controlled by tapping a small portion of the pump power to another photodetector, where its output voltage regulates the power supply voltage.

So far, we have discussed only single-stage EDFA configurations. There are also dual-stage EDFA system configurations, in which two EDF sections are placed in tandem. These EDFA systems are designed to provide very high gains (G > 40 dB) and high output powers (> 20-dBm), particularly for DWDM cable TV networks. Figure 6.11 shows three possible dual-stage EDFA system configurations. In the first EDFA system configuration [Figure 6.11(a)] unidirectional forward pumping is employed in for both stages. In the second EDFA system configuration [Figure 6.11(b)], unidirectional forward pumping is employed for the first stage and a unidirectional backward pumping is employed in the second stage. In the third EDFA system configuration [Figure 6.11(c)], unidirectional backward pumping is employed for the first stage and unidirectional forward pumping is employed for the second stage. A midway OI and/or a narrow optical bandpass filter are typically employed with this EDFA design. The role of the midway OI is to suppress the buildup of an ASE peak at 1530-nm. The backward-propagating ASE from the second stage is prevented by the midway OI from entering into the first stage. The result is an improved EDFA gain and noise performance. The optimal location of the midway OI must be determined using numerical simulations. The combination of both the midway OI and optical filter is the most efficient way to suppress the amplifier self-saturation caused by ASE. Additional possible EDFA configurations, which have not been shown here, include bidirectional forward and backward pumping for each stage, and unidirectional backward pumping for each stage.

6.3 Amplifier Noise and CNR Calculation

Understanding noise generation in optical amplifiers and its impact on amplifier performance is a critical issue in lightwave communication networks. First, we will review how the new noise terms for optical amplifiers are generated. Then, the optical and electrical methods to measure amplifier noise figure will be discussed. Third, the calculation of the CNR for SCM lightwave systems will be explained.

6.3.1 Optical Fiber Amplifier Noise

In lightwave communication systems, optical amplifiers are used to amplify the propagating signals as in-line amplifiers, as power booster amplifiers (to the laser transmitter), or as pre-amplifiers (to the receiver). In order to determine the SNR or the CNR of the signal in these communication systems, one must analyze the noise contributions of the optical amplifier. Let us assume an optical amplifier with gain G so that the output power is related to the input power by $P_{out} = GP_{in}$. Furthermore, let us assume that we have an optical receiver following the optical amplifier with an average DC photocurrent given by

$$I_R = \left[\frac{q \cdot \eta}{h\nu}\right] G \cdot P_{in} = R \cdot G \cdot P_{in} \tag{6.21}$$

where q is the electronic charge, η_D is the quantum efficiency of the photodetector, and we have assumed 100% coupling efficiency between the amplifier and the photodetector. Using Equation (6.16) for a single-stage amplifier, the photocurrent at the receiver due to the ASE is given by

$$I_{ASE} = R \cdot P_{ASE} = 2R \cdot n_{sp} \cdot h\nu[G-1]\Delta\nu \tag{6.22}$$

The current variance in the photodetector due to the ASE shot noise is given by

$$\sigma_{ASE}^2 = \left\langle i_{ASE}^2 \right\rangle = 2qI_{ASE} = 4q^2 \cdot \eta \cdot n_{sp}[G-1]\Delta\nu \tag{6.23}$$

Of course, there is also a shot-noise component due to the photocurrent I_R. Let us define a related quantity to the ASE power called the single-sided spectral density of the ASE, which is nearly constant and can be written as

$$S_{sp}(\nu) = \frac{P_{ASE}}{2\Delta\nu} = n_{sp}[G-1]h\nu \tag{6.24}$$

The effect of the ASE is to add noise fluctuations to the amplified power, which are converted to current fluctuations during the photodetection process. The primary contributions to the optical receiver noise come from the beating of the spontaneous emission with the signal, which is called a signal-spontaneous beat noise, and with itself, which is called a spontaneous-spontaneous beat noise. Using Equation (6.24), the variance of the photocurrent due to the signal-spontaneous beat noise is given by

$$\sigma_{S-ASE}^2 = \left\langle i_{S-ASE}^2 \right\rangle = 4I_R[R \cdot S_{ASE}] = 4\frac{(q\eta)^2}{h\nu} n_{sp} \cdot G[G-1]P_{in} \tag{6.25}$$

where a factor of 2 in Equation (6.25) is used because one needs to consider a double-sided ASE spectral density, and another factor of 2 is used to include both light polarizations. Notice that the signal-spontaneous (S-ASE) beat noise is independent of the amplifier optical bandwidth (Δv). Consequently, one cannot remove the ASE without eliminating the desired signal at the same time. In contrast, the spontaneous-spontaneous beat noise is proportional to the amplifier bandwidth. This means that a narrow bandpass optical filter can be used to remove this ASE beat noise. The variance of the spontaneous-spontaneous (ASE-ASE) beat noise is given by

$$\sigma^2_{ASE-ASE} = \left\langle i^2_{ASE-ASE} \right\rangle = 4\left[R \cdot S_{ASE} \right]^2 \Delta v = 4\left[q \cdot \eta \cdot n_{sp}(G-1) \right]^2 \cdot \Delta v \qquad (6.26)$$

where the factor of 4 in Equation (6.26) is used for the same considerations as in Equation (6.25). When the amplifier gain G is high (i.e., small input power levels), the current variance of both the S-ASE and ASE-ASE beat noise is proportional to G^2, while the ASE shot noise is only proportional to G. Consequently, the ASE shot noise contribution can be neglected compared with the ASE-ASE and S-ASE beat noise contributions in the photodetector. We will use this approximation in the calculation of the SNR in the next section.

6.3.2 CNR and Noise-Figure Calculation

Now we are ready to analyze the CNR of fiber-optics communication system, particularly for cable TV applications. Let us assume an SCM communication system with an in-line single-stage optical amplifier, with m modulation index per RF channel, and DC photocurrent of I_R. Then, using the CNR definition in chapter 5, the CNR can be written as

$$CNR = \frac{\left\langle i^2_R \right\rangle}{\left[\sigma^2_{shot} + \sigma^2_{th} + \sigma^2_{RIN} + \sigma^2_{S-ASE} + \sigma^2_{ASE-ASE} \right]B} \qquad (6.27)$$

where $\langle i^2_R \rangle = (mI_R)^2/2$ is the mean-square signal photocurrent. The photodetector shot-noise variance is given by the following equation

$$\sigma^2_{shot} = \left\langle i^2_{shot} \right\rangle = 2q(I_R + I_{ASE}) = 2q^2\eta\left[\frac{G \cdot P_{in}}{hv} + 2n_{sp}(G-1)\Delta v \right] \qquad (6.28)$$

From a practical aspect, it is convenient to express the CNR given by Equation (6.27) in terms of the amplifier noise figure. The amplifier noise figure is defined as

$$F = \frac{SNR_{in}}{SNR_{out}} = \frac{CNR_{in}}{CNR_{out}} \qquad (6.29)$$

where "in" and "out" refer to the SNR or the CNR measured at the output of the photodetector with and without the optical amplifier, respectively. Without the optical amplifier, the photodetector is essentially shot-noise limited. Consequently, the CNR_{in} can be written as

$$CNR_{in} = \frac{(mI_R)^2}{4qI_R} = \frac{m^2 \cdot \eta \cdot P_{in}}{4h\nu} \qquad (6.30)$$

In the presence of the optical amplifier, the photodetector is primarily limited by the shot noise and the signal-spontaneous beat noise. The ASE-ASE beat noise and the ASE shot noise contributions can be neglected compared with the signal-ASE beat noise contribution. The contribution to the inverse CNR caused by the signal-ASE beat noise can be written as

$$CNR_{S-ASE}^{-1} = \frac{8h\nu \cdot n_{sp}}{m^2 \cdot P_{in}} \left[1 - \frac{1}{G} \right] \qquad (6.31)$$

The resultant CNR at the photodetector with the optical amplifier can be calculated according to

$$CNR_{out}^{-1} = CNR_{shot}^{-1} + CNR_{S-ASE}^{-1} = \frac{4h\nu}{m^2 \cdot P_{in} \cdot G} \left[1 + 2n_{sp}(G-1) \right] \qquad (6.32)$$

Substituting Equations (6.30) and (6.32) into Equation (6.29), we get

$$F = 2n_{sp} \left[1 - \frac{1}{G} \right] + \frac{1}{G} \qquad (6.33)$$

For a high-gain operation ($G \gg 1$), the amplifier noise figure can be approximated by $2n_{sp}$. If we further assume an ideal optical amplifier with a complete population inversion, then $n_{sp} = 1$. This means that in the quantum noise limit, the amplifier has a noise figure of 3-dB [$= 10 \cdot \log_{10}(2)$]. Furthermore, neglecting the right-hand side term $1/G$ from Equation (6.33), the optical amplifier noise figure can be written as

$$F = 2n_{sp} \left[1 - \frac{1}{G} \right] \qquad (6.34)$$

Substituting the noise figure expression given by Equation (6.34) in Equation (6.27) for the CNR, we obtain [30]

$$CNR = \frac{m^2}{2B\left[RIN + \frac{2q}{I_R} + \left(\frac{N_R}{I_R}\right)^2 + 2h\nu\left(\frac{F}{P_{in}}\right) + \Delta\nu\left(\frac{F \cdot h\nu}{P_{in}}\right)^2\right]} \qquad (6.35)$$

where B is the photodetector electrical bandwidth, RIN is the overall communication system RIN, and N_R is the photodetector noise equivalent current. Equation (6.35) is very useful expression since it describes the CNR dependence in terms of two primary measurable parameters, namely, the amplifier noise figure (F) and its optical input power. Furthermore, Equation (6.35) suggests that the transmitted CNR of the RF channel after the in-line EDFA is primarily governed by the signal-ASE and ASE-ASE beat noise at low optical input levels (< -10-dBm). In this input power regime, the amplifier noise figure approaches the quantum noise limit of 3-dB. If the noise figure is nearly unchanged when the EDFA is operating in saturation (optical input power $\geq$ 0-dBm), the AM CNR becomes limited by the receiver's thermal and shot noise at a given detected optical power. However, if the amplifier noise figure increases monotonically with the optical input power, then it sets the upper limit on the AM CNR as seen from Equation (6.35). As we will see later, robust SCM transmission of AM-VSB video channels often requires the EDFAs to operate in saturation.

6.3.3 Noise-Figure Measurement

The noise figure of an optical amplifier can be measured using either the optical or electrical methods. The optical measurement method uses an optical spectrum analyzer and/or an optical power meter. It is easily done, and provides accurate results in the small-signal regime (< -10 dBm). On the other hand, the electrical method, which uses an electrical spectrum analyzer, is a more accurate method when the amplifier operates in saturation. However, it is more complicated to measure.

The optical method is based on Equation (6.16) for the ASE power and Equation (6.34) for the amplifier noise figure in the small-signal regime. Substituting Equation (6.34) in Equation (6.16), one obtains

$$P_{ASE}(L) = F \cdot G \cdot h\nu \cdot \Delta\nu \qquad (6.36)$$

Thus, the noise figure can be calculated according to Equation (6.36) by measuring amplifier small-signal gain G ($G \gg 1$) and ASE power. The ASE power can be measured by turning the input optical signal off and placing an optical bandpass filter with bandwidth $\Delta\nu$, centered at the signal wavelength. As the input signal power is increased, the amplifier saturates and ASE power becomes compressed. The compressed ASE power cannot be measured using this method, since when the input signal is turned off the compressed ASE becomes uncompressed. By adding the polarization nulling technique, one can still use the optical method to measure the compressed ASE power. The polarization-nulling technique uses a polarization controller and/or a polarizer between the optical amplifier and the optical spectrum analyzer to null the input signal [31]. However, there are two problems with the optical method when the amplifier is operating in saturation. First, residual spectral components of the ASE signal may still exist, creating a measurement uncertainty. Second, the measured noise figure may be overestimated due to a phenomenon called polarization hole burning (PHB) [32]. The magnitude of the compressed ASE noise measured in the "orthogonal" polarization may be significantly different than in the "parallel" polarization, particularly when the optical amplifier is operating in a deep saturation. Since only the compressed ASE with "parallel" polarization generates the signal ASE beat noise, the measured "orthogonal" ASE may not be very reliable under these conditions. Furthermore, interferometric noise effect caused by multiple optical reflections can add uncertainty to the optical measurement.

The electrical method is based on using Equations (6.34) and (6.35), requiring the use of an electrical spectrum analyzer, particularly when the amplifier is saturated. For an SCM lightwave communication system, the CNR of the transmitted RF channel is measured on an electrical spectrum analyzer both with and without the optical amplifier as the optical input power is increased. In the large-signal regime, the measured CNR of the RF channel is limited by the shot and thermal noise at the photodetector. The RIN in Equation (6.35) is the overall system RIN noise, including RIN from multiple reflections. Using Equation (6.34), the noise figure (F) can be easily be calculated at various optical input power levels to the amplifier. It should be pointed out that for DWDM cable TV networks, the EDFAs are typically operating in saturation, where it is advantageous to use the electrical method. The noise characteristics of EDFAs operating in the saturation regime are discussed in the next section.

6.4 EDFA Requirements for Cable TV Networks

6.4.1 EDFA's Noise-Figure Requirement

As we learned earlier, the CNR of the transmitted RF channel through a high-power in-line EDFA is limited by the EDFA noise in the small-signal regime. Moreover, the transmitted CNR may be further reduced in a cascaded EDFA link, which is often required to transport the AM video channels, due to EDFA noise buildup. Thus, selecting the proper EDFA system design configuration is essential to meet the overall transmission requirements. The selection process is best illustrated by the following comparison. Two in-line 980-nm pumped-EDFAs with different characteristics but with the same small-signal noise figure have been selected. The experimental setup for the multichannel AM-VSB video lightwave transmission system consisted of a 1552-nm EM-DFB laser transmitter with a built-in power amplifier, 75-km of a standard SMF with an in-line 980-nm pumped EDFA, and an optical receiver [30]. The EM-DFB laser transmitter has built-in mechanisms for a predistortion linearization and the suppression of SBS. The SBS effect converts the transmitted optical signal in the fiber to a backward-scattered one, and thus limits the maximum optical power that can be launched into the standard SMF [33]. Two types of 980-nm pumped EDFA were used as in-line amplifiers, which are labeled EDFA type A and type B. EDFA type A is a single-stage single-pump amplifier with a 980-nm unidirectional forward pumping. EDFA type B is a dual-stage amplifier with a 980-nm unidirectional backward pumping for the first stage, and unidirectional forward pumping for the second stage. The saturated output power levels of both EDFA type A and B are 14.2-dBm and 16.2-dBm, respectively. Both EDFA types were designed to achieve simultaneously high output power with low NF and overall flat gain. Seventy-one AM-VSB channels (55.25- to 505.25-MHz) simulated by CW carriers from a multitone generator, were used to modulate the 1550-nm EM-DFB laser with an output power of +16-dBm. The AM-VSB video channels were transmitted through a standard SMF and in-line EDFAs, and then detected at 0-dBm optical power level at the optical receiver. Frequency-domain simulations were used to analyze the two different types of in-line EDFAs. The algorithm for the EDF was based on the quasi-analytical steady-state model, which was developed by Saleh and Jopson, for a two-level EDFA for which ESA can be neglected [34, 35]. This model gives accurate results for low-gain amplifiers (gain $\leq$ 20-dB) pumped at 980-nm or 1480-nm. However, at higher gains, the three-level model EDF, which was discussed in Section 6.1.2, was used. The various EDFA parameters used in the simulations are summarized in Table 1. The Erbium-doping profile of the EDF was identical for the two EDFA types. In addition, the isolator forward pump loss and backward loss were assumed to be infinite for both EDFA types. The optical spectra in the 100-nm band of each in-line EDFA in the AM video lightwave link were measured at various optical input power levels up to +5.9-dBm. For example, the measured and calculated optical spectra of EDFA type A at an optical input power of −13-dBm is shown in Figure 6.12 [30]. Notice that the ASE peak is located around 1530-nm. Excellent agreement is obtained between the measured and calculated optical spectra for both types of EDFAs based on the frequency- domain simulations at various optical input powers. The calculated gain for both EDFAs, which is reduced by a similar amount as the optical input power, is increased. EDFA type B has about 2-dB higher gain than EDFA type A at the same optical input power level. Since the saturated gain of both EDFAs remains high ($G \gg 1$), the noise figure can still be approximated by the

Table 6.1 Simulation parameters for 980 -nm pumped EDFA types "A" and "B".

EDFA Parameter	EDFA "A"	EDFA "B"
Pump Power (mW)	100	90
Pump Wavelength (nm)	980	980
Fiber Length (m)	30	1st Stage = 25, 2nd Stage = 30
WDM Pump Loss (dB)	0.6	0.6
WDM Signal Loss (dB)	1	0.5
Isolator Forward Signal Loss (dB)	0.8	0.65

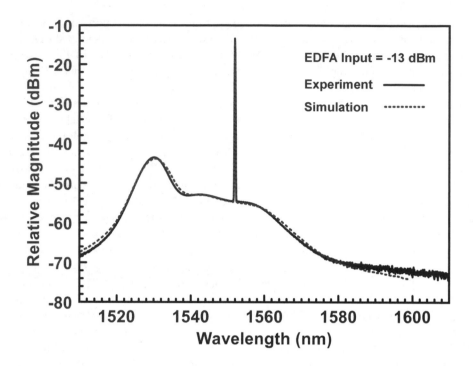

Figure 6.12 Measured and calculated optical spectrum of EDFA type A with an optical
input power of –13-dBm. After Ref. [30] (© 1997 IEEE).

$2n_{sp}$ parameter. The noise figure was measured at the optical receiver with and without the
in-line EDFA according to Equation (6.29) at various optical input power levels. Figure 6.13
shows the measured and calculated noise figure for both EDFA types as a function of the
optical input power levels.   The noise figure was calculated from the simulated gain profiles
and n_{sp} for each EDFA using Equation (6.34). The noise figure can also be calculated from
the measured CNR of the transmitted analog channels using Equation (6.35) with the listed

Table 6.2 System parameters for CNR calculation.

Parameter	Specification
Modulation Index per Channel (m)	3.2%
Electrical Noise Bandwidth of AM Channel	4-MHz
Optical Bandwidth of the EDFA	40 nm
Receiver Photocurrent	0.85-mA
Receiver Noise Equivalent Current	7-pA/√Hz
RIN	–164 dBc/Hz

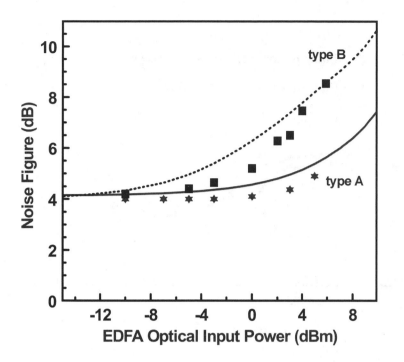

Figure 6.13 Measured (symbols) and calculated (lines) noise figure versus the EDFA
optical input power level for EDFA type A and B. After Ref. [30]
(© 1997 IEEE).

parameters in Table 6.2. Notice that although the small-signal noise figure of these EDFAs is nearly the same (= 4 dB), the noise figure of EDFA type B dramatically increases to 8.5-dB at high optical input power (= +6 dBm). In contrast, the noise figure of EDFA type A remains essentially unchanged even at these high optical input levels.

This noise figure behavior of these EDFAs can be explained as follows. The lowest EDFA noise figure is achieved when the pump propagates in the same direction as the signal, as is the case for EDFA type A. There is a much stronger inversion depletion of the Er^{3+} ions at high optical input levels in the backward-pumped first stage of EDFA type B compared with EDFA type A. In addition, the midway isolator or filter as shown in Figure 6.11 (c) can be eliminated, since the ASE peak is probably suppressed when the EDFA operates in saturation. Consequently, a single-stage EDFA with unidirectional forward pumping is advantageous for analog video transmission over cable TV networks.

The noise figure of gain-saturated EDFAs has been the investigation subject of many studies [30, 36–38]. The reason is that the theoretical noise figure (F) based on Equation (6.34)

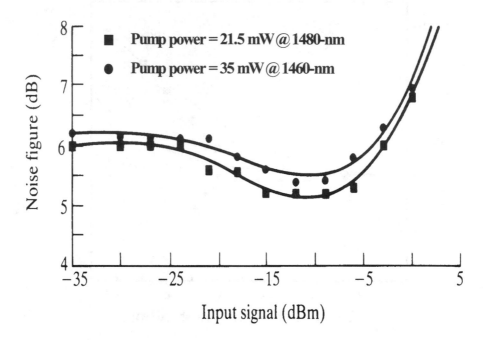

Figure 6.14 Measured and calculated saturated noise figure versus the optical input power for pump powers of 21.5-mW (circles) and 35-mW (square) in a both-end 1480-nm pumped EDFA. After Ref. [37] (© 1991 Optical Society of America).

is predicted to be approximately equal to $2n_{sp}$ if the EDFA saturated gain remains high. This may seem contradictory to various experimental results (i.e., Figure 6.13), which show that the noise figure is actually increasing. To resolve this apparent dilemma, one should remember that the noise figure derivation in the previous sections assumed a small-signal regime in which one has a linear amplification. However, in the saturated-gain regime, the EDFA operates as a nonlinear amplifier, where no rigorous definition has been developed. It has been proposed to define the saturated noise figure (SNF) parameter as $2n_{sp}^{min}(1 + X)$, where X is an excess noise parameter due to saturation [36]. This definition is based on a simplified model with the following three assumptions. First, there is no ASE self-saturation, which means that signal-induced saturation is the main mechanism. Second, the pumping power is much larger than the amplifier saturating power, which is aimed at power amplifiers. Third, the pump power is uniformly distributed along the fiber, which is a reasonable approximation only when the fraction of the pump power absorbed in the EDFA is small. However, in most power amplifiers, this fraction can be large, say 60%, leading to potentially inaccurate noise figure predictions [19].

The noise figure of gain-saturated EDFA can sometimes exhibit a complex behavior in highly performing doped fibers. Figure 6.14 shows, for example, the noise figure behavior as a function of the optical input power to a both-end 1480-nm pumped EDFA [36]. In this example, the small-signal gain is 39 dB for a signal wavelength of 1550-nm. Notice that the saturated noise figure decreases by 1-dB compared with the small-signal value, exhibiting a minimum at an optical input power level near −10-dBm. At higher input power levels, the saturated noise figure increases to 7-dB at an optical input power of 0-dBm. The dip in the noise figure, which occurs at high-gain amplifiers, can be explained by the redistribution of the ASE power along the EDF due to the saturating signal. For sufficiently high optical input power, the backward traveling ASE, which degrades the noise figure, is reduced near the fiber ends. Thus, a higher medium inversion is achieved, and a lower noise figure is obtained.

6.4.2 EDFA's CNR Requirement

Based on these results, what is the required noise figure of a high-power in-line amplifier for an AM video lightwave links?

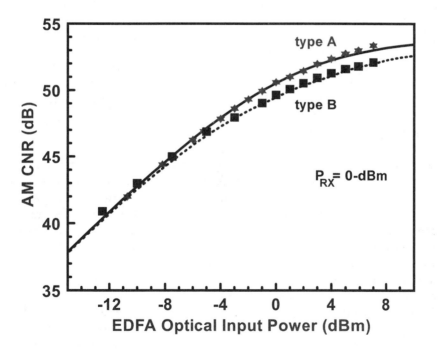

Figure 6.15 Measured (symbols) and calculated (lines) AM CNR versus the EDFA optical input power levels (in dBm) for EDFA types A and B. After Ref. [30] (© 1997 IEEE).

To answer this question, let us examine the following measurement. The AM CNR was first measured for SMF lengths up to 50-km without using the in-line EDFAs. The worst- case measured AM CNR at 325.25-MHz (channel 41) was 53.75-dB with low CSO and CTB distortions (≤ −65 dBc). Then, the measurements were repeated with type A or type B in-line EDFAs in the lightwave link with the same detected optical power of 0-dBm at the receiver.　Figure 6.15 shows the worst-case measured and calculated AM CNR at channel 4 as a function of the EDFA's optical input power level (dBm) for the two EDFAs. Both the calculated noise figures and the AM CNR results are in good agreement with the measured results. Notice that the AM CNR difference between the two EDFAs is increased to 1.7-dB at an optical input power level of +6 dBm. Negligible degradation in both the CTB and CSO distortions in the AM video lightwave links was observed using these in-line EDFAs. The experimental results in Figures 6.13 and 6.14 suggest that an in-line amplifier with a noise figure of 5-dB or less is needed to achieve CNR greater or equal to 50-dB for analog video lightwave trunking links [39, 40].

To test this requirement, an AM-VSB video lightwave trunking link with a 39 dB link budget was constructed using two similar in-line EDFAs, where the optical input power for the first in-line EDFA was kept at +3-dBm and with 0-dBm at the receiver. With +3-dBm optical input power to the second in-line EDFA, the measured AM CNR was 50.4-dB using the single-stage EDFA, representing a CNR loss of 3.4-dB. In contrast, the measured AM CNR was only 48-dB using the dual-stage EDFA, or a CNR loss of 5.8-dB under the same operating conditions. Therefore, single-stage EDFAs are advantageous in AM video lightwave links, particularly when cascaded EDFAs are required to achieve the CNR requirement of the designed link. The performance characteristics of cascaded EDFA-based transmission links for both analog and QAM channels are further discussed in Chapter 10.

6.4.3 Gain-Flattened EDFAs

So far, we have discussed the EDFA requirements for single-wavelength cable TV networks. As we learned in Chapter 1, HFC cable TV access networks are evolving toward DWDM networks for targeted delivery of analog and digital video/data services. To support the delivery of digital QAM channels over DWDM networks, EDFAs with very broad bandwidths (tens of nanometers) and with flat gain (within 1-dB) over the amplifier bandwidth are required. Gain differences among digital channels transmitted at different wavelengths, particularly over a cascaded EDFA link, may result in a large difference in the received optical power, causing an unacceptable BER. In addition, gain equalization is also desirable for analog video transmission to suppress the CSO distortions induced by gain tilt (see Section 10.5) [41]. Let us review the recent research and development in this area. The main approaches to gain equalization of EDFAs can be divided into passive (or static) methods and active methods. The passive methods include the use of long-period gratings, which have a grating periodicity in the order of hundreds of microns and act as low-loss wavelength selective filters [42]. The problem with the passive methods is that they produce a flat gain only for a predetermined gain level, and produce an undesirable gain tilt when the gain level is changed. Furthermore, there is no way to adapt the filter transmission profile when the EDFA operating conditions are changed. The active methods for EDFA gain equalization, which are based on active optical filtering, can produce a complex filter transmission response necessary to even out the EDFA gain for different input signals and pump powers. Early work has focused on using integrated-optic LiNbO$_3$ waveguide devices, where the acousto-optic coupling occurs between the TE and the TM polarization modes through a surface acoustic wave.

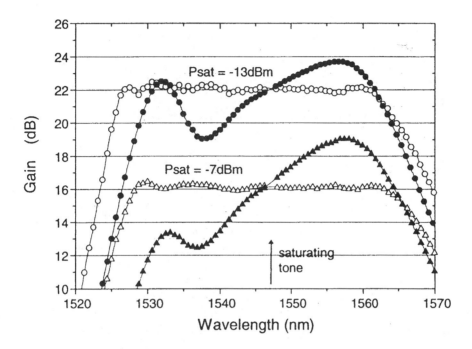

Figure 6.16 EDFA gain profiles with (filled circles) and without (open circles) equalization at −7-dBm and −13-dBm saturation levels. After Ref. [44] (© 1998 IEEE).

To obtain polarization-independent operation, complex polarization diversity configurations were required, resulting in various problems relating to the use of two independent acousto-optic polarization converters [43]. Recent works have demonstrated dynamic EDFA gain equalization using an all-fiber active gain-flattening filter, which consist of two AOTFs in tandem [18, 44]. Each AOTF was driven by three RF signals at different amplitudes and frequencies, which generates sinc-function notch profiles at variable wavelengths and rejection ratio, to produce acousto-optic mode conversion from the fundamental spatial mode to the cladding modes. Figure 6.16 shows the measured gain profile with (open circles) and without (filled circles) gain equalization for two saturating signal powers of −7-dBm and −13-dBm [44]. Gain variations over 5-dB were reduced to less than 0.7-dB with gain bandwidth over 35-nm.

References

1. G. P. Agrawal, *Fiber-Optic Communication Systems*, John Wiley & Sons (New York, 1992).
2. F. P. Kapron, "Transmission Properties of Optical Fibers," *Optoelectronic Technology and Lightwave Communication Systems*, Ed. Chinlon Lin, Van Nostrand-Reinhold (New Jersey, 1989).
3. D. Marcuse, *Theory of Dielectric Optical Waveguides*, Academic Press, New York, 2nd Ed. (1991).
4. J. Straus and B. Kawasaki, "Passive Optical Components," *Optical Fiber Transmission*, Ed. E. E. Basch, Howard W. Sams & Co., Indianapolis (1987).
5. B. Nyman, "Passive Components," *OFC Tutorial* (1998).
6. N. Takato, T. Kominato, A. Sugita, K. Jinguji, H. Toba, and M. Kawachi, "Silica-Based Integrated Optic Mach-Zehnder Multi/Demultiplexer Family with Channel Spacing of 0.01-250-nm," *IEEE Journal of Lightwave Technology* **8**, 1120–1127 (1990).
7. C. Dragone, "An N x N Optical Multiplexer Using a Planar Arrangement of Two Star Couplers," *IEEE Photonics Technology Letters* **3**, 812–815 (1991).
8. C. Dragone, C. A. Edwards, and R. C. Kistler, "Integrated Optics N x N Multiplexer on Silicon," *IEEE Photonics Technology Letters* **3**, 896–899 (1991).
9. M. K. Smit and C. Van Dam, "PHASAR-Based WDM Devices: Principles, Design and Applications," *IEEE Journal of Selected Topics in Quantum Electronics* **2**, 236–250 (1996).
10. W. I. Way, *Broadband Hybrid Fiber/Coax Access System Technologies*, Academic Press (San Diego, 1999).
11. M. R. Amersfoort, C. R. de Boer, B. H. Verbeek, P. Demeester, A. Looyen, J. J. G. M. vander Tol, and A. Kuntze, "Low-loss Phase-Array Based 4-Channel Wavelength demultiplexer Integrated with Photodetectors," *IEEE Photonics Technology Letters* **6**, 62–64 (1994).
12. J.-L. Archambault and S. G. Grubb, "Fiber Gratings in Lasers and Amplifiers," *IEEE Journal of Lightwave Technology* **15**, 1378–1390 (1997).
13. K. O. Hill, B. Malo, F. Bilodeau, D. C. Johnson, and J. Albert, "Bragg Gratings Fabricated in Monomode Photosensitive Optical Fiber by UV Exposure Through a Phase Mask," *Applied Physics Letters* **62**, 1035–1037 (1993).

14. W. H. Loh, R. I. Laming, N. Robinson, A. Cavaciuti, F. Vaninett, C. J. Anderson, M. N. Zervas, and M. J. Cole, "Dispersion Compensation over Distances in Excess of 500 km for 10-Gb/s Systems Using Chirped Fiber Gratings," *IEEE Photonics Technology Letters* **8**, 944–946 (1996).

15. D. Pastor, J. Capmany, and J. Marti, "Reduction of Dispersion Induced Composite Triple beat and Second-Order Intermodulation in Subcarrier Multiplexed Systems Using Fiber Grating Equalizers," *IEEE Photonics Technology Letters* **9**, 1280–1282 (1997).

16. C. M. Miller and J. W. Miller, "Wavelength-Locked, Two-Stage Fiber Fabry-Perot Filter for Dense Wavelength Division Demultiplexing in Erbium-Doped Fiber Amplifier Spectrum," *IEE Electronics Letters* **28**, 216–217 (1992).

17. K. Hirabayashi, H. Tuda, and T. Kurokawa, "New Structure of Tunable Wavelength-Selective Filters with a Liquid Crystal for FDM Systems," *IEEE Photonics Technology Letters* **3**, 741–743 (1991).

18. H. S. Kim, S. H. Yun, I. K. Kwang, and B. Y. Kim, "All-Fiber Acousto-Optic Tunable Notch Filter with Electronically Controllable Spectral Profile," *Optics Letters* **22**, 1476–1478 (1997).

19. E. Desurvire, *Erbium-Doped Fiber Amplifiers Principles and Applications*, John Wiley & Sons (New York, 1994).

20. W. J. Miniscalco, "Erbium-Doped Glasses for Fiber Amplifiers at 1500-nm," *IEEE Journal of Lightwave Technology* **9**, 234–250 (1991).

21. A. Mori, Y. Ohishi, M. Yamada, H. Ono, Y. Nishida, K. Oikawa, and S. Sudo, "1.5-µm Broadband Amplification by Tellurite-Based EDFAs," Technical Digest, *OFC Conference*, paper PD-1, Dallas, Texas (1997).

22. J. L. Zyskind, E. Desurvire, J. W. Sulhoff, and D. J. DiGiovanni, "Determination of Homogeneous Linewidth by Spectral Gain Hole-Burning in an Erbium-Doped Fiber Amplifier with GeO_2:SiO_2 Core," *IEEE Photonics Technology Letters* **2**, 869–871 (1990).

23. C. R. Giles and E. Desurvire, "Propagation of Signal and Noise in Concatenated Erbium-Doped Fiber Optical Amplifiers," *IEEE Journal of Lightwave Technology* **9**, 147–154 (1991).

24. H. Asano, S. Takano, M. Kawaradani, M. Kitamura, and I. Mito, "1.48-µm High-Power InGaAs/InGaAsP MQW LD's for Er-Doped Fiber Amplifiers," *IEEE Photonics Technology Letters* **3**, 415–417 (1991).

25. S. Ishikawa, "Recent Progress in Reliable 980-nm Pump Laser Diodes," *Technical Digest, OFC Conference*, paper ThC3 (1995).

26. X. F. Shum, R. Parke, G. Harnagel, R. Lang, and D. Welch, "1.2 W Single-Mode Fiber-Coupled MOPA at 980 nm," Technical Digest, *OFC Conference*, paper PD13 (1995).

27. S. O'Brien, R. Lang, R. Parke, J. Major, D. F. Welch, and D. Mehuys, "2.2-W Continuous-Wave Diffraction-Limited Monolithically Integrated Master Oscillator Power Amplifier," *IEEE Photonics Technology Letters* **9**, 440–442 (1997).

28. H. M. Presby, "Near 100% Efficient Fiber Microlenses," OSA Proceeding on Conference on Optical Fiber Communications, paper PD24 (1992).

29. H. Kuwahara, M. Sazaki, and N. Tokoyo, "Efficient Coupling from Semiconductor Lasers into Single-Mode-Fibers with Tapered Hemispherical Ends," *Applied Optics* **19**, 2578 (1980).

30. S. Ovadia, "CNR Limitations of Er-Doped Fiber Amplifiers in AM-VSB Video Lightwave Trunking Systems," *IEEE Photonics Technology Letters* **9**, 1152–1154 (1997).

31. J. Aspell, J. Federici, B. M. Nyman, D. L. Wilson, and D. S. Shenk, "Accurate Noise Figure Measurements of Erbium-Doped Fiber Amplifiers in Saturation Conditions," Technical Digest, OFC conference, paper ThA4 (1992).

32. D. W. Hall, R. A. Hass, W. F. Krupke, and M. J. Weber, "Spectral and Polarization Hole Burning in Neodymium Glass Lasers," *IEEE Journal of Quantum Electronics* **19**, 1704 (1990).

33. X. Mao, R. W. Tkach, A. R. Chraplyvy, R. M. Jopson, and R. M. Derosier, "Stimulated Brillouin Threshold Dependence on Fiber Type and Uniformity," *IEEE Photonics Technology Letters* **4**, 66–69 (1992).

34. A. A. M. Saleh, R. M. Jopson, J. D. Evankow, and J. Aspell, "Modeling of Gain in Erbium-Doped Fiber Amplifiers", *IEEE Photonics Technology Letters* **2**, 714–717 (1990).

35. R. M. Jopson and A. A. M. Saleh, "Modeling of Gain and Noise in Erbium-Doped Fiber Amplifiers", *SPIE proceedings of Fiber Laser Sources and Amplifiers III*, vol. **1851**, 114–119 (1991).

36. W. I. Way, A. C. VonLehman, M. J. Andrejco, M. A. Saifi, and C. Lin, "Noise Figure of Gain-Saturated Erbium-Doped Fiber Amplifier Pumped at 980-nm," Proceedings of Topical Meeting on Optical Amplifiers and Applications, 1990, paper TuB3, 134.

37. J. Marcerou, H. Fevrier, J. Hervo, and J. Auge, "Noise Characteristics of the EDFA in Gain Saturation Regimes," OSA Proceeding of Topical Meeting on Optical Amplifiers and Their Applications, paper THE1-1, 162–165 (1991).

38. R. G. Smart, J. L. Zyskind, J. W. Suhoff, and D. J. DiGiovanni, "An Investigation of the Noise Figure and Conversion Efficiency of 0.98 μm Pumped Erbium-Doped Fiber Amplifiers Under Saturated Conditions," *IEEE Photonics Technology Letters* **4**, 1261–1264 (1992).

39. W. I. Way, M. M. Choy, A. Yi-Yan, M. Andrejco, M. Saifi and C. Lin, "Multichannel AM-VSB Television Signal Transmission using an Erbium-Doped Optical Fiber Power Amplifier," *IEEE Photonics Technology Letters* **1**, 343–345 (1989).

40. S. Ovadia and C. Lin, "Performance Characteristics and Applications of Hybrid Multichannel AM-VSB/M-QAM Video Lightwave Transmission Systems," *IEEE Journal of Lightwave Technology* **16**, 1171–1186 (1998).

41. C. Y. Kuo and E. E. Bergmann, "Second-Order Distortion and Electronic Compensation in Analog Links Containing Fiber Amplifiers," *IEEE Journal of Lightwave Technology* **10**, 1751–1759 (1992).

42. A. M. Vensarkar, P. J. Lemaire, J. B. Judkins, V. Bhatia, T. Erdogan, and J. E. Sipe, "Long-Period Fiber Gratings as Band-Rejection Filters," *IEEE Journal of Lightwave Technology* **14**, 58–65 (1996).

43. F. Tian, C. Harizi, H. Herrmann, V. Reimann, R. Ricken, U. Rust, W. Sohler, F. Wehrmann, and S. Westenhofer, "Polarization-Independent Integrated Optical, Acoustically Tunable, Double-Stage Wavelength Filter," *IEEE Journal of Lightwave Technology* **12**, 1192–1197 (1994).

44. H. S. Kim, S. H. Yun, H. K. Kim, N. Park, and B. Y. Kim, "Actively Gain-Flattened Erbium-Doped Fiber Amplifier over 35 nm by Using All-fiber Acousto-optic Tunable Filters," *IEEE Photonics Technology Letters* **10**, 790–792 (1998).

CHAPTER 7

RF DIGITAL QAM MODEMS

The purpose of this chapter is to explain the design and operation principles of RF cable modulators and demodulators (modems) for transporting digital video and data signals over the HFC network using the quadrature amplitude modulation (QAM) format based on a well-known standard. The principles of design and operation of QAM modems are essential for the understanding of cable modems and digital set-top boxes, which are discussed in Chapter 8. We shall start with a brief review of all the building blocks of QAM modems. In Section 7.2, the MPEG transport framing is discussed, which enables the transmission of MPEG packets. The next four sections will explain each of the four layers in the forward-error-correction (FEC) coding scheme. The FEC layers include Reed-Solomon coding, Trellis coded modulation, interleaving, and randomization. The building blocks, operation, and design of M-ary QAM modulators are discussed in Sections 7.7 and 7.8. Three key building blocks of the QAM receiver, namely, adaptive equalizer, carrier, and timing recovery are explained in Sections 7.9 and 7.10. Two important parameters to test the fidelity of the QAM transmitter are introduced in Section 7.11. Finally, the BER of M-ary QAM signals in a Gaussian channel are derived in Section 7.12.

7.1 RF QAM Modem Building Blocks

Radio frequency (RF) digital transmitters and receivers constitute one of the most important systems in HFC networks to transport digital video and/or data signals from the cable TV headend to the subscriber and back to the headend. In particular, the QAM modulation format was selected as the standard format for cable TV access networks by well-known standard organizations DOCSIS [1], DAVIC [2], and ITU-T J.83 Annex B [3]. Figure 7.1 shows a simplified block diagram of a QAM transmitter and receiver operating over an HFC network. The QAM transmitter portion of the modem consists of a Reed-Solomon (R-S) encoder, a convolutional interleaver, a randomizer or scrambler, a Trellis encoder (TCM), a symbol mapper, an M-ary QAM modulator with differential encoder, and an RF up-converter. The QAM receiver portion of the modem consists of an RF tuner, a QAM demodulator, a symbol demapper, a TCM decoder, a convolutional deinterleaver, an R-S decoder, and a derandomizer (descrambler). All these QAM modem building blocks are explained in the following sections based on the DOCSIS 1.1 protocol [1], which is the primary standard used in the U.S. Figure 7.2 illustrates the FEC section of the QAM modem.

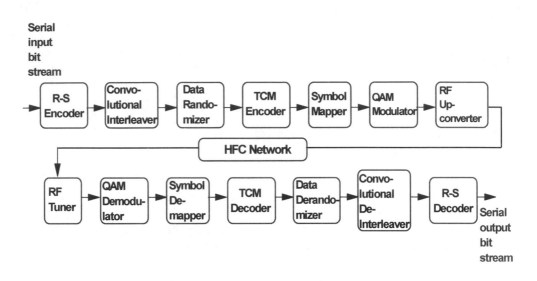

Figure 7.1 A simplified block diagram of RF QAM modem operating over HFC network.

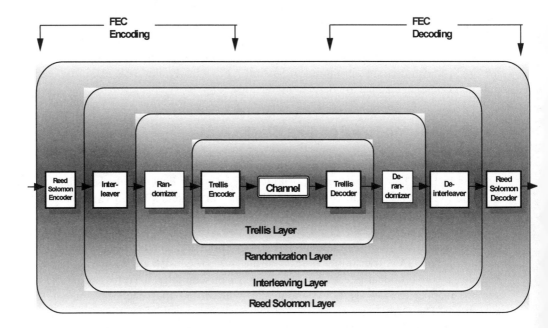

Figure 7.2 Four-layer block diagram of the forward-error-correction code.
After Ref. [4].

The FEC section uses various types of error-correcting algorithms and techniques used to transport data reliably over the cable channel. It consists of four processing layers as follow:

- **Reed-Solomon (R-S) Coding** – This layer provides block encoding and decoding to correct up to three symbols within an R-S block.
- **Interleaving** – This layer evenly disperses the symbols in time, preventing a burst of symbol errors from being sent to the R-S decoder.
- **Randomization** – This layer randomizes the data on the channel to allow effective QAM demodulator synchronization.
- **Trellis Coding** – This layer provides convolutional encoding and the possibility of using soft-decision Trellis decoding of random channel errors.

Before the data are FEC encoded in the QAM transmitter and after the data are FEC decoded in the receiver can be applied, MPEG stream transport framing is needed. The following section explains this important processing layer.

7.2 MPEG Transport Framing

The MPEG transport framing, which is the outermost layer of processing, enables robust

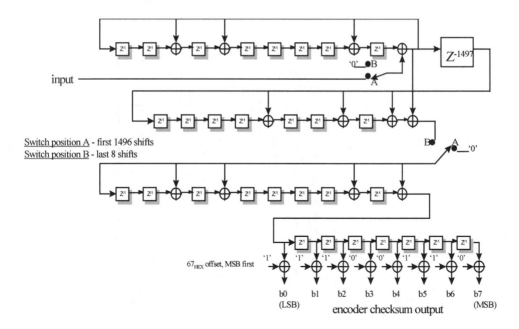

Figure 7.3 Checksum generator for the MPEG-2 sync byte encoder using LFSR. After Ref. [4].

transmission of the MPEG packets to the receiver [4]. This is accomplished by delivering the MPEG packet synchronization byte at the receiver output, and setting up the MPEG packet error flag for uncorrected errors that occurred during the transmission. The MPEG transport-framing block receives an MPEG-2 transport data stream consisting of a continuous stream of fixed length 188 byte packets (n =1504 bits), which is serially transmitted with the most-significant bit (MSB) first. The first byte of the MPEG packet is a sync byte, which has a constant value of 47_{HEX}, and it is used for both synchronization and additional error correction above that provided by the R-S code in the FEC. The rest of the MPEG payload is data (k =1496 bits).

The problem of additional error detection and synchronization is solved using a parity checksum. Parity checksum is a coset of finite-impulse-response (FIR) parity check linear block code, which is substituted for this sync byte. Figure 7.3 shows the parity checksum generator structure for the MPEG-2 sync byte encoder using linear feedback shift register (LFSR), where all additions are assumed to be modulo 2. The parity checksum is computed by sending the 1496-bit MPEG-2 payload through the LFSR (Figure 7.3), which is described by the following polynomials:

$$f(x) = \left[1 + b(x) \cdot X^{1497}\right] / g(x)$$

$$g(x) = 1 + X + X^5 + X^6 + X^8 \tag{7.1}$$

$$b(x) = 1 + X + X^3 + X^7$$

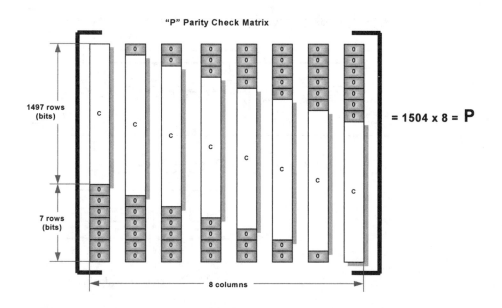

Figure 7.4 Structure of the parity check matrix P. After Ref. [4].

The parity checksum generator works as follows: before the MPEG-2 packet payload is shifted into the LFSR, the LFSR is initialized to zero. The input MPEG-2 packets are first divided by $g(x)$, and the output is split. Part of the output is delayed by 1497 bits and multiplied by $b(x)$, and then added to the non-delayed part to form $f(x)$. After the 1496 data bits are received, the encoder input is set to zero, and the switch position is changed from A to B. The eight parity bits are sequentially output by eight additional shifts of the LFSR content. 0x67 offset is added to the parity checksum byte at the MPEG-2 sync byte encoder to improve the auto-correlation properties, and to produce the 0x47 value during the parity checksum decoding at the receiver when a valid codeword is presented. A valid checksum is identified at the MPEG-2 decoder using a parity-check matrix, P, which is shown in Figure 7.4. The parity-check matrix consists of eight columns of 1497-bit vector, which we will call vector C. Vector C is calculated according to the $f(x)$ polynomial given by Equation 7.1. It is downshifted by one bit position as one proceeds from the leftmost column of the matrix to the right. The unoccupied bit positions in each data column of the parity-check matrix are filled with zeros. The syndrome generator at the decoder can be implemented using either FIR or infinite-impulse-response (IIR) filters. For example, Figure 7.5 shows an IIR filter implementation of the syndrome generator for the MPEG-2 sync byte at the decoder [5]. Note that the checksum at the MPEG-2 sync decoder is calculated always on the previous 187 bytes. Packet synchronization at the MPEG decoder is achieved as follows: the received MPEG-2 vector, which is called vector R, includes 187 bytes of MPEG-2 data followed by the checksum byte, yielding a total of 1504 bits. Multiplying (modulo 2) vector R by the parity-check matrix P yields an eight-bit-length vector S. A valid checksum is indicated when S = [0100, 0111] = 47_{HEX}.

The MPEG-2 decoder computes a sliding checksum on the serial data stream, using the detection of a valid codeword to detect the start of a packet. Once packet synchronization

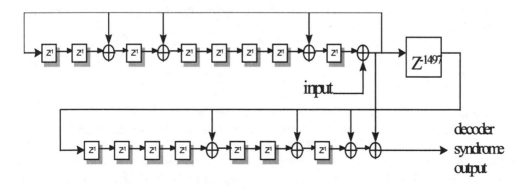

Figure 7.5 Syndrome generator for the MPEG-2 sync decoder. After Ref. [4].

has been achieved, the absence of a valid codeword at the beginning of the MPEG packet indicates a packet error. The error flag of the previous packet can then be set. To provide a standard MPEG-2 data stream, the normal sync word is reinserted in place of the parity checksum as the data is output from the decoder.

It should be pointed out that if other than MPEG-2 transport protocols such as ATM are used, this outer processing layer is removed or bypassed. The FEC layer is essentially transparent to all other transport protocols. The ATM header-error-control byte (HEC) typically provides the proper packet framing and error detection. Isochronous ATM streams are therefore carried transparently without overhead for MPEG or quasi-MPEG packet encapsulation.

7.3 Reed-Solomon Codes

The *nonbinary* Reed-Solomon (R-S) codes are a special subclass of the Bose-Chadhuri-Hocquenghem (BCH) codes, which are a large class of powerful random-error correcting cyclic codes [5]. A very useful feature of the R-S codes is that they achieve the largest possible code minimum distance for any linear code with the same encoder input and output lengths. For nonbinary codes, the distance between two code words is defined as the number of nonbinary symbols in which the sequences differ. For R-S codes, the minimum distance is given by $d_{min} = n-k+1$, where n is the total number of code symbols and k is the number of encoded data symbols in the encoded data block. Furthermore, the R-S code is capable of correcting up to T symbol errors, which is given by $T = (d_{min}-1)/2 = (n-k)/2$, corresponding to no more than $2T$ parity-check symbols. R-S codes are particularly effective for burst error correction; they are effective for channels with memory. Relatively long R-S codes can be efficiently implemented in a digital receiver with typical hard-decision decoding algorithms. For a T error-correcting R-S code with an alphabet of 2^m symbols, $n = 2^m-1$, and $k = 2^m-1-2T$, where m = 2,3,..., the R-S decoded-symbol-error probability can be approximated by

$$P_E \cong \frac{1}{n} \sum_{j=T+1}^{n} j \binom{n}{j} p^j (1-p)^{n-j} \tag{7.2}$$

where p is the symbol-error probability with no R-S coding.

For MPEG-2 transmission, the transport stream is R-S encoded using a (128,122) code over the so-called *Galois* field (GF) (128). This means that each R-S block consists of 122 information symbols and 6 parity symbols and thus has the capability of correcting up to T = 3 errors per R-S block. The same R-S code is used for both 64-QAM and 256-QAM formats, but with different Trellis coding. In addition, the FEC frame format is different for each modulation format. For the DAVIC standard, R-S T=8 (204,188) coding over GF(256) is used for both 64-QAM and 256-QAM modulation formats.

Figure 7.6 shows the packet frame format for 64-QAM modulation. For 64-QAM, an FEC frame consists of six symbols sync trailer, which is appended to the end of 60 R-S blocks, with each block containing 128 symbols. Each R-S and sync symbol consists of 7 bits.

Thus, each FEC frame consists of a total of 53,760 data bits and 42 frame sync trailer bits. For 256-QAM, an FEC frame consists of a 40-bit sync trailer, which is appended to the end of an 88 R-S block FEC frame. Each symbol in the R-S blocks is a 7-bit symbol. The total number of data bits in a 256-QAM FEC frame is 78,848 bits.

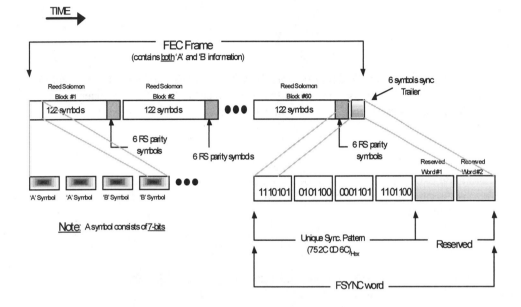

Figure 7.6 FEC frame packet format for a 64-QAM modulation. After Ref. [4].

For the ITU-T J.83 Annex B standard, the R-S encoder is implemented as follows: a systematic encoder is utilized to implement R-S T = 3 (128,122) code over the GF (128). The primitive polynomial used to form the field over GF(128) is:

$$p(X) = X^7 + X^3 + 1 \tag{7.3}$$

The generator polynomial used by the encoder is:

$$g(X) = (X+\alpha)(X+\alpha^2)(X+\alpha^3)(X+\alpha^4)(X+\alpha^5)$$
$$= X^5 + \alpha^{52}X^4 + \alpha^{116}X^3 + \alpha^{119}X^2 + \alpha^{61}X + \alpha^{15} \tag{7.4}$$

where α is the root of the primitive polynomial, namely $p(\alpha) = 0$. Notice that all the coefficients of $g(X)$ are from GF(128). The message polynomial input to the encoder consists of 122, 7-bit symbols, and is described

$$m(X) = m_{121}X^{121} + m_{120}X^{120} + \ldots + m_1 X + m_0 \tag{7.5}$$

This message polynomial is first multiplied by X^5; then, it is divided by the generator polynomial $g(X)$ to form a remainder, described by the following equation:

$$r(X) = \frac{X^5 m(X)}{g(X)} = r_4 X^4 + r_3 X^3 + r_2 X^2 + r_1 X + r_0 \tag{7.6}$$

This remainder constitutes five parity symbols, which are then added to the message polynomial to form a 127-symbol code word that is an even multiple of the generator polynomial. Then, the generated code word can be described by the following polynomial:

$$c(X) = m_{121}X^{126} + m_{120}X^{125} + m_{119}X^{124} + \ldots + r_4 X^4 + r_3 X^3 + r_2 X^2 + r_1 X + r_0 \tag{7.7}$$

A valid code word will have roots at the first through fifth powers of α.

The last symbol of the transmitted R-S block is formed by extending the parity symbol (c*) as follows:

$$c^* = c(\alpha^6) \tag{7.8}$$

In other words, the code word is evaluated at the sixth power of α. Thus, the extended code word would appear as follows:

$$c^{'} = X \cdot c(X) + c^* = m_{121}X^{127} + m_{120}X^{126} + \ldots + r_1 X^2 + r_0 X + c^* \tag{7.9}$$

The structure of the R-S block, which illustrates the order of transmitted output symbols from the R-S encoder, is shown as follows:

$$m_{121} \quad m_{120} \quad m_{119} \quad \ldots \quad r_1 \quad r_0 \quad c^* \text{ (order sent is left to right)}$$

For 64-QAM, the first four symbols of the FSYNC word contain the 28-bit "unique" synchronization pattern (1110101 0101100 0001101 1101100) or (75 2C 0D 6C)$_{HEX}$. The remaining two symbols are reserved for future applications. The FSYNC word will be inserted by the encoder and will be detected at the decoder. The decoder circuits search for this pattern and determine the location of the frame end when it is found.

For 256-QAM, the 40-bit frame sync trailer is divided as follows: 32 bits are the "unique" synchronization pattern (0111 0001 1110 1000 0100 1101 1101 0100) or (71 E8 4D D4)$_{HEX}$, 3 bits are a control word that determines the size of the interleaver employed, and 5 bits are a reserved word that is set to zero.

Note that there is no synchronization relationship between the transmitted R-S block and transport data packets. Thus, MPEG-2 synchronization must be obtained independently from R-S frame synchronization. This keeps the FEC and transport layers decoupled and independent. The downstream 64-QAM and 256-QAM parameters for ITU-T J.83 Annex B and A are summarized in Tables 7.3 and 7.4 (Section 7.12).

7.4 Interleaver/Deinterleaver

The concept of a convolutional interleaver has been proposed by Ramsey [6] and Forney [7]. Interleaving the input symbols before transmission and deinterleaving them after reception allows the QAM receiver to spread the burst errors in time, allowing the receiver to handle them as if they were random errors. Thus, the QAM receiver transforms a channel with memory to a memoryless channel, enabling random-error-correcting codes such as an R-S code to be used effectively. Convolutional interleaving is employed in both 64-QAM and 256-QAM modulation formats. A convolutional interleaver is typically characterized by the number of its shift registers, which is also called the "depth" I (symbols), and by the symbol delay increment J per register. Due to memory cost and end-to-end delay for the transmitted symbols, it is advantageous to limit the interleaver (I, J) values in certain applications.

The concept of convolutional interleaving is illustrated in Figure 7.7. The interleaving commutator position increments at the R-S symbol frequency, with a single symbol output from each position. With a convolutional interleaver, the R-S code symbols are sequentially shifted into the bank of 128 registers (the width of each register is 7 bits, which matches the R-S symbol size). Each successive register has J symbols more storage than the preceding register. The first interleaver path has a zero time delay, while the second path has J symbol

Table 7.1 Convolutional interleaver characteristics for 64-QAM and 256-QAM.

I	J	Burst Protection 64/256-QAM	Latency 64/256-QAM
8	16	5.9 μs/4.1 μs	0.22 ms/0.15 ms
16	8	12 μs/8.2 μs	0.48 ms/0.33 ms
32	4	24 μs/16 μs	0.98 ms/0.68 ms
64	2	47 μs/33 μs	2.0 ms/1.4 ms
128	1	95 μs/66 μs	4.0 ms/2.8 ms
128	2	189 μs/132 μs	8.0 ms/5.6 ms
128	4	377 μs	16.0 ms/11.2 ms

period of delay, and the third path 2·J symbol periods of delay, and so on, up to the 128th path. Thus, the last interleaver register has 127·J symbol periods of delay. At the receiver, the time-delay process through each of the deinterleaver paths is reversed such that the total net delay for each R-S symbol is the same through the interleaver and deinterleaver.

Burst noise in the channel can cause a series of erroneous symbols or multiple errors in each symbol. These errors are spread over many R-S blocks by the deinterleaver such that the resultant symbol errors per block are within the range of the R-S decoder correction capability. For 64-QAM, in the absence of all other impairments, ignoring any error propagation from the Trellis decoder, the burst tolerance as measured at the input of the deinterleaver is 95-μsec. To accomplish this, J is set equal to 1 in the interleaver.

For 256-QAM, the interleaver continues to employ a depth of I = 128, but here J may be selected by a 3-bit field in the frame sync trailer to be an integer value between 1 and 8. Table 7.1 shows the burst noise protection and latency convolutional interleaver with different I and J values. The burst noise protection, which is sometimes called erasure correction (EC) of a convolutional interleaver, is given by [2]:

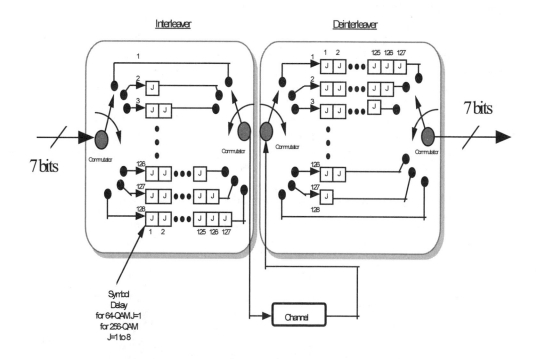

Figure 7.7 Interleaver and deinterleaver functional block diagram. After Ref. [4].

$$EC(\mu s) = \frac{(I \cdot J + 1)}{R_{RS}} \left[\frac{T \cdot I}{N} \right] \tag{7.10}$$

where N is the R-S block size, and R_{RS} is the R-S symbol rate. The end-to-end delay or latency is given by:

$$Latency(\mu s) = \frac{I \cdot J \cdot (I - 1)}{R_{RS}} \tag{7.11}$$

From Equations (7.10) and (7.11), larger interleaver size (I·J) requires larger memory to perform the convolutional interleaving operation, causing a larger the end-to-end delay. The ITU-T J.83 Annex B standard for downstream QAM channels provides a burst tolerance up to 95-μs and 66-μs for 64-QAM and 256-QAM, respectively, for an $I = 128$, $J = 1$ interleaver setting. The performance of the interleaver in the QAM receiver to combat burst noise will be discussed in Chapter 8.

7.5 Trellis-Coded Modulation (TCM)

The discussion on trellis-coded modulation (TCM) is divided into three parts. First, we introduce the concept of convolutional coding with punctured pattern. Second, the Viterbi decoding process in the QAM receiver is reviewed. Then, the use of TCM in both 64-QAM and 256-QAM modulation formats for downstream transmission is discussed.

7.5.1 Punctured Convolutional Coding

Trellis coding is employed as the inner coding scheme, while R-S coding is employed as the outer coding scheme as shown in the QAM modem block diagram in Figure 7.1. The ITU-T J.83 Annex B standard for downstream QAM transmission, which is used in the U.S, employs a concatenated coding scheme. The combination of modulation with Trellis coding is termed *trellis-coded modulation* (TCM), which was first published and analyzed by Ungerboeck [8, 9]. It allows the introduction of redundancy to improve the SNR by increasing the symbol constellation without increasing the symbol rate or bandwidth.

The TCM coding provides significantly larger "free distance" or minimum Euclidean distance (d_{free}) than the minimum distance between uncoded modulation signals at the same information rate, bandwidth, and power. The convolutional encoder consists of a shift register with $k \cdot K$ stages, where K is called the *constraint length* of the code. The constraint length represents the number of k-bit shifts in which the encoder output is influenced by a single bit. At any given time, k bits enter the shift register, and the contents of the last k stages of the shift register are dropped. n linear combinations of the shift register contents are computed to generate an encoded output waveform. The code rate of the convolutional encoder is $k/n < 1$, where $n-k$ is the number of redundant bits. Figure 7.8 shows, for example, a 1/2-rate convolutional encoder with constraint length of 7, followed by a puncturing unit. The convolutional encoder connections between the shift register and each of the n-modulo-2 adders may be characterized by generating polynomials, as we learned in Section 7.2 on R-S codes. Each polynomial has a degree of $K-1$ or less, and its coefficients are either 1 or 0, depending on whether a connection exists or does not exist between the shift register and the particular modulo-2 adder. The generating polynomials are sometimes written in octal notation. For example, the generating polynomials for the convolutional encoder shown in Figure 10.5 are given by

$$g_1(X) = 1 + X^2 + X^3 + X^5 + X^6$$
$$g_2(X) = 1 + X + X^2 + X^3 + X^6 \qquad\qquad (7.12)$$

or 133 and 171 octal.

Punctured coding is an attractive method for implementing rate-selectable convolutional encoders using existing hardware. To create a punctured code, one first encodes the data using a $1/n$ convolutional encoder and then deletes some of the channel symbols at the encoder output, which is called "puncturing." The puncturing pattern is represented by "0" or "1" symbols, where a "1" symbol indicates that a channel symbol is to be transmitted and "0" indicates a channel symbol to be deleted. Figure 7.8 shows, for example, how to create

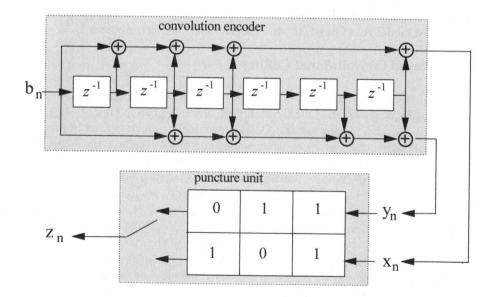

Figure 7.8 3/4-rate punctured convolutional encoder with constraint length K = 7 using a 1/2-rate convolutional encoder.

3/4-rate code from the 1/2-rate code using the puncturing pattern. Each of the columns of the puncturing pattern corresponds to a bit input to the encoder, while each bit in the pattern corresponds to an output-channel bit. The puncturing pattern can also be written in a matrix format as [P1, P2] = [011, 101]. Thus, the puncturing unit outputs 4 bits for every 3 bits input, resulting in an overall 3/4-rate convolutional encoder. Using the basic 1/2 code, this method allows the implementation of any rate $(n-1)/n$ that is slightly less efficient than the optimum code for that rate.

7.5.2 Viterbi Decoding

At the QAM receiver, the received bit stream needs to be decoded to recover the original data with an error correction code. The state of a $1/n$ convolutional encoder is defined as the content of the rightmost $K-1$ shift-registers. The output of the encoder is determined by the input bits and the state of the encoder. There are three common methods to characterize convolutional encoders. The first method is called a state diagram, which is a graphical representation of all the possible states and the possible transitions from one state to another. The problem with the state diagram is that it does not show the time evolution of the encoder transitions. The second method is called a tree diagram, which shows the time evolution of each successive input bit, where each tree branch describes an output branch word. The

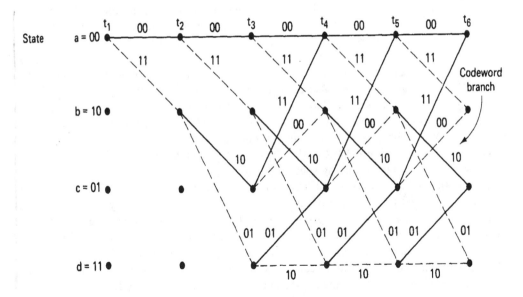

Figure 7.9 Trellis diagram for 1/2-rate convolutional code with K = 3, where the solid and dashed lines indicate the output generated by input bits "0" and "1", respectively. After Ref. [10].

drawback of the tree-diagram method is that the number of tree branches increases as 2^M, where M is the number of bits in the input sequence. The third method, which is also the most useful method, is called a trellis diagram, where the nodes of the trellis characterize the encoder states, and solid and dashed lines represent outputs generated by bits "0" and "1", respectively. At any given time, the trellis diagram requires 2^{M-1} nodes to represent the 2^{M-1} possible encoder states. Figure 7.9 illustrates the concept of a time-expended trellis diagram of 1/2-rate convolutional code with $K = 3$ ($d_{free} = 5$). The generating polynomials for this encoder are $g_1(X) = 1 + X + X^2$ and $g_2(X) = 1 + X^2$. Notice that the trellis diagram at each stage after the initial transition consists of four nodes, corresponding to four different states labeled a, b, c, and d. At each node in the trellis there are two incoming paths and two outgoing paths after the second stage. The optimum decoding method is maximum-likelihood decoding, where the decoder attempts to find the closest "valid" sequence to the received bit stream. The most important algorithm for maximum-likelihood decoding is the Viterbi algorithm [11,12]. The Viterbi decoder recognizes at any given time that certain trellis paths cannot possibly belong to the maximum-likelihood path, and thus it discards them. When binary convolutional code (BCC) with $k = 1$ and constraint length K is decoded using the Viterbi algorithm, there are 2^{K-1} surviving paths at each stage, and 2^{K-1} add compare operations. For BCC in which k bits each time are shifted into the encoder, there are $2^{k(K-1)}$ surviving paths at each stage, and 2^k paths merge at each node. Since there is only one surviving path at each node, namely, the minimum-distance path, the required number of computations at each stage increases exponentially with both k and K. Thus, in practice, the Viterbi algorithm decoder is typically implemented only for small K (< 10) and k.

The Viterbi decoder can be implemented using the so-called *hard-decision* or *soft-decision* decoding. In the hard-decision mode, the output of the QAM demodulator is quantized to either "1" or "0", and fed to the Viterbi decoder. If the demodulator output is quantized to more than two levels based on a Gaussian distribution, the decoding is called a soft-decision decoding. For example, eight levels (3-bits) of soft-decision Viterbi decoding reduces the required E_b/N_o by about 2-dB in a Gaussian channel compared to a hard-decision (1-bit) Viterbi decoding at the same BER [13]. The cost of this improvement is the larger required memory size and also the decoding speed at the receiver.

An essential part of the TCM scheme, which was developed by Ungerboeck, is the set-partitioning concept [9]. Set partitioning divides a signal set successively into smaller subsets while increasingly maximizing the smallest free distance within each subset. The set partitioning can be carried out to the limit where each subset contains only a single point. However, in many practical cases, such set partitioning is not necessary.

The performance of TCM coding can be expressed in terms of coding gain relative to the uncoded modulation set. Thus, the coding gain (G) for a given TCM code is given by

$$G = \frac{\left[d_{min}^2 / P_{av} \right]_{coded}}{\left[d_{min}^2 / P_{av} \right]_{uncoded}}$$

(7.13)

where d_{min} is the minimum Euclidean distance, and P_{av} is the average power of the modulation. Table 7.2 shows, for example, the coding gains for different trellis-coded M-ary. QAM modulation formats [9]. N_{fed} represents the number of signal sequences with the minimum distance that diverge at any state and after one or more transitions remerge at that state. The asymptotic coding gain represents the maximum possible coding gain for a given M-ary QAM modulation over the uncoded modulation. Notice that for a trellis-coded 64-QAM modulation, most of the coding gain (4.56-dB) can be achieved using only 16 states as indicated in Table 7.2. Viterbi decoding is also more tolerant than sequential decoding to metric table and receiver AGC errors. The inherent parallelism of the Viterbi decoder makes it easy to implement in hardware, an important consideration for high-speed decoding.

An important design parameter for a Viterbi decoder is the path memory length. An ideal Viterbi decoder would keep every possible path in memory and delay a final decision about the first bits of the packet until the very end. However, various simulations have shown that several constraint lengths back, the paths usually merge into a single maximum-likelihood candidate. Thus, a hardware implementation of a Viterbi decoder (unpunctured), which typically retains 4–5 constraint lengths of decoded data, is sufficient to achieve nearly the same uncorrected error rate as an ideal decoder. For a $K = 7$ code, a 32-bit path memory is a good match to a 32-bit word size for a digital signal processor (DSP) implementation. This is fast and efficient since 32-bit operations on a 32-bit DSP are no more costly than smaller operations.

Table 7.2 Coding gains for different trellis-coded M-ary QAM modulations. After Ref. [9] (© 1987 IEEE).

Number of states	Code rate	G (16-QAM/ 8-QAM) (dB)	G (32-QAM/ 16-QAM) (dB)	G (64-QAM/ 32-QAM) (dB)	Asymptotic coding gain (dB)	N_{fed}
4	1/2	3.01	3.01	2.80	3.01	4
8	2/3	3.98	3.98	3.77	3.98	16
16	2/3	4.77	4.77	4.56	4.77	56
32	2/3	4.77	4.77	4.56	4.77	16
64	2/3	5.44	5.44	4.23	5.44	56
128	2/3	6.02	6.02	5.81	6.02	344
256	2/3	6.02	6.02	5.81	6.02	44

7.5.3 TCM for 64/256-QAM Modulation

For a 64-QAM modulation, the input to the trellis-coded modulator is a 28-bit sequence of four 7-bit R-S symbols, which are labeled in pairs of "A" symbols and "B" symbols. A block diagram of a trellis-coded modulator for 64-QAM modulation is shown in Figure 7.10. The trellis-coded modulator includes a punctured 1/2-rate BCC that is used to introduce the redundancy into the least-significant bits (LSBs) of the trellis group. The convolutional encoder is a 16-state nonsystematic 1/2-rate encoder with the generating polynomials $g_1 = 010$ 101 and $g_2 = 011\ 111$, or (25,37) in octal notation. Equivalently, one convolutional encoder can be described in terms of the generator matrix $[1 \oplus D^2 \oplus D^4, 1 \oplus D \oplus D^2 \oplus D^3 \oplus D^4]$. The puncturing matrix for 4/5-rate is [P1, P2] = [0001; 1111], which produces a single serial bit stream. All 28 bits are assigned to a trellis group, where each trellis group forms 5 QAM symbols, as shown in Figure 7.10. Each trellis group is further divided into two groups: two upper or most-significant bits (MSBs) uncoded bit streams, and one lower or LSB-coded bit stream. Of the 28 bits in the trellis group, the four LSBs of the trellis group are fed to the BCC for coding, producing five coded bits. The remaining 10 bits are sent to the mapper

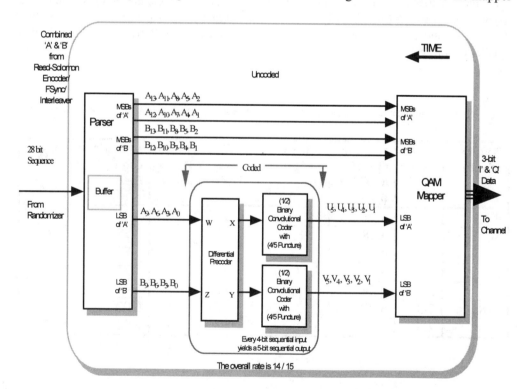

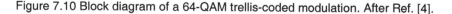

Figure 7.10 Block diagram of a 64-QAM trellis-coded modulation. After Ref. [4].

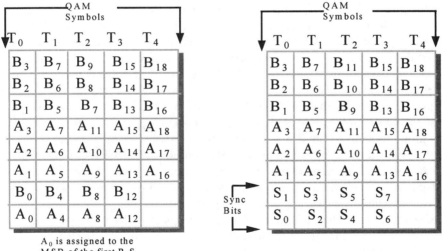

Figure 7.11 Trellis group for a 64-QAM modulation. After Ref. [4].

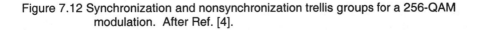

Figure 7.12 Synchronization and nonsynchronization trellis groups for a 256-QAM modulation. After Ref. [4].

uncoded. This will produce an overall output of 15 bits. Thus, the overall TCM rate for 64-QAM is 14/15. The trellis group is formed from R-S symbols as follows: for the "A" symbols, the R-S symbols are read, from MSB to LSB, A_{10}, A_8, A_7, A_5, A_4, A_2, A_1 and A_9, A_6, A_3, A_0, A_{13}, A_{12}, A_{11}. The four MSBs of the second symbol are fed into the BCC, one bit at a time, LSB first. The remaining bits of the second symbol and all the bits of the first symbol are fed into the mapper, uncoded, LSB first one bit at a time. The four bits sent to the BCC will produce five coded bits labeled U_1, U_2, U_3, U_4, U_5. The same process is done for the "B" bits. The process can be seen in Figure 7.10. With 64-QAM, four R-S symbols conveniently fit into one trellis group, and in this case the sync word is embedded in every position of the trellis group.

For a 256-QAM modulation, the TCM is using the same 1/2-rate BCC and the same 4/5 punctured matrix as a 64-QAM modulation. However, in this case all the sync information is embedded only in the trellis group LSBs. There are two different types of trellis groups in 256-QAM, namely the nonsynchronization and the synchronization groups, which will be labeled as nonsync and sync groups, respectively. Each trellis group will produce five QAM symbols; the nonsync group will contain 38 data bits while the sync group will contain 30 data bits and 8 sync bits. Figure 7.12 shows both the nonsync Trellis group and the sync trellis group. Since there are 88 R-S blocks per FEC frame, there will be a total of 2,076 trellis groups per frame. Of these trellis groups, 2,071 are nonsync trellis groups and five are sync trellis groups. The five sync trellis groups come at the end of the frame. The frame sync trailer is aligned to the trellis groups. The trellis group is further divided into two groups: one upper or MSB uncoded bit stream and one lower or LSB coded bit stream. The first bit of the nonsync trellis group is assigned to the MSB of the first R-S symbol in the FEC frame. The output from each BCC is the five parity bits labeled U_1 through U_5 and V_1 through V_5, respectively. To form trellis groups from R-S code words the R-S code words are serialized beginning with the MSB of the first codeword. The MSB of the first symbol is placed in position A_0 of the nonsync trellis group shown in Figure 7.12. Bits are then placed in trellis group locations, from R-S symbols, in the following order: A_0 B_0 A_1 ... B_3 A_4 B_4 ... B_{16} B_{17} B_{18}. For the sync trellis groups, the bits from the serialized R-S symbols begin at location A_1 instead of A_0; from each trellis group, four bits are coded by the BCC producing five coded bits. Each of the last five sync trellis groups in an FEC frame contain 8 sync bits, $S_0 S_1 ... S_7$ in the frame sync trailer as shown in Figure 7.12.

Of the 38 input bits to the TCM modulator, two groups of 4 bits of the differentially precoded bit stream in a trellis group are separately encoded by the 1/2-rate BCC. Each BCC produces 5 coded bits, and the remaining bits are sent to the QAM mapper uncoded. This produces a total output of 40 bits, resulting in a 19/20-rate TCM for 256-QAM modulation.

7.5.4 Differential Precoder

The differential precoder shown in Figure 7.13 performs the 90° rotationally invariant trellis coding. Rotationally invariant coding is employed in both the 64- and 256-QAM modulators. The key for robust modem design is to have very fast recovery from a carrier phase slip. In a nonrotationally invariant design, a carrier phase slip will require a major re-synchronization of the FEC, leading to a burst of errors at the FEC output.

The differential precoder allows the information to be carried by the change in phase, rather than by the absolute phase. For 64-QAM, the 3rd and the 6th bits of the 6-bit symbols are differentially encoded, and for 256-QAM, the 4th and 8th bits are differentially encoded. The differential precoder is described by the following coupled set of equations:

$$X_j = W_j + X_{j-1} + Z_j \left[X_{j-1} + Y_{j-1} \right]$$

$$Y_j = Z_j + W_j + Y_{j-1} + Z_j \left[X_{j-1} + Y_{j-1} \right]$$

(7.14)

If you mask out the 3rd and the 6th bits in 64-QAM as in Figure 7.10 as well as the 4th and 8th bits in 256-QAM, the remaining bits have 90° rotational invariance, which is inherent in the labeling of the symbol constellation.

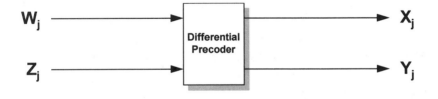

Figure 7.13 Differential precoder for TCM as shown in Figure 7.10.

7.6 Randomizer/Derandomizer

The randomizer is the third layer of processing in the FEC block diagram. The randomizer provides an even distribution of the symbols in the constellation, resulting in a flat QAM spectrum, which enables the QAM demodulator to maintain proper lock. The randomizer adds a pseudo-random noise (PN) sequence of 7-bit symbols to the transmitted signal to assure a random transmitted sequence.

For both 64- and 256-QAM, the randomizer is initialized during the FEC frame trailer and is enabled at the first symbol after the trailer. Thus, the trailer itself is not randomized. Initialization is defined as preloading to the all "1" state, for the randomizer structure shown in

Figure 7.14. The generating polynomial of the randomizer, which uses a linear feedback shift register specified by a GF (2^7), is defined as follows:

$$g(X) = X^3 + X + \alpha^3 \tag{7.15}$$

where $\alpha^7 + \alpha^3 + 1 = 0$. The corresponding randomizer structure is shown in Figure 7.14.

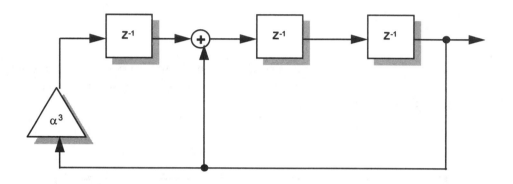

Figure 7.14 A QAM randomizer structure (7-bit byte scrambler).

7.7 M-ary QAM Modulator Design and Operation

7.7.1 Baseband Shaping Filter

One of the main tasks of a baseband digital communication system is to cancel the effect of the so-called *intersymbol interference* (ISI) [10]. Due to distortions in the transmitter, in the receiver, or in the channel, the received signals overlap with one another in time. In other words, the end of one received pulse interferes in the detection process with the beginning of the following pulse, generating ISI. Nyquist [14], who has investigated this problem, showed that the theoretical minimum system bandwidth needed to detect R_s symbols/s without generating ISI is $R_s/2$ Hz. This condition is satisfied for an ideal filter with a rectangular transfer function H(f) with a Nyquist bandwidth $W = R_s/2 = 1/2T_s$. The impulse response of this filter, which is given by the inverse Fourier transform, is $h(t) = \mathrm{sinc}(t/T_s)$. However, such a brick-wall filter is very difficult to implement since it requires an infinite time delay and would be very sensitive to small timing errors. The most commonly used filter in digital systems, which smoothly approximates the Nyquist filter, is called a Nyquist *raised-cosine* (RC) filter, whose transfer function can be written as

$$H(f) = \begin{cases} 1 & for \quad |f| < R_s(1-\alpha)/2 \\ \cos^2\left[\dfrac{\pi}{2}\dfrac{|f| - R_s(1-\alpha)/2}{\alpha R_s}\right] & for \quad R_s(1-\alpha)/2 \le |f| \le R_s(1+\alpha) \\ 0 \, for \quad |f| > R_s(1+\alpha) \end{cases} \tag{7.16}$$

where $W = R_s(1 + \alpha)/2$ is the −6-dB bandwidth, and α is the so-called bandwidth roll-off factor, which represents the fractional excess bandwidth over the Nyquist bandwidth. The corresponding impulse response $h(t)$ of this filter is given by:

$$h(t) = R_s \cdot \operatorname{sinc}[R_s t] \cdot \frac{\cos[\pi \alpha R_s t]}{1 - 4(\alpha R_s t)^2} \tag{7.17}$$

where sinc$(x) = \sin(\pi x)/\pi x$. This filter is very popular since its impulse response h(t) goes through zero when it is sampled every T_s, and thus it has no ISI. It should be pointed out that $h(t)$ is a *noncausal* filter since its impulse response starts from t = −∞. However, $h(t-t_0)$, which is a delayed version of $h(t)$ is causal and can be generated by a real filter. Notice from Equation (7.17) that $h(t)$ with a smaller roll-off factor (α) has large oscillating amplitudes between positive and negative values.

A closely related filter to the Nyquist RC filter is the square-root-raised-cosine (SRRC) shaping filter. It is well known that when a communication system is operating in the presence of additive white Gaussian noise (AWGN), then, the BER is minimized by optimum partitioning of the transmitter and receiver filters. These transmit and receiver filters are called a *matched* filter pair since their transfer function product must yield the RC filter as described by Equation (7.16). Thus, the frequency response of each filter is simply the square root of $H(f)$ or the SRRC filter, whose impulse response is given by [10]

$$h_{SRRC}(t) = \sqrt{R_s} \cdot \frac{\sin[\pi R_s(1-\alpha)t] + (4\alpha R_s t) \cdot \cos[\pi R_s(1+\alpha)t]}{\pi R_s t\left[1 - (4\alpha R_s t)^2\right]} \tag{7.18}$$

Figure 7.15 shows, for example, the frequency and amplitude response of an SRRC filter with 160 taps and with a 12% roll-off factor. Notice the side-lobs in the SRRC frequency response, which are caused by the truncated length of an ideal filter. There are different in-band and out-of-band amplitude variation requirements on a digitally implemented SRRC filter, particularly in the transmitter. For example, the MCNS DOCSIS 1.1 protocol requires that the amplitude variations generated by SRRC filters in an upstream QAM transmitter be within ±0.3-dB for any frequency within $f_C \pm R_S/4$, and less than −30-dBc for any frequency within $f_C \pm 5R_S/8$, where f_C is the channel center frequency [1].

SRRC Frequency Response (dB) **SRRC Impulse Response**

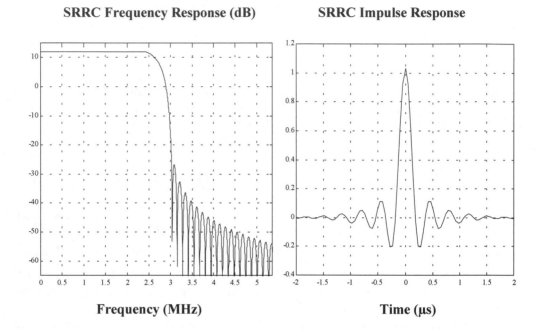

Frequency (MHz) **Time (µs)**

Figure 7.15 Frequency and impulse response of 160-taps SRRC filter with 12% roll-off factor.

7.7.2 Building Blocks of M-ary QAM Modulator

The task of an M-ary QAM modulator is to transform a baseband bit sequence to a QAM constellation with M possible amplitude and phase levels within a given bandwidth at IF frequency. There are various ways to digitally implement a conventional M-ary QAM modulator. Figure 7.16 shows, for example, a block diagram of a digitally implemented conventional QAM modulator. The serial input bit stream, which is typically a binary non-return-to-zero (NRZ) sequence, has $\log_2 (M)$ bits per symbol. The input symbols are first demultiplexed into two branches, one in-phase (I branch) and one in-quadrature-phase (Q branch) by a serial-to-parallel converter, where each branch has $k = \log_2(M)/2$ of the bits per symbol. These bits need to be mapped to 2^K possible signal amplitudes, which can be done in several ways. The preferred mapping is the one in which adjacent signal amplitudes differ by one binary bit. Such mapping is based on the fact that random errors due to a Gaussian noise in the demodulation of the desired signal will most likely affect the adjacent amplitudes to the transmitted signal amplitude. The result is only a single bit error. This mapping scheme is known as Gray coding [10]. Since M-ary QAM constellations have $\pi/2$ symmetry, the inherent $\pm \pi/2$ phase ambiguity in the QAM modulator needs to be resolved. This means that the QAM receiver can lock the carrier for every $\pi/2$ phase error. Differential encoding, as was explained in Section 7.4.4, encodes the next symbol based on the current input symbol and the previous encoded symbol.

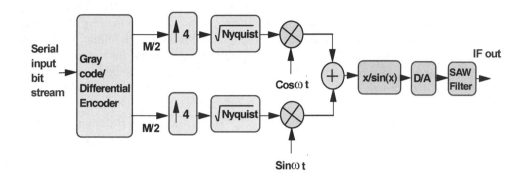

Figure 7.16 Simplified block diagram of a digitally implemented QAM modulator. After Ref. [15].

In many digital applications, the QAM modulator is required to operate at a given symbol rate, which is independent of the input symbol rate. For example, Figure 7.16 shows a 1x4 interpolation filter, which has four sample points per symbol and provides the needed sampling rate up-conversion. In the QAM receiver (Section 7.7), a 1x4 decimation filter is used. The advantage of using interpolation or decimation filters in the QAM modulator or demodulator is the elimination of the need to use high-speed digital multipliers and adders. Then, each of the differentially encoded baseband bit sequences is filtered using an SRRC shaping filter to combat ISI, which will be generated at the QAM receiver.

To form the QAM constellation, the I and Q branches of the M-ary QAM modulator are multiplied by $\sin(2\pi f_c t)$ and $\cos(2\pi f_c t)$, where f_c represents the QAM carrier frequency. Mathematically, the M-ary QAM signal can be written as

$$S(t) = I(t) \cdot \cos(2\pi f_c t) + Q(t) \cdot \sin(2\pi f_c t) \tag{7.19}$$

where $I(t)$ and $Q(t)$ are the 2^K amplitude and phase levels of the M-ary QAM constellation. Notice that there is a $\pi/2$ phase shift between the I and Q branches. In practical applications, the quadrature oscillators can be implemented using the so-called d*irect digital frequency synthesizer* (DDFS) [16]. The DDFS can generate frequencies in steps of $\Delta f = F_s/2^N$, where F_S is the modulator sampling frequency and N is the number of bits used in the DDFS phase accumulator. The QAM carrier frequency f_c can be generated by the DDFS by $f_c = W \cdot \Delta f$, where W is a constant word that is added to the DDFS accumulator at F_S rate [17].

After combining the I and Q branches, a x/sin(x) correction filter is applied before the digital-to-analog converter (DAC). In an ideal DAC, the output voltage is instantaneous for each sample. Thus, the output voltage between samples has to be interpolated to make the analog output continuous in both time and amplitude. If the Nyquist sampling criterion is used and the interpolation filter eliminates all the frequencies above the folding frequency, then the output is an exact replica of the original input signal, except for the quantization errors. For practical reasons, natural sampling instead of impulse sampling is used in the D/A converter. This means that the DAC output is typically held constant for the full duration between samples. The frequency transfer function of the interpolation filter causes a sin(x)/x error, typically less than 4-dB up to $f/2F_S$. The x/sin(x) filter is designed to compensate for the frequency roll-off [18].

When the M-ary QAM modulator is designed, the magnitude of all the discrete and spurious noise outside the QAM channel is particularly important. This is because when QAM channels are stacked right next to one another, a large amount of spurious and discrete noise of one QAM channel appears as an in-band noise, which will degrade the BER of the transmitted QAM channel. To overcome this possible limitation, a surface-acoustic-wave (SAW) filter is used at the IF frequency of the QAM modulator. The SAW filters have high insertion loss, typically about 20-dB, and they may be expensive. Despite that, SAW filters are used since they are FIR filters with a very flat in-band group delay. In North America, the IF frequency of QAM modulator is typically 43.75-MHz or 44-MHz, while in Europe the IF frequency is 36.125-MHz. The IF output of the QAM modulator is typically upconverted to the desired RF frequency in the 54 MHz to 860 MHz range.

7.8 M-ary QAM Receiver Design and Operation

The task of the QAM receiver is the inverse of the QAM transmitter, namely, to convert the received RF signal to a digital baseband bit sequence. According to Figure 7.1, the first building block of the QAM receiver is the RF tuner. Since the RF tuner is also an essential part of the digital set-top box, it will be discussed separately in Chapter 8 on subscriber home terminals. Figure 7.17 shows a simplified block diagram of a digitally implemented M-ary QAM receiver [19, 20]. The RF tuner output is an analog QAM signal typically at an IF of 44-MHz. The analog front end of the QAM receiver is shown, for example, in Figure 7.18 [19]. There are two widely used methods to digitize and down-convert the analog IF signal before it is presented to the QAM demodulator. In the first method, the IF signal is down-converted to a secondary IF frequency equal to the QAM symbol rate ($\cong$ 5 MHz) before it is digitized. The down-conversion process typically requires the use of a SAW filter, a programmable gain amplifier (PGA), a voltage-controlled oscillator (VCO) or a numerically controlled oscillator (NCO), and a mixer. Then, the mixer output is fed to an ADC with a typical 10-bit resolution. The PGA provides the necessary adjustable gain (up to 20-dB) before it is input to the sample-and-hold circuit in the ADC converter. The ADC samples the input signal at a higher sampling rate than the symbol rate, typically four times, in order to allow all digital quadrature down-conversion and digital filtering. Notice that the error signals generated from the tracking and acquisition loops can be used for example to drive two VCOs to adjust the input carrier frequency and timing accuracy as shown in Figure 7.18. Another implementation of the front-end section of the QAM receiver is to replace the two VCOs with fixed-frequency oscillators and adjust the input carrier frequency in the quadrature oscillators. In the second method, which is called direct IF sampling, the 44-MHz IF frequency is passed through a SAW filter and PGA before it is directly digitized and sampled typically by a 10-bit ADC to produce a digital IF signal. The ADC is locked to a very stable reference crystal. The digitally sampled IF signal is first converted to a baseband signal, normally using a multirate filter, and then it is separated into its I and Q

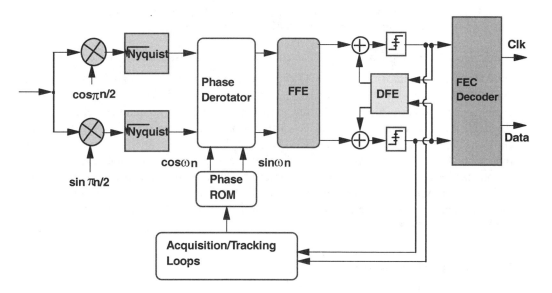

Figure 7.17 Simplified block diagram of a digitally implemented M-ary QAM receiver. After Ref. [19].

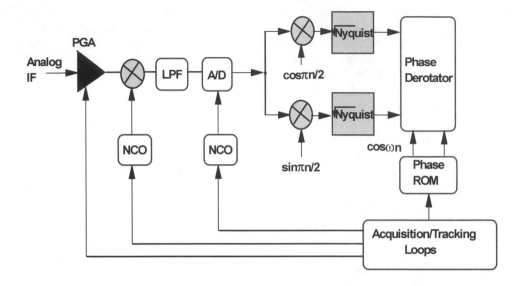

Figure 7.18 Block diagram of the analog front end of the QAM receiver. PGA =
programmable gain amplifier. After Ref. [19].

components by multiplication with the quadrature oscillators, namely, the corresponding $\sin(2\pi f_c n)$ and $\cos(2\pi f_c n)$. The I and Q components of the baseband signals are filtered by dual SRRC filters. Since the received sampled signal may have a large amount of phase errors, a decision-directed phase-locked loop (PLL), which is sometimes called a *phase derotator* loop, is used to estimate the carrier phase magnitude and direction for proper compensation. The block in Figure 7.17, which is labeled acquisition and tracking loops, represents the QAM carrier frequency and timing recovery loops. Since the adaptive equalization, carrier recovery, and timing recovery are essential parts of the QAM receiver design and operation, they are discussed in more detail in the following sections. The operation of the FEC decoder block is more complicated than that of the FEC encoder block, and it will not be discussed here.

7.9 Adaptive Equalizer

One of the primary components of a QAM receiver is its adaptive equalizer (AE). The AE is designed to adapt its coefficients to minimize the ISI resulting from linear amplitude and phase distortions in the channel, which cannot be removed by the SRRC filters. In the following section, we will review the state-of-the-art adaptive equalization methods and practices that are employed in QAM receivers used in cable modems and digital set-top terminals.

In order to compensate for a channel's linear distortions, one will need to estimate the actual channel impulse response and adjust accordingly an equalizer filter in the receiver to equalize the channel. This problem was first proposed and addressed by R. W. Lucky [21–23]. The adaptation process adapts the coefficients of the equalizer filter to minimize the effect of ISI at the receiver output, generated by the channel linear distortions. The equalizer adaptation is driven by the magnitude of an error signal from a decision device called a slicer, which indicates to the equalizer the direction its coefficients must change in order to more accurately represent the data symbols at the slicer input. An ideal equalizer will have a transfer function inverse to that of the transmission channel. There are three main approaches to solve the equalization problem as follows: (A) linear equalization, (B) decision-feedback equalization, and (C) transmitter precoding. First, let us explain the linear equalization approach.

The linear transversal equalizer (LTE), which represents the simplest equalizer, is constructed using a tapped delay line, a series of tap weights or coefficients c_n, and a slicer. The impulse response and the frequency response of such an equalizer can be written as [24]

$$g_k(t) = \sum_{n=-N}^{n=N} c_{nk} \cdot \delta(t - nT) \tag{7.20}$$

$$G_k(t) = \sum_{n=-N}^{n=N} c'_{nk} \cdot e^{-j2\pi nTf} \tag{7.21}$$

where $\{c_{nk}\}$ are the $(2N+1)$ equalizer tap coefficients at the kth sampling time. It was also assumed that the LTE taps are symmetrically located around the main tap (zero-time delay), and the delay time between adjacent taps is a symbol period T. Such an equalizer is called a symbol-spaced equalizer. In practice, the LTE filter is approximated by a finite-duration impulse response (FIR) filter. If the output of the equalizer is sampled at $1/T < 2W$, where 2W is the channel bandwidth, then the frequencies in the received signal above the folding frequency $1/T$ are aliased into frequencies below $1/T$. The equalizer compensates for the distorted signal caused by the aliasing effect. In contrast, if the time-delay between adjacent taps τ satisfies the condition $\tau \le 1/2W \le T$, then no aliasing occurs, and the equalizer compensates for the "real" channel distortion. Such an equalizer is called a *fractionally spaced equalizer*. Typically, $\tau = T/2$ is selected.

Let us assume that the impulse response of the transmission channel is $h(t)$, and the equalizer output signal $x(t)$ is sampled when $t = mT$; then the equalizer output signal is given by

$$x(mT) = \sum_{n=-N}^{N} c_{nm} \cdot h(mT - n\tau) \tag{7.22}$$

where n = 0, ±1, ...,±N, and τ is the time delay between adjacent taps. Now, we may apply the condition that Equation (7.22) is equal to zero for $m \neq 0$, and equal to 1 for $m = 0$. This condition is called the zero-forcing (ZF) condition. Thus, one obtains (2N+1) linear equations for the coefficients of the *zero-forcing equalizer* [24]. It should be noted that a finite-length ZF-LTE does not completely eliminate ISI. However, as the number of taps N is increased, the ISI is reduced to an acceptable level. The main drawback of the ZF equalizer is that it ignores the presence of an additive noise, which may result in a noise enhancement. For example, suppose that a deep null caused by multipath reflections occurs in the channel frequency response. The ZF will provide a large gain to equalize the total response, and in the process also will enhance the noise in the channel, which in turn reduces the transmitted SNR in the receiver.

The noise enhancement problem of the ZF equalizer can be minimized using an AE with an adaptation algorithm based on the minimum-mean-square-error (MMSE) criterion. Let us assume that all the signals are wide-sense stationary. This assumption will become clear later. Suppose that the desired sample output of the equalizer at $t = mT$ is the transmitted values $\{d_m\}$. Then, the mean-square error (MSE) between the actual output sample $x(mT)$ and the desired values $\{d_m\}$ is given by

$$MSE = E\left\{\left[x(mT) - d_m\right]^2\right\} = E\left\{\left[\sum_{n=-N}^{N} c_{nm}h(mT - n\tau) - d_m\right]^2\right\}$$

$$= \sum_{n=-N}^{N}\sum_{k=-N}^{N} c_{nm}c_{km}E\left[h(mT - n\tau)h(mT - k\tau)\right] - 2\sum_{k=-N}^{N} c_{km}E\left[h(mT - k\tau)d_m\right] + E(d_m^2)$$

(7.23)

where the expectation is taken with respect to the desired values $\{d_m\}$ and the channel noise. The MMSE condition is obtained by requiring that the derivative of Equation (7.23) with respect to the equalizer coefficients $\{c_{nm}\}$ be equal to zero. Thus, the necessary conditions for the MMSE are given by

$$\sum_{n=-N}^{N} c_{nm}E\left[h(mT - n\tau)h(mT - k\tau)\right] = E\left[h(mT - k\tau)d_m\right]$$

(7.24)

for $k = 0, \pm 1, \pm 2, ...,\pm N$. Now, we have (2N+1) linear equations for the equalizer coefficients that depend on the statistical properties of the channel noise.

In general, the solution for the equalizer coefficients [Equation (7.24)] is usually obtained by an iterative process. However, since the channel characteristics change with time, adaptive channel equalization is needed. Thus, the adaptive equalizer must adapt its coefficients in response to the channel time variations to reduce ISI. Combining the adaptive equalizer with the *least mean square* (LMS) algorithm provides an approximate solution to Equation (7.24). The LMS algorithm, which is also known as a *stochastic gradient or steepest-descent* algorithm, does not provide an exact solution to the problem of minimizing the MSE [25, 26]. This approximation is the result of not requiring that the channel characteristics be known or stationary. The LMS approach is to substitute time-average estimates for the ensemble estimates in Equation (7.24) and to solve for the equalizer coefficients. The LMS algorithm works as follows: the equalizer output is driven by an error signal, which is given by

$$e_m = d_m - \sum_{n=-N}^{N} c_{nm} h(mT - n\tau)$$

(7.25)

Thus, the error signal is simply the difference between the desired and actual equalizer output. The equalizer coefficients are adjusted during each symbol or fraction of a symbol period to track more accurately the data symbols at the slicer input. Since the equalizer output is typically noisy, a threshold decision device, which is known as a slicer, is used to quantize the output and make decisions. Since the MSE as given by Equation (7.23) has a quadratic dependence on the equalizer coefficients, it has a unique global minimum. Thus, the equalizer coefficients are iteratively adjusted to minimize the MSE by steeply descending along the MSE surface. The n equalizer coefficient at the $(m+1)$ sampling time is given by

$$c_{n(m+1)} = c_{nm} - \frac{\mu}{2} \frac{\partial \left[e_m^2 \right]}{\partial c_{nm}}$$

(7.26)

for $n = 0, \pm 1, \ldots, \pm N$, where μ is the adaptation step size that controls the size of the change in c_{nm} at each update. The division by 2 in Equation (7.26) was done to avoid a factor of 2 in the subsequent adaptation algorithm. Notice from Equation (7.26) that we have taken the gradient with respect to the equalizer coefficients of the error-signal square instead of the MSE of e_m. This is because the iterative process rapidly attenuates the channel noise, and thus provides a good approximation. Substituting Equation (7.25) in Equation (7.26), one obtains

$$c_{n(m+1)} = c_{nm} + \mu e_m h^*(mT - n\tau)$$

(7.27)

where h^* represents the complex channel impulse response input to the equalizer. Equation (7.27) is the LMS algorithm. Initially, the adaptive equalizer can be trained by a known training sequence over the channel. This allows the equalizer to initially adjust its coefficients. Once the initial adjustment is completed, it can operate in a decision-directed mode, where the slicer output is sufficiently reliable to allow the equalizer to adapt to the channel characteristics for any input.

The convergence rate of adaptive equalizers is also an important feature in their design. When fast converging algorithms for adaptive filters are needed, the *recursive-least-squares* (RLS) and/or the *Kalman* filter algorithms are typically used [26–28]. These equalizer adaptation algorithms have been recognized to provide the fastest convergence rates. In fact, the RLS algorithm converges (in MSE) in about $2N$ iteration, where N is the number of equalizer taps. This convergence rate of the RLS algorithm is about an order of magnitude faster than that of the LMS algorithm [26]. Several RLS algorithms were developed where the number of operations per iteration grows linearly rather than quadratic with N. These efficient algorithms are called fast RLS algorithms [27]. The potential disadvantage of the RLS algorithm over the LMS algorithm is its numerical stability, particularly for fixed-point arithmetic.

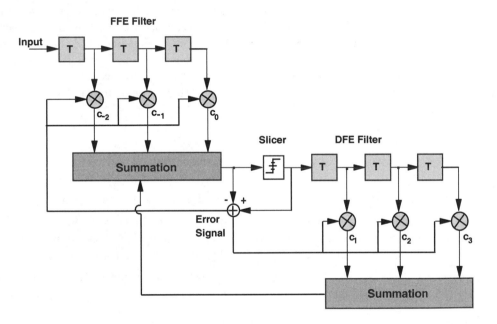

Figure 7.19 Block diagram of adaptive equalizer with feed-forward and decision-feedback filters.

Equalization techniques that do not require a training sequence for the initial adjustment of the equalizer coefficients are called *blind* or *self-recovering* equalization. Blind equalization allows the equalizer to converge without any preliminary carrier phase recovery. The various blind equalization methods will briefly be described here, and the reader is encouraged to check the literature [28–31]. Three different classes of blind equalization techniques have been developed after the pioneering work by Sato [29]. The first class of algorithms is based on the LMS adaptation algorithm, where the most widely used algorithm in practice is the so-called *Godard* or *constant-modulus algorithm* (CMA) for the adaptive equalizer [30]. The CMA is designed to minimize deviations of the blind equalizer from a constant modulus. The advantage of this algorithm is that it does not require knowing the carrier phase. However, it tends to converge slowly compared with a known training sequence. The second class of algorithms estimates the channel characteristics based on the second- and higher-order statistics of the received signal and then computes the equalizer coefficients [31, 32]. The main drawback of this algorithm class is the large amount of computational complexity involved in the estimation of the higher-order moments of the received signal. The third class of blind equalization algorithms is based on the maximum-likelihood criterion. This algorithm, which jointly estimates the channel impulse response and data sequence similar to the Viterbi algorithm approach, is suitable only for low symbol rates because of its computational complexity.

The adaptive linear equalizers described so far may perform poorly in transmission channels with large linear distortions such as cable TV channels with multipath echoes. A decision-feedback equalizer (DFE) is a nonlinear filter that uses the slicer decisions on the previously detected symbols to eliminate the ISI of the currently input symbol to the slicer. Figure 7.19 shows a block diagram of adaptive equalizer filter with DFE and feed-forward equalizer (FFE) filters. The first filter in Figure 7.19 is an FFE, which is identical to the form of linear equalizer that we have discussed earlier, showing only three coefficients. Each block T represents a symbol-spaced or fractional-spaced time delay between adjacent taps. Notice that the output from all the taps is summed and fed forward to the slicer. In contrast, the output from the DFE is subtracted from the output of the FFE before it is fed to the slicer. For the equalizer shown in Figure 7.19, the output of the slicer is given by

$$x(mT) = \sum_{n=-(N-1)}^{n=0} c_{nm} \cdot h(mT - n\tau) - \sum_{n=1}^{n=M} a_{nm} \cdot \hat{d}_{m-n} \qquad (7.28)$$

where the FFE is anticausal with N coefficients $\{c_{nm}\}$ and the DFE is causal with M coefficients $\{a_{nm}\}$. The desired slicer output d_{m-n} has been replaced by the estimated decision outputs. The MSE criterion can be applied to Equation (7.28) and solved for the DFE coefficients, which are given by [24]

$$a_{km} = \sum_{n=-(N-1)}^{n=0} c_{nm} \cdot h(kT - n\tau) \qquad (7.29)$$

for $1 \leq k \leq M$. Since the DFE coefficients in Equation (7.29) are expressed in terms of the FFE coefficients, one can substitute this solution in Equation (7.24), and apply the MSE criterion to minimize the $\{c_{nm}\}$ coefficients. A simple extension of the stochastic gradient algorithm can be applied here to enable a joint adaptation of both the FFE and DFE filters. Although DFE is able to combat linear channel distortions, it is not the optimum equalizer since it does not necessarily minimize the probability of error. The ability of the DFE to cancel residual ISI caused by past symbols provides more freedom in the selection of the FFE coefficients. In particular, the FFE does not need to approximate the inverse of the channel impulse response and thus avoids excessive noise enhancement. However, the DFE may be sensitive to an error propagation, namely, there is a greater probability for more incorrect decisions from the slicer following the first one. Fortunately, the error propagation is not usually catastrophic. In practice, both adaptive FFE and DFE are used to combat linear and nonlinear distortions over cable TV channels.

The third approach to the equalization problem is to use transmitter precoding. This approach will be explained later in Chapter 11. The performance of adaptive equalizers used in consumer home terminals such as CMs and digital STBs will be discussed in Chapter 8.

7.10 Carrier and Timing Recovery

So far, it has been assumed that both the carrier frequency and phase and the symbol timing are known at the receiver. The QAM receiver performs demodulation of the transmitted signal where its timing reference is usually independent from the transmitter. Extracting the carrier frequency and phase as well as the symbol timing from the received signal are referred to as a carrier and a timing recovery, respectively, which are essential to the operation of the QAM receiver, particularly over cable TV networks. This is because the QAM receiver has to operate in the simultaneous presence of multiple impairment environments, which exist over the cable TV plants. The following subsections briefly explain the operation principles of some of the currently used methods for carrier and timing recovery in QAM receivers.

7.10.1 Carrier Recovery in QAM Receivers

In this section we will assume that the symbol timing is known in order to extract the carrier frequency. Carrier recovery is an essential part of the QAM receiver, even if differential coding is employed. When a QAM signal is detected at the receiver, the sampled QAM constellation points are spinning around the origin with a fixed phase error. Thus, the goal of the carrier recovery loop is to minimize the phase error, extract the carrier frequency and phase, and allow the QAM receiver to proceed with the demodulation process.

The carrier recovery methods can be classified into feedforward and feedback methods. In the feedforward method, the carrier recovery circuit generates a signal at four times the carrier frequency ($4f_c$), which is then refined using a bandpass filter (BPF) or a PLL. For QPSK constellations, the phase error of the detected signal changes only by multiple of 2π, and the signal would contain a discrete component at $4f_c$. This method suffers from self-noise or pattern jitter, and is suitable only for QPSK modulation.

In general, the feedback methods for carrier recovery can further be divided into two main categories: (A) Costas loops [33] and times-N loops [33, 34], and (B) decision-directed loops [35, 36]. Figure 7.20 shows a block diagram of a fourth-order Costas loop for a carrier recovery [33]. Let us assume that θ_e is the phase error between the received baseband signal and the VCO, which is used to generate the local carrier at f_c. The received signal is split four ways, where each branch is multiplied by the VCO signal with 0°, 45°, 90°, and 135° phase shifts, respectively. The LPF at each branch allows one to keep only the lowest order terms in f_c. It can be shown (after some manipulations) that the estimated phase error output from the phase detector, which drives the VCO to match the phase of the received signal, is proportional to $\sin(4\theta_e)$.

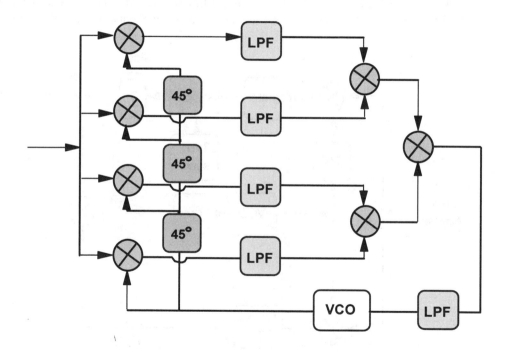

Figure 7.20 Block diagram of a fourth-order Costas loop for M-ary QAM carrier recovery.

Another popular method for a carrier recovery, which is stochastically equivalent to a hard-limited Costas loop, is called the demodulation-remodulation method. In the demodulation part, the two-way split received signal is multiplied by the VCO output, where a phase shift of 90° has been introduced between the two branches. The resultant baseband signals, after passing through the LPF and the slicer, remodulate the locally generated carrier where their phases are compared with the received signal to generate the phase error control signal for the VCO. This method, which uses two-phase detectors, may require a careful balance of the delays between the demodulation and remodulation signal paths. The primary drawback of this method is that it estimates the phase error continuously, and thus the control signal to the VCO can be corrupted by self-noise or pattern noise even with no added noise. For QAM constellations higher than QPSK, the carrier acquisition speed may be a limiting factor, requiring a larger loop bandwidth.

To overcome these issues, let us discuss the second category of feedback loops for carrier recovery, namely, the *decision-directed* carrier recovery loop [35–37]. Figure 7.21 shows a generic block diagram of a decision-directed carrier recovery loop. The received signal is multiplied by the quadrature carriers, which are obtained from the VCO output. The resultant signal is sampled and passed through a slicer, where a decision is made on the symbol every T seconds. The error signal is filtered by the LPF to reject the double-frequency term and to drive the VCO. Notice that the second quadrature component of the sampled signal is

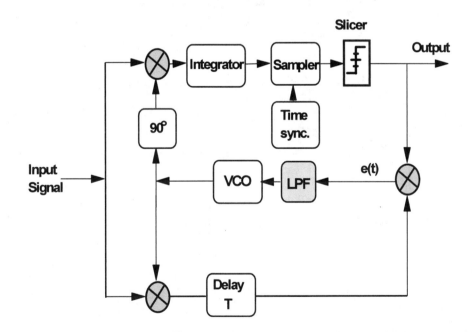

Figure 7.21 Basic block diagram of a decision directed carrier recovery loop.

delayed by T seconds to allow the slicer to make a decision. In practice, this method provides a smaller error variance compared with either the Costas or times-N carrier recovery loops if the receiver is operating at error rates below 0.01. A decision-directed carrier recovery loop can be implemented with or without an adaptive equalizer. Without an adaptive equalizer, the receiver makes a decision based on the most likely transmitted constellation point given the received symbol. This is usually the minimum Euclidean distance between the transmitted and received constellation points. These decisions are then used to generate a phase-error control signal to adjust the erroneous phase of the received constellation points.

The use of an adaptive equalizer in conjunction with a decision-directed loop creates several practical difficulties that need to be overcome. Placing an adaptive equalizer inside the decision-directed loop would cause long symbol delays, and would impair the loop's ability to track phase jitter. This problem is typically solved by placing the adaptive FFE before the carrier recovery loop.

7.10.2 Timing Recovery in QAM Receivers

The primary task of the timing recovery loop is to generate a clock signal at the symbol rate R_s or a multiple of R_s from the received signal. For state-of-the-art QAM receivers, it is not practical to transmit a separate clock signal with the data; the clock signal must be extracted from the received modulated signal using a additional circuitry in the receiver.

In general, the timing recovery methods can be categorized into two classes, which are called the forward-acting [33] and the feedback methods [38–40]. The forward-acting timing-recovery method extracts a timing tone directly from the received signal. The timing tone, which has an average frequency exactly equal to the symbol rate, is used to synchronize the receiver to the detected digital signal. Since the timing tone has some amount of a

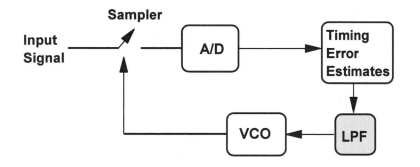

Figure 7.22 A generic block diagram of an MMSE feedback-timing-recovery loop.

phase jitter or a timing jitter, a PLL is typically used to reduce the timing jitter to an acceptable level. In the feedback-timing-recovery methods, a feedback loop is used in conjunction with the PLL to adjust the sampled phase in an iterative way and drive the VCO in order to minimize the timing error.

The most popular-timing-recovery method in the forward-acting category is the spectral-line method. In this method, the timing error τ_k is determined by the zero crossing of the timing tone and PLL. The modulated signal is passed through a timing tone detector such as an envelope détector, LPF, and a PLL, where the resultant signal is used to drive a VCO. The feedback-timing-recovery category can be divided into two different techniques, namely, the MMSE and the baud-rate timing recovery. The operating principles and the advantages and disadvantages of each method are explained as follows: Figure 7.22 shows a basic block diagram of an MMSE timing-recovery loop. The MMSE timing recovery adjusts the timing error to minimize the expected square error between the slicer output (estimated symbols) and the corrected data symbols. Notice that we assumed that the received signal is in baseband and it is sampled every symbol period (T). The MMSE timing-

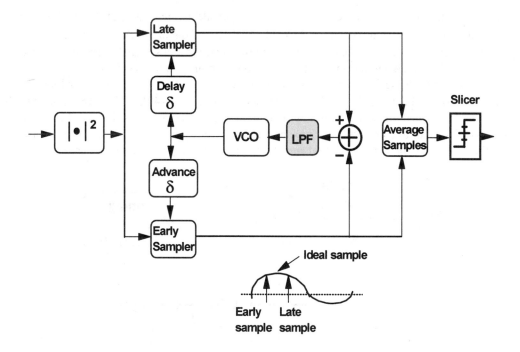

Figure 7.23 A block diagram of an early-late timing recovery loop. After Ref. [40].

recovery method, which is sometimes called the LMS timing recovery, can be implemented using various techniques. The MMSE can be implemented using the stochastic gradient algorithm to minimize the timing error in the same way that it was used for the adaptive equalizer (Section 7.8). The main problem with this approach is that it operates on the input to the slicer, which has been equalized and demodulated. However, both the carrier recovery and the equalizer assumed that the timing recovery has been achieved.

Other timing recovery techniques, which are closely related to the MMSE method, have been reported in the literature. One of the well-known techniques is the *early-late timing recovery* as shown in Figure 7.23 [35]. In this technique, the received baseband waveform is first squared to make all the peaks positive, and then it is sampled at two different times, once before and once after the sampling instant at the same amount of δ. The sampling instant is adjusted by the VCO until the early and late samples are equal. This technique assumes that the received waveform peaks are at the correct sampling points and the waveform around the peak is symmetric. The averaged-timing-error signal is used to drive the VCO, whose output is the desired clock signal. The drawback of this technique is that the waveform peaks of the received signal (in the time domain) do not occur every sampling period for QAM sequences.

7.11 MER and EVM

Now that we understand the operation of the various building blocks of a QAM transmitter and receiver, we are ready to introduce two important parameters, which are related to one another, and which provide one with a very practical tool to test the fidelity of a QAM transmitter and to calibrate the CNR measurement of a QAM receiver under test.

7.11.1 MER and EVM Definition

The first parameter is the SNR due to any impairment. It is equivalent to the SNR for Gaussian noise generated by inter-symbol interference (ISI) that would degrade the transmitter or receiver by the same amount as the impairment. This parameter is also called the modulation-error ratio (MER), which is typically measured on a standard QAM test receiver such as the Agilent 89441A Vector Signal Analyzer, and is defined as follows:

$$MER(dB) = -10 \cdot \log \left\{ \frac{\displaystyle\sum_{j=1}^{N} (\delta I_j^2 + \delta Q_j^2)}{\displaystyle\sum_{j=1}^{N} (I_j^2 + Q_j^2)} \right\} \tag{7.30}$$

where I and Q are the ideal QAM constellation data points, δI and δQ are the errors in the received data points relative to the ideal QAM constellation points due to the impairment, and N is the number of points that were captured in the data sample. N in Equation (7.30) is the data sample size, and typically should be much greater than the number of points in the constellation in order to capture a representative sample. In other words, it is a measurement of the constellation-cluster variance due to any impairment and transmitter imperfections measured relative to the ideal constellation point locations.

The related parameter to the MER is the error-vector magnitude (EVM), which is defined as

$$EVM = 100\% \cdot \sqrt{\frac{\frac{1}{N}\sum_{j=1}^{N}(\delta I_j^2 + \delta Q_j^2)}{S_{max}^2}} \tag{7.31}$$

where S_{max} is the magnitude of the vector to the outermost state of the M-ary QAM constellation. Since the numerator is the same in each case, the EVM and MER are related as follows

$$\frac{EVM^2}{MER} = \frac{\frac{1}{N}\sum_{j=1}^{N}(I_j^2 + Q_j^2)}{S_{max}^2} = \frac{P_{rms}}{P_{max}} = 1/R \tag{7.32}$$

where R is the ratio of the peak power to the average power of the M-ary QAM constellation. Thus, Equation (7.32) defines the relationship between the MER and the EVM.

7.11.2 MER and EVM Test Procedure

To accurately measure the MER or EVM of a QAM transmitter, a calibrated reference QAM receiver is needed. Such an instrument is, for example, the Agilent 89441A Vector Signal Analyzer [41]. This instrument can analyze a broad range of performance parameters, including both the MER and the EVM from QPSK to 256-QAM signals. The Agilent 89441A can operate either in continuous or burst data mode. However, for simplicity, we will explain the MER test procedure for a QAM transmitter operating in a continuous mode. The measurement of MER by the Agilent 89441A is done by capturing a number of symbols, analyzing them, and calculating the result. Because of the randomizer, each QAM constellation point should be equally likely. In a typical measurement, the HP89441A calculates the MER of each point, but weights them all equally since they are all equally likely to occur. The Agilent 89441A should be set, for example, to sample (result length) N = 512 symbols per measurement and then average over at least 10 measurements. It is recommended that a larger number of sample points, particularly for 256-QAM constellations, should be specified.

The Agilent 89441A dynamically tracks the carrier-phase loop, baud loop, and AGC-gain loop. The Agilent 89441A does not perform real-time signal processing so there really is no tracking loop bandwidth. The instrument will determine the best overall carrier phase, baud timing, and AGC gain for the overall sample and will hold those parameters fixed for the entire sample size. The effective carrier, baud, and AGC loop bandwidth will be determined by the size of the data sample. For example, when demodulating 64-QAM signals at 5-Mbaud and using a sample size (result length) of 512 points with 1 point per symbol (512 symbols), the duration of the sample data is 0.2-Mbaud·512 = 0.2-µs·512 ≅ 102-µs. This corresponds to a bandwidth of 10-kHz. In other words, if the carrier phase changes are slow compared to 10-kHz, the change during the sample time will be small and will have small

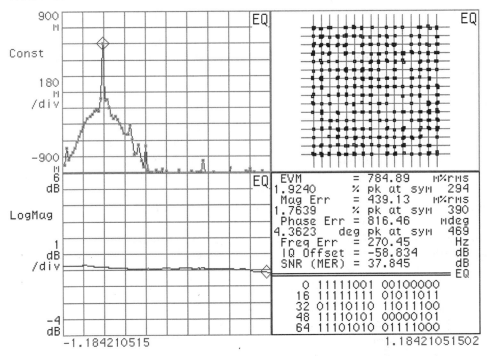

Figure 7.24 Measured MER (37.845 dB) for 256-QAM equalized constellation (99 taps) using Agilent 89441A instrument with a result length = 512 symbols.

small effect on the MER. If the change is fast compared to 10-kHz, the change during the sample time will be significant. A new set of parameters is calculated for each set of samples so the "tracking bandwidth" is in effect determined by the time duration of the result length. When set up to receive ITU-T J.83B type M-ary QAM signals, the Agilent 89441A instrument appears to behave as a well-damped second-order loop with a carrier bandwidth of about 10-kHz. Below 10-kHz, the phase perturbations are reduced at 12-dB per octave. It seems to track these parameters but with loops that are narrower than typically would be used in a QAM receiver. Thus, the Agilent instrument displays the degradations that would affect a real QAM receiver.

Figure 7.24 shows an example of a four-window display of a 256-QAM signal demodulated by the Agilent 89441A. A 256-QAM constellation with the measured EVM and MER parameters is shown in the top and bottom right side window of Figure 7.24. Notice that some of the 256-QAM constellation points are missing from the data sample due to the limited data sample size ($N = 512$). The impulse response and its Fourier transform are also shown in the top and bottom left side of Figure 7.24. The Agilent 89441A receiver contains an adaptive FIR filter, where the number of symbol-space taps ranges from 3 to 99 (odd values). The adaptation process minimizes the EVM and therefore maximizes the MER. Consequently, the equalized MER will be larger than the unequalized MER. Thus, it is recommended to specify the performance of a QAM transmitter with and without the adaptive equalizer using the Agilent 89441A instrument. The number of equalizer taps that are used in the Agilent 89441A can be determined according to the following equation [41]:

$$Taps = \left[(L-1) \cdot (po \operatorname{int} s\,/\,symbols) \right] + 1 \qquad\qquad (7.33)$$

where L is the FFE filter length. For a symbol-spaced equalizer, one point per symbol can be used. The main tap of the FFE filter is located at $(L-1)/2$ up to a setting of 31-taps, and at the 15th tap up to 75 taps. From 76 to 99 taps, the main tap is located at $L/5$, as shown in the Figure 7.24 example. For longer FIR filters, the center tap is selected closer to the beginning of the filter in order to accommodate multipath measurements.

In order to measure the MER and the EVM, the following parameters have to be set on the Agilent 89441A instrument:

- Demodulation format– the QAM mode
  – Digital demodulation (QPSK–16-QAM)
  – Video demodulation (64-QAM–256-QAM)
- Nominal symbol rate
- Measurement filter – Root raised cosine filter[1]

---

[1] The combination of a root-raised cosine filter in the Agilent 89441A with a root-raised cosine in the CM transmitter yields in an overall raised cosine response that matches the reference filter.

- Reference filter – Raised Cosine
- Roll-off factor – the α parameter
- Enable the adaptive equalizer and setup its length (number of symbols)
- QAM channel frequency and channel bandwidth
- Select data result length (N) – should be much larger than the QAM constellation size
- Set to continuous sweep mode

The MER of the tested QAM transmitter (MER_T) can be estimated using the following relation:

$$10^{-MER_T/10} = 10^{-MER_M/10} - 10^{-MER_I/10} \tag{7.34}$$

where MER_M and MER_I are the measured MER of the QAM transmitter and of the Agilent 89441A instrument, respectively. For example, the measured 256-QAM MER in Figure 7.24 is 37.845-dB, while the calibrated instrument MER is 42-dB. Thus, the corrected 256-QAM MER is 39.95-dB. Once the MER or the EVM parameters have been measured for a given QAM transmitter, they can be used in a similar way to estimate the MER or EVM of an unknown QAM receiver.

7.12 BER of M-ary QAM Signals in AWGN Channel

It can be shown that rectangular M-ary QAM constellations in which $M = 2^k$ for an even k are equivalent to two-phase amplitude modulation (PAM) constellations, each having $L = \sqrt{M} = 2^{k/2}$ constellation points on each of the QAM carriers [35]. Thus, the probability of error for M-ary QAM signals in AWGN channel is easily determined from the probability of error for PAM signals. Recalling that L-ary PAM signals are represented geometrically by one-dimensional constellation points located at $\pm d$, $\pm 3d$,..., $\pm(L-1)d$, where the Euclidean distance between adjacent points is 2d. The average symbol energy of the L-ary PAM constellation is given by

$$E_s = \left[\frac{L^2 - 1}{3} \right] \cdot d^2 \tag{7.35}$$

Let us assume an AWGN channel with zero mean and variance $\sigma_n^2 = N_0/2$, where N_0 is the noise spectral density and all the PAM amplitude levels are equally likely. The probability of error for the L-ary PAM is the probability that the noise magnitude exceeds half the distance between adjacent levels (d). The exception is for the two outer levels $\pm (L-1)\cdot d$, where an error can occur only in one direction. Thus, it can be shown that P_L is given by

$$P_L = 2\left[1 - \frac{1}{L}\right] Q\left[\sqrt{\frac{6E_s}{(L^2-1)N_0}}\right] \tag{7.36}$$

where $Q(\bullet)$ is defined as

$$Q(x) = \frac{1}{\sqrt{2\pi}}\int_x^\infty e^{-u^2/2}\,du = \frac{1}{2}\,erfc\left(\frac{x}{\sqrt{2}}\right) \tag{7.37}$$

,

where erfc(x) is the well-known complementary error function. Recalling that the relationship between the energy per symbol to the energy per bit is $E_s = E_b\cdot\log_2(L)$, the uncoded BER for L-ary PAM signals assuming Gray coding is given by

$$P_b^{PAM} = \frac{2}{\log_2 L}\left[1 - L^{-1}\right]\cdot Q\left[\sqrt{\frac{6(\log_2 L)E_s}{(L^2-1)N_0}}\right] \tag{7.38}$$

Following Proakis's approach, the probability of error for the M-ary QAM signal can be determined from the probability of error for the PAM signals, since one can demodulate the PAM signals on each of the quadrature components of the receiver. Thus, the probability for a correct decision for an M-ary QAM constellation is given by

$$P_c = \left[1 - P_{\sqrt{M}}\right]^2 \tag{7.39}$$

where $P_{\sqrt{M}}$ is the probability of error of $\sqrt{M}$-ary PAM constellations with 50% of the average power in each of the QAM carriers. Thus, the probability of symbol error for the M-ary QAM constellation is given by

$$P_{QAM} = 1 - \left[1 - P_{\sqrt{M}}\right]^2 = 2P_{\sqrt{M}} - P_{\sqrt{M}}^2 \tag{7.40}$$

where $P_{\sqrt{M}}$, has been adjusted for M-ary QAM, and is given by

$$P_{\sqrt{M}} = 2\left[1 - M^{-1/2}\right]\cdot Q\left[\sqrt{\frac{3E_s}{(M-1)N_0}}\right] \tag{7.41}$$

If $P_{\sqrt{M}} \ll 1$, then Equation (7.40) can be approximated by

$$P_{QAM} \approx 2P_{\sqrt{M}} = 4\left[1 - M^{-1/2}\right] \cdot Q\left[\sqrt{\frac{3E_s}{(M-1)N_0}}\right]$$ (7.42)

Assuming Gray coding, the uncoded M-ary QAM BER for rectangular constellations is given by

$$P_b^{QAM} = \frac{4}{\log_2 M}\left[1 - M^{-1/2}\right] \cdot Q\left[\sqrt{\frac{6(\log_2 \sqrt{M})E_s}{(M-1)N_0}}\right]$$ (7.43)

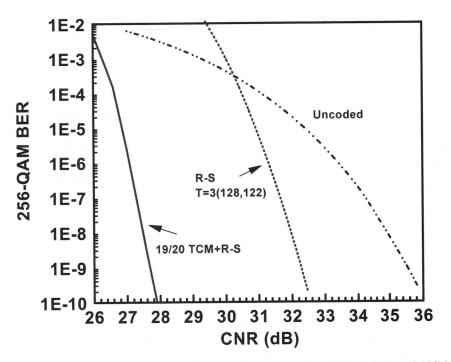

Figure 7.25 Theoretical 256-QAM uncoded, R-S T=3 (128,122) coded, and 19/20 TCM+R-S coded BER versus the CNR (in dB) for AWGN channel.

To calculate the coded R-S 64-QAM or 256-QAM BER, substitute Equation (7.43) into Equation (7.1) for the R-S symbol error rate. Figure 7.25 shows, for example, the theoretical 256-QAM uncoded, R-S T = 3 (128,122) coded, and 19/20 TCM and R-S T = 3 (128,122) coded BER versus the CNR, assuming the presence of only AWGN. It was assumed in Figure 7.25 that the channel bandwidth is equal to the symbol rate bandwidth, which means that the CNR is equivalent to E_s/N_0. Notice that the concatenated R-S T = 3 (128,122) with 19/20 rate TCM reduces the required CNR by more than 7-dB at BER of 10^{-9}.

The transmission parameters for 64-QAM and 256-QAM modulation formats, which are based on DOCSIS 1.1 protocol [1] and ITU-T J.83B standard [42], are summarized in Table 7.3. Notice the particular values that are used for the downstream symbol rate for both 64-QAM and 256-QAM. In order to understand where these values came from, one needs to understand some historical background. Many cable TV headends receive their analog and digital video feed from various satellite transponders, which are operating at QPSK modulation with an information rate of 26.97035 Mb/s (See Appendix C). Thus, to convert the incoming multiplexed video stream into 64-QAM, one needs to match the channel information rate. Thus, the symbol rate for 64-QAM signals can be calculated as follows:

$$\left[26.97035Mb/s \cdot \left(\frac{128}{122}\right) \cdot \left(\frac{15}{14}\right) \cdot \left(\frac{53,802}{53,760}\right)\right]/6 = 5.056941 - Mbaud \tag{7.44}$$

where we have taken into account the FEC overhead as well as the MPEG sync bytes. For a 256-QAM modulation, there are no satellite transponders that are transmitting at an information rate of 38.8107 Mb/s. However, the Grand Alliance, which initially developed the standard for digital HDTV, has selected its information rate such two HDTV channels can fit exactly into a single 256-QAM channel. Thus, the symbol rate for 256-QAM modulation format can be calculated as follows:

$$\left[38.8107Mb/s \cdot \left(\frac{128}{122}\right) \cdot \left(\frac{20}{19}\right) \cdot \left(\frac{78,888}{78,848}\right)\right]/8 = 5.360537 - Mbaud \tag{7.45}$$

In comparison, Table 7.4 shows the corresponding ITU-T J.83 Annex A standard for 64-QAM and 256-QAM modulation formats. This standard has been developed for 8-MHz channel bandwidth, which is mostly used in European countries. Notice that the FEC coding approach in this standard has been to use R-S block code with a larger T = 8 rather than use R-S code with a shorter T concatenated with TCM. The advantage of the concatenated code in Annex B is the addition of about 2-dB coding gain at BER = $1 \cdot 10^{-9}$ compared with

Table 7.3 Downstream cable TV transmission parameters for 64-QAM and 256-QAM modulation formats based on ITU-T J.83 Annex B standard [41].

PARAMETER	256-QAM	64-QAM
Channel bandwidth	6-MHz	6-MHz
Total symbol rate	5.360537 MBaud	5.056941 MBaud
Total channel bit rate	42.884 Mb/s	30.341644 Mb/s
Information rate (includes MPEG sync bytes)	38.8107 Mb/s	26.97035 Mb/s
Bits per symbol	8 bits	6 bits
R-S coding	T = 3 (128,122)	T = 3 (128,122)
TCM coding	19/20 rate	14/15 rate
FEC sync overhead	78,888/78,848 (40-bit sync word)	53,802/53,760 (42-bit sync word)
Interleaver depth	I = 128, J = 1	I = 8,16,32,64,128 J = 16,8,4,2,1
Roll-off factor	$\alpha = 0.12$	$\alpha = 0.18$
IF center frequency	44 or 43.75 MHz	44 or 43.75 MHz
Theoretical CNR (only AWGN at coded BER = $1 \cdot 10^{-9}$)	27.9-dB	21.8-dB

Table 7.4 Downstream 64/256-QAM transmission parameters based on ITU-T J.83 Annex A standard [41].

PARAMETER	256-QAM	64-QAM
Channel bandwidth	8-MHz	8-MHz
Total symbol rate	6.952 MBaud	6.952 MBaud
Total channel bit rate	55.616 Mb/s	41.712 Mb/s
Information rate (includes MPEG sync bytes)	50.981 Mb/s	38.236 Mb/s
Bits per symbol	8 bits	6 bits
R-S coding	T = 8 (204,188)	T=8 (204,188)
FEC sync overhead	188/187	188/187
Interleaver depth	I = 204, J = 1	I = 12, J = 17
Roll-off factor	$\alpha = 0.15$	$\alpha = 0.15$
IF center frequency	36.125 MHz	36.125 MHz
Theoretical CNR (only AWGN at coded BER = $1 \cdot 10^{-9}$)	29.9-dB	23.8-dB

Annex A for a Gaussian channel. The drawback is the more complex receiver design, and the need for a deeper convolutional interleaver in the presence of burst and impulse noise.

References

1. CableLabs *Data-Over-Cable Service Interface Specifications (DOCSIS)*, Radio Frequency Interface Specifications SP-RFIv1.10I-990311 (1999). See also http://www.cablemodem.com.
2. Digital Audio-Visual Council (DAVIC) 1.2 Specification Part 8, Lower Layer Protocols and Physical Interfaces, Revision 4.2 (1997).
3. IEEE 802.14a High-Capacity Physical Layer Specification, IEEE 802.14a Hi-Phy Task Group, Draft 1 (1999).
4. DVS-031, *Digital Video Transmission Standard for Cable Television*, Society for Cable Telecommunication Engineers (1998).
5. S. Lin and D. J. Catello, *Error-Correction Coding for Digital Communications*, Prentice Hall, Englewood Cliffs, New Jersey (1983).
6. J. L. Ramsey, "Realization of Optimum Interleavers," *IEEE Transactions on Information Theory* **IT16**, 338-345 (1970).
7. G. D. Forney, "Burst-Correcting Codes for the Classic Bursty Channel," *IEEE Transactions on Communication Technology* **COM19**, 772–781 (1971).
8. G. Ungerboeck, "Trellis-Coded Modulation with Redundant Signal Sets, Part I; Introduction," *IEEE Communication Magazine* **25**, 5–11 (1987).
9. G. Ungerboeck, "Trellis-Coded Modulation with Redundant Signal Sets, Part II; State of the art," *IEEE Communication Magazine* **25**, 12–21 (1987).
10. B. Sklar, *Digital Communications*, Prentice Hall, Englewood Cliffs, New Jersey (1988).
11. G. D. Forney, Jr., "The Viterbi Algorithm," *Proceedings of IEEE*, 268–278 (1973).
12. A. J. Viterbi, "Error Bounds for Convolutional Codes and Asymptotically Optimum Decoding Algorithm," *IEEE Transactions on Information Theory*, **IT-13**, 260–269 (1967).
13. L. H. Charles Lee, *Convolutional Coding: Fundamentals and Applications*, Artech House, Boston (1997).
14. H. Nyquist, "Certain Topics of Telegraph Transmission Theory," *Transactions of American Institute of Electrical Engineers* **47**, 617–644 (1928).
15. H. Graham, "Modulator Having Direct Digital Synthesis for Broadband RF Transmission," U.S. Patent Number 5,412,352, May (1995).
16. H. T. Nicholas III, H. Samueli, "A 150-MHz Direct Digital Frequency Synthesizer in 1.25-μm CMOS with −90-dBc Spurious Performance," *IEEE Journal of Solid State Circuits*, **26**, 1959-1969 (1991).

17. B. Daneshrad and H. Samueli, "A Carrier and Timing Recovery Technique for QAM Transmission on Digital Subscriber Loops," *Technical Digest of International Conference on Communications*, Geneva, Switzerland, 1804–1808 (1993).

18. M. Kolber, "DSP for Analog Engineers," *General Instrument Journal*, 29–37 (1999).

19. H. Samueli and C. P. Reames "System for, and Method of, Processing Quadrature Amplitude Modulated Signals," U.S. Patent Number 5,754,591, May (1998).

20. B. C. Wong and H. Samueli, "A 200 MHz All-Digital QAM Modulator and Demodulator in 1.2-μm CMOS for Digital Radio Applications," *IEEE Journal of Solid State Circuits* **26**, 1970–1979 (1991).

21. R. W. Lucky, "Automatic Equalization for Digital Communications," *Bell System Technical Journal* **44**, 547–588 (1965).

22. R. W. Lucky and H. R. Rudin, "An Automatic Equalizer for General Purpose Communication Channels," *Bell System Technical Journal* **46**, 2179 (1967).

23. R. W. Lucky, J. Salz, and E. J. Weldon Jr., *Principles of Data Communication*, McGraw-Hill, New York (1968).

24. E. A. Lee and D. G. Messerschmitt, *Digital Communication*, 2nd Ed., Kluwer Academic Publishers (1996).

25. B. Widrow, J. M. McCool, M. G. Larimore, and C. R. Johnson Jr., "Stationary and Non-stationary Learning Characteristics of the LMS Adaptive Filter," *Proceedings of the IEEE* **64**, 1151–1162 (1976).

26. S. Haykin, "Adaptive Filter Theory," Prentice Hall, Second Edition (1991).

27. S. U. H. Qureshi, "Adaptive Equalization," *Proceedings of the IEEE* **73**, 1349–1387 (1985).

28. D. D. Falconer and L. Ljung, "Application of Fast Kalman Estimation to Adaptive Equalization," *IEEE Transactions on Communications*, **COM-26**, 1439–1446 (1978).

29. Y. Sato, "A Method of Self-Recovering Equalization for Multilevel Amplitude Modulation," *IEEE Transactions on Communications*, **COM-23**, 679–682 (1975).

30. D. N. Godard, "Self-Recovering Equalization and Carrier Tracking in Two-Dimensional Data Communication Systems," *IEEE Transactions on Communications* **COM-28**, 1867–1875 (1980).

31. D. Hatzinakos and C. L. Nikias, "Blind Equalization Using a Tricepstrum-Based Algorithm," *IEEE Transactions on Communications*, **COM-39**, 669–682 (1991).

32. L. Tong, G. Xu, and T. Kailath, "Blind Identification and Equalization Based on Second-Order Statistics," *IEEE Transactions on Information Theory*, **IT-40**, 340–349 (1994).

33. J. A. C. Bingham, *The Theory and Practice of Modem Design*, J. Wiley & Sons, New York (1988).

34. A. J. Rustako, L. J. Greenstein, R. S. Roman, and A. A. M. Saleh, "Using Times-Four Carrier Recovery in M-QAM Digital Radio Receivers," *IEEE Journal of Selected Areas on Communications*, **SAC-5**, 524–533 (1987).

35. J. G. Proakis, *Digital Communications*, 3rd Ed., McGraw-Hill, New York (1995).

36. J. G. Proakis, P. R. Drouilhet Jr., and R. Price, "Performance of Coherent Detection Systems Using Decision-Directed Channel Measurement," *IEEE Transactions on Communication Systems*, **CS-12**, 54–63 (1964).

37. I. Horikawa, T. Murase, and Y. Saito, "Design and Performance of 200 Mbits/s 16-QAM Digital Radio System," *IEEE Transactions on Communications*, **COM-27**, 1953–1958 (1979).

38. K. H. Mueller and M. Muller, "Timing Recovery in Digital Synchronous Data Receivers," *IEEE Transactions on Communications*, **COM-24**, 516–531 (1976).

39. B. Daneshrad, and H. Samueli, "A Carrier and Timing Recovery Technique for QAM Transmission on Digital Subscriber Loops," Technical Digest of *IEEE International Conference on Communications*, 1804–1808 (1993).

40. W. Webb and L. Hanzo, *Modern Quadrature Amplitude Modulation: Principles and Applications for Fixed and Wireless Communications*, IEEE Press, New York (1995).

41. HP 89440/AGILENT 89441A Operator's Guide, Hewlett-Packard Co. (1995).

42. Telecommunication Standardization Sector of International Telecommunication Union (ITU-T) Recommendation J.83: Digital Multi-Programme Systems for Television, Sound and Data Services for Cable Distribution (4/97).

CHAPTER 8

SUBSCRIBER HOME TERMINALS

In this chapter we shall discuss the architecture and performance of the most important sub-scriber home terminals that are currently employed in HFC networks, namely, digital set-top boxes (STBs) and cable modems (CMs). Chapter 8 will start with a brief overview of the basic building blocks of a digital STB with a detailed discussion on each of the building blocks in the following sections. In Section 8.2 the operating principles of the RF cable TV tuner are reviewed, while in Section 8.3 the operation and coding scheme of the out-of-band (OOB) receiver are reviewed. Section 8.4 will discuss the building blocks and operation of the upstream QAM transmitter based on what was learned in Chapter 7 on RF QAM mo-dems. After the QAM signal has been demodulated, the demultiplexing and decoding of the MPEG video and audio streams are discussed in Section 8.5. One of the key building blocks of the digital STB is the conditional access and control block, which is discussed in Section 8.6. Another important STB building block is the processing and display of video images and graphics content on the TV screen to enable the viewing of the different STB applica-tions, which is reviewed in Section 8.7. The requirements of the STB hardware resources such as its central processing unit (CPU) and memory are reviewed in Section 8.8. Section 8.9 will introduce the architecture of an advanced digital STB platform with a built-in DOCSIS CM. The last section will review the different transmission impairments of 64/256-QAM signals over HFC networks such as phase noise, multipath echoes, burst noise, and AM hum modulation. Understanding these impairments enables one to develop a link budget for the digital STB operating in a cable TV network.

8.1 Digital Set-Top Box Building Blocks

As was discussed in Chapter 1, the interactive digital STB acts as a broadband gateway to the home to enable the delivery of existing analog cable TV channels, MPEG-2 compressed audio and video channels using different TV-based interactive applications, and Internet-based applications [1–3]. (See Chapter 12.) Figure 8.1 shows a simplified block diagram of an interactive digital STB for use in HFC networks [4], which consists of four main sections. The RF front-end of the digital STB consists of an A/B switch, diplex filter, RF cable TV tuner, and an OOB tuner. The A/B switch is designed to switch between two possible input ports to the STB. The diplex filter, as we learned in Section 2.4.2, provides the necessary isolation, which is typically better than 65-dB, between the downstream and the upstream

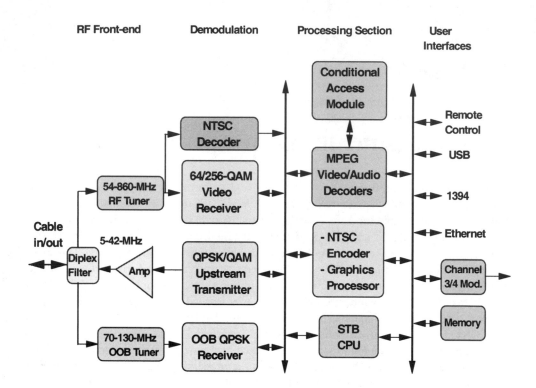

Figure 8.1 Simplified block diagram of an interactive digital set-top box. After Ref. [3]
(© 1999 IEEE).

portions of the cable TV RF spectrum. The downstream cable TV tuner is discussed in Sections 8.2 and 8.3. The second section of the digital STB consists of a 64/256-QAM receiver based on the ITU–T J.83 Annex A/B standard, an OOB QPSK receiver, an upstream QPSK/QAM RF transmitter, and an NTSC decoder. The central section of the digital STB has a variety of processing capabilities, depending on the type of multimedia information. It consists of a conditional access module or a security processor, an MPEG transport stream demultiplexer and decoder, an NTSC encoder, a graphics processor, a high-performance CPU, and different types of memory modules such as flash, read-only memory (ROM), and synchronous-dynamic-random-access memory (SDRAM). The back section of the digital STB platform provides different types of output interfaces for the multimedia information including an RF channel 3 or 4 NTSC modulator, composite video and audio outputs, a universal serial bus (USB), an IEEE 1394 or FireWire [5], a 10-base-T Ethernet, and a telephone modem. The operating principles and performance requirements of each of these building blocks are presented in the following sections. The performance requirements of most of these building blocks have been addressed by the following two standard organizations:

(A) The digital-video subcommittee (DVS) of the Society of Cable Telecommunication Engineers (SCTE)

(B) CableLabs OpenCable specification, which often refers to some of the DVS documents.

The various relevant documents of both DVS and OpenCable will be referenced throughout this chapter.

8.2 Cable TV RF Tuner

The STB's cable TV RF tuner is designed to downconvert the received RF cable TV channel, which may be either AM-VSB or QAM channel format, to an intermediate frequency (IF). If the downconverted signal is an M-ary QAM signal, then it is sent to the QAM receiver in the STB. If the downconverted signal is an AM-VSB channel, it is demodulated by the NTSC demodulator and then decoded by the NTSC decoder to a baseband video stream. The operating principle of cable TV tuners is based on the superheterodyne detection process, which was invented by E. H. Armstrong during World War I [6]. Cable TV tuners can be divided into two main types, namely, single and dual-conversion tuners. Dual-

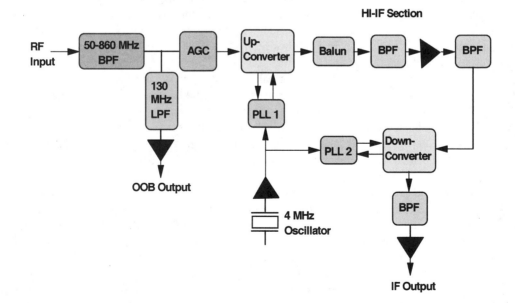

Figure 8.2 Simplified block diagram of a dual-conversion cable TV tuner. After Ref. [7].

conversion tuners, which have two frequency conversion stages between the input and output ports, are typically used in digital STBs in order to achieve the spurious and nonlinear distortion performance necessary for robust operation in an HFC network [7]. Figure 8.2 shows a block diagram of a typical dual-conversion superheterodyne cable TV tuner. The dual-conversion tuner can be divided into three main sections as follows: (A) upconverter section, (B) high-IF section, and (C) downconverter section. In the upconverter section, the input AM-VSB or M-QAM signals from 54-MHz to 860-MHz are first upconverted to a fixed IF, which is the sum of the first synthesized phase-locked loop (PLL) frequency and the tuned channel visual carrier (f_{in}). The first synthesized PLL frequency is obtained by taking the difference between the first local oscillator (LO) frequency (f_{LO}) and the tuned channel visual carrier. The subscriber, using an infrared remote control, selects the tuned channel frequency. The first LO frequency, which is typically between 1.1-GHz and 2-GHz, is selected such that the frequency difference between f_{LO} and the tuned channel visual carrier is above the input signal range. Part of the input RF signal is split before the first upconversion stage in the tuner, then filtered by 130-MHz low-pass filter (LPF), and amplified for the OOB receiver. The operation of the OOB receiver is discussed in the next section. In addition, a DC or low-frequency (< 15-MHz) control signal is used to select the proper port of the A/B switch. Note that the upconverter section essentially determines the noise figure (NF) and distortion performance of the tuner as well as the phase noise degradation of the signal.

Bipolar semiconductor technology was traditionally used for low cost synthesized PLL circuits. However, the wireless industry has generated a large market for very low-power synthesizers using bipolar complementary metal oxide semiconductor (BiCMOS) ICs having power consumption of less than 100-mW. Furthermore, replacing two synthesized PLLs using bipolar ICs with one dual-synthesizer BiCMOS IC lowers the power consumption even further with additional cost benefits.

In the high-IF section of the tuner, a balun transformer is typically used to connect directly between the unbalanced transmission line (coaxial cable) input to the tuner and the balanced load at the tuner output [8]. Using balun transformers eliminates potential problems from unbalanced-to-balanced operation and possible impedance mismatches, which may cause ringing, ghosting, and unwanted radiation. The fixed frequency IF signal is then filtered by a 6-MHz BPF, and amplified before it is downconverted by the second stage to a 44-MHz center frequency.

The traditional narrow-band high-frequency amplifier design uses an air-coil between the supply and the collector, which is often hand inserted in the tuner. New tuner designs use a quarter-wavelength transmission line to eliminate the need for the air-coil. Using both voltage feedback and constant base current-source-emitter resistors helps to achieve a noise figure performance of less than 2.5 dB with a low-cost bipolar transistor [7].

In practice, the high-IF section is typically implemented using ceramic BPFs. The ceramic filters use high dielectric materials to realize their small size with the required frequency response [9]. Although these filters cannot easily achieve the desired image attenuation in a single device, a pair of ceramic filters with an amplifier in between can achieve the required response. In fact, better than 0.5-dB frequency flatness over a 6-MHz channel bandwidth was achieved in General Instrument's dual-conversion tuner using ceramic filters [7]. Each filter has a typical image rejection of 40 dB, a 1-dB bandwidth of 35-MHz, and an insertion loss of 1.5 dB.

The third section of the tuner downconverts the high frequency IF signal to a fixed 44-MHz frequency. In practice, both the upconverter and downconverter sections of the cable TV tuner are implemented using Gallium-Arsenite (GaAs) integrated circuits, requiring only 5 V [10]. This approach simplifies the cable TV tuner design, improves its reliability, and reduces the cost. Section 8.10.1 discusses the tuner characteristics and performance requirements such as phase noise and noise figure in a digital STB for robust QAM transmission.

8.3 Out-of-Band (OOB) Receiver

An active OOB data channel provides continuous communication from the cable TV headend to the digital STB at the subscriber's home via the cable TV HFC distribution network. The digital STB typically remains powered up even when it is in the "off" state. The OOB channel remains active independent of the tuned TV channel, whether the received channel is analog or digital and whether the digital STB box is turned "on" or "off." Thus, whenever the digital STB is connected to the coaxial cable and AC power, the OOB channel is active for downstream communication.

There are two standards, which are currently being used in STB, for the transmission of the OOB signals. The OOB transmission standards are specified in SCTE DVS-167 [11] and SCTE DVS-178 [12] documents. DVS-167 specifies the following two transmission rates: (I) 1.544-Mb/s for grade A, and (II) 3.088-Mb/s for grade B (optional). The OOB transmission rate according to DVS-178 is 2.048-Mb/s. This information rate is nominally divided into two parts as follows: (1) 1.544-Mb/s is used for the addressable data stream, application program downloads, program guides, and future services not yet conceived, and (2) 0.461-Mb/s is reserved for future applications.

In general, any service directly linked to an individual MPEG service or MPEG transport stream, which is sometimes called transport multiplex, is carried in-band within the specific multiplex. Global applications, those requiring continuous communications, or those services available simultaneously with any of the available television services, are carried in the OOB channel. At the cable TV headend, the desired parts of each input data stream, which originated from various sources, are multiplexed into a 1.544-Mb/s MPEG-2 transport stream for output. Then, null MPEG packets are added to this stream to pad it up to the transmitted stream rate of 2.005-Mb/s. The MPEG transport stream is then QPSK modulated, error encoded, and upconverted to an RF frequency in the 70-MHz to 130-MHz band.

Table 8.1 Out-of-band cable TV transmission specifications according SCTE   DVS-178 standard.

Parameter Name	Specifications
Transport Protocol:	MPEG-2 transport stream (MPEG-TS) compliant
Modulation:	QPSK, differential coding for 90°-phase invariance
Symbol Rate:	1.024-MBaud
Symbol Size:	2 bits per symbol
Channel Spacing (BW):	1.8-MHz
Transmission Frequency Band:	71 to 129-MHz
Carrier Center Frequency:	75.25^1 MHz $\pm$ 0.01%
Data Rate:	2.048-Mb/s $\pm$ 0.01%
Forward Error Correction:	96,94 Reed-Solomon block code, T = 1, 8-bit symbols
FEC Framing:	Locked to MPEG-TS, two FEC blocks per MPEG packet
Interleaving:	Convolutional (96,8)
Nominal Information Rate:	2.005-Mb/s (132.8 b/s margin)
Frequency Response:	Raised cosine filter, α = 0.5 (receiver only)

As was explained in Section 8.2, the 70–130 MHz frequency band is split at the front end of the RF tuner and directed to the OOB receiver. Earlier practice has assumed one of several default frequencies such as 75.25-MHz or 72.75-MHz, in order to allow the QPSK receiver to downconvert to a secondary IF equal to the symbol rate. The current digital STB practice has an agile OOB receiver that can tune to any RF frequency in the 70- to 130-MHz band. As with the downstream QAM receiver, the received IF is oversampled, typically four times, separated into "I" and "Q" components, and Nyquist filtered to produce a complex baseband data stream. Then, a simple symbol-spaced feedforward and feedback equalizer with typically two to four taps is used to remove the ISI generated by the presence of various impairments such as multipath echoes over the cable TV network. The FEC scheme for the OOB channel, which is similar to downstream QAM channels, is composed of four layers as follows: (A) frame synchronization, (B) de-interleaving, (C) Reed-Solomon (R-S) decoding, and (D) de-randomization. Table 8.1 summarizes the physical attributes of the OOB chan-

---

[1] Other possible OOB carrier center frequencies are 72.75 MHz and 104.2 MHz.

nel, which is commonly used in the cable TV industry [12]. The OOB FEC scheme according to the DVS-178 standard is described in more detail in the following sections.

8.3.1 OOB Randomizer

The randomization of the MPEG-2 transport stream in the OOB receiver is similar to the randomization done in the QAM receiver. The randomizer circuit performs exclusive-OR operation on the input MPEG transport sequence with the randomizer's pseudorandom number (PN) generator output sequence. The randomization frame consists of two MPEG packets with the randomizer PN generator reset at the start of every second MPEG-TS packet. MPEG-TS sync bytes are inverted on alternate packets to improve receiver synchronization performance. The randomizer is a 13-bit counter implemented as a linear feedback shift register (LFSR) as shown in Figure 8.3. Binary arithmetic and taps are placed at the output of stages 13, 11, 10, and 1. Stages 2 through 9 are loaded with a seed value of "0." The corresponding generating polynomial is defined as

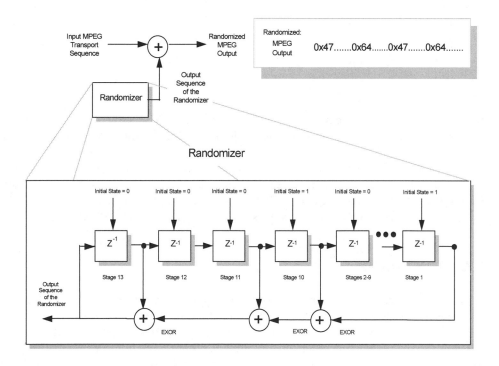

Figure 8.3 Block diagram of OOB randomizer. After Ref. [12].

$$g(X) = X^{13} + X^{11} + X^{10} + X + 1 \tag{8.1}$$

The same circuit is used for de-randomizing the received MPEG-2 transport stream packets. The sync symbol of the first MPEG-TS packet in a frame remains 0x47 after randomization because the first randomizer output byte after reset is "0x00." The second MPEG-2 sync byte is changed by the randomizer but will be returned to the MPEG-TS standard 0x47 by the de-randomizer at the receive site.

8.3.2 OOB Reed-Solomon Coding

The forward error correction (FEC) code in the OOB transmission system is a Reed-Solomon (R-S) block code [13]. No convolutional coding is required for the relatively robust QPSK transmission on cable TV networks. The R-S code is T = 1 (96,94) over GF(256), which is capable of one symbol error correction every R-S block of 96 symbols. The GF(256) is constructed based on the following primitive polynomial over GF(2):

$$p(X) = X^8 + X^4 + X^3 + X^2 + 1 \tag{8.2}$$

and the generating polynomial for the R-S code is defined as

$$g(X) = (X - \alpha)(X - \alpha^2) = X^2 + \alpha^{26}X + \alpha^3 \tag{8.3}$$

where α is a primitive element in GF(256). The OOB FEC frame consists of two R-S blocks with two parity symbols in each R-S block, which equals one MPEG transport packet.

  Mapping from FEC frame to MPEG-TS packets is done as follows: the first 94 bytes are unaltered and used directly as received. The next two bytes are the parity bytes obtained from the R-S polynomial calculation. Two blocks of 96 bytes are sent for every 188-byte MPEG packet received. The FEC frame is reset at the start of each MPEG transport stream packet.

8.3.3 OOB Interleaver

Interleaving the coded R-S symbols before transmission and de-interleaving after the reception causes multiple burst errors that occur during transmission to be spread out in time, and thus to be handled by the receiver as if they were random errors. Separating the R-S symbols in time enables the random error-correcting R-S code to be useful in a bursty-noise environment. Using a convolutional interleaver with a depth of I = 8 symbols, the R-S T = 1 (96,94) decoder can correct an error burst of 8 symbols, which corresponds to a burst noise protection of 32 µsec. Interleaving is synchronized to the R-S blocks, and thus to MPEG-2 transport stream packets. MPEG-2 transport stream sync bytes always pass through commutator branch 1 of the interleaver and hence are not delayed through the interleaver.

8.3.4 OOB QPSK Mapping

The OOB modulator uses a differential encoding scheme to resolve the 90° phase ambiguity in the detection of the QPSK signal at the demodulator. The OOB QPSK demodulator should be capable of handling both forms of differential coding as listed in Table 8.2. Also, a means of selecting the appropriate form of decoding for the user's system must be present in the QPSK demodulator.

Table 8.2 Differential Coding Scheme for OOB QPSK Signal.

I Data	Q Data	Default Carrier Phase Changes	Alternate Carrier Phase Changes
0	0	No Change	No Change
0	1	−90° CW	+90° CW
1	0	+90° CW	−90° CW
1	1	180°	180°

8.4 RF QAM Transceiver

The RF QAM transceiver in the digital STB consists of a downstream QAM receiver and an upstream QPSK/QAM burst transmitter. The design and operation of the QAM receiver was already discussed in Chapter 7. The upstream RF transmitter in either the digital STB or CM enable subscribers via the cable TV network to transmit data packets to the CMTS or other burst receivers, which are located at the cable TV headend. The discussion here will focus on the upstream burst transmitter in the STB. RF upstream transmission requirements of CMs will be discussed in Chapter 11 when the physical layer of DOCSIS protocol is explained [14]. Figure 8.4 shows a simplified block diagram of an RF upstream QAM transmitter. For most applications, the data generated at the subscriber's home must be transmitted in short packets. The small asynchronous transfer mode (ATM) protocol cell structure is well suited for this application [15]. The input data packets to be transmitted are first placed in a small burst first-in-first-out (FIFO) buffer (not shown in Figure 8.4) in order to de-couple the input rate from the transmission data rate. In other word, the FIFO buffer allows the data packets to be input while a burst is actively being transmitted. The FEC coding scheme that follows the FIFO buffer consists of a randomization layer and an R-S coding layer, which provide the necessary coding gain for the AWGN channel for combating burst noise (see the discussion in following sections).

Before we discuss the preamble prepend block, let us introduce the burst packet format. For network connectivity reasons, it is convenient to transmit the data packet to the headend using an ATM packet [15]. Table 8.3 shows the upstream packet format according to the SCTE DVS-178 standard [12]. Each ATM packet is concatenated with a 28-bit unique word (UW), a one-byte packet sequence counter, and eight R-S parity bytes. The 28-bit UW, which can be written as (I, Q), is used to identify the start of the data packet for robust sync detection by the return-path receiver. The packet sequence byte consists of a message number (three bits), and a sequence number (five bits). The message number is used to associate upstream cells with a particular protocol data unit (PDU). It is incremented every time the first cell of a new PDU is sent. The sequence number, which has a field length of five bits, is used to identify the order of the cells within a PDU. It starts at zero for each new message number and is used by the headend return-path demodulator to detect missing cells for the RF modem report-backs. The preamble, which is composed of a UW and a packet sequence number, is added to each FEC-encoded ATM packet by the preamble prepend block shown in Figure 8.4. Similar to the downstream QAM receiver, the return-path transmitter also uses differential encoding to enable phase invariant reception at the cable TV headend. Two

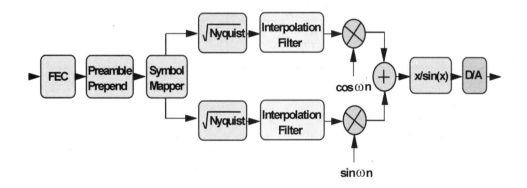

Figure 8.4 Block diagram of an upstream QAM burst transmitter.

Table 8.3 Packet format for an upstream transmitter in a digital set-top box.

Parameter	Specification
Unique Word	28 bits (1100 1100 1100 1100 1100 1100 0000)
Packet Sequence	1 byte
ATM Data	53 bytes
R-S Parity	8 bytes

modes of differential decoding can be defined (see Table 7.5) to accommodate different system local oscillators. The output data from the differential encoder feeds the Nyquist pulse-shaping filters, which are implemented using SRRC filters with a roll-off factor of 25% or 50% (α = 0.25 or 0.5). For robust upstream transmission, particularly at high symbol rates (> 1-Mbaud), the amplitude ripple of the filter within the passband as well as the maximum attenuation of the stopband ripple needs to meet particular specifications. For example, the in-band amplitude ripple is specified to be within ± 0.3-dB and stopband attenuation greater than 30-dB according to DOCSIS 1.1 protocol [14]. Interpolation filters are used after the SRRC filters to increase the sampling rate of the baseband signal up to the symbol rate. Then, the output from the interpolation filters is multiplied by the quadrature oscillators using for example DDFS [16]. As was discussed in Section 7.4, an x/sin(x) correction filter is applied to the digitally shaped data burst before it is converted to an analog waveform using for example a 10-bit D/A converter. Table 8.4 summarizes the upstream transmitter specifications according to SCTE DVS-178 and DVS-167 standards. The DVS-178 and DVS-167 specifications represent upstream differentially encoded QPSK transmitters in digital STBs from General Instrument and Scientific Atlanta, respectively. Notice that both

Table 8.4 Upstream RF transmitter parameters for a digital STB.

Parameter Name	DVS-178 Specification	DVS-167 Specification
Modulation Type	Differentially-Encoded QPSK	Differentially-Encoded QPSK
Access Scheme	Polling and ALOHA (programmable)	TDMA
Data Transmission Rate	256 kb/s	256 kb/s (grade A) 1.544 Mb/s (grade B) 3.088 Mb/s (grade C)
Bits per Symbol	2	2
Channel Spacing	192-kHz	200-kHz, 1 MHz, 2-MHz
Transmit Filter Shape	SRRC, α = 0.5	SRRC, α = 0.3
FEC Code	R-S T = 4 (62,54)	R-S T = 3 (59,53)
RF Output Power Range	+24 dBmV to +60 dBmV	+25 dBmV to +53 dBmV
Spurious Output Level (Idle state)	< −30 dBmV (in-band), <−65 dBmV (out-of-band)	> 30 dB (out-of-band)
Frequency Range	8 to 40-MHz	8 to 26.5 MHz

upstream transmitters according to DVS-167 and DVS-178 do not utilize the full 35-MHz upstream band. The latest advanced digital STB from Motorola/General Instrument, which is called DCT5000+, provides significantly higher upstream symbol rates according to DOCSIS 1.1 [14]. In fact, the DOCSIS 1.1 protocol, which will be discussed in more detail in Chapter 11, specifies upstream symbol rates up to 2.56-Mbaud using 16-QAM format, and an upstream frequency up to 42-MHz.

8.4.1 RF Upstream FEC

A block code FEC is used to allow both correction of some transmission errors and detection of packets that cannot be corrected. For many applications, upstream packets that cannot be corrected can be retransmitted. Block or convolutional interleaving is not appropriate since their function is to spread out error bursts over many FEC blocks. These upstream transmissions often consist of a single FEC block. The upstream FEC in the digital STB transmitter consists of a R-S code and a randomizer, and it is described in the next two subsections. Note that this upstream FEC is different than the upstream FEC for CMs as specified by DOCSIS 1.0/1.1 (see Section 11.3.1).

8.4.1.1. Upstream R-S Coding

The FEC code in the return-path transmission link is a R-S T = 4 (62,54) code over the GF(256) field. Each R-S symbol consists of eight bits. This FEC code is capable of correcting four symbol errors for R-S block of 62 symbols. The following primitive polynomial over GF(256) is used:

$$p(X) = X^8 + X^7 + X^2 + X + 1 \tag{8.4}$$

The generating polynomial for this FEC code is

$$g(X) = (X - \alpha^{120})(X - \alpha^{121})(X - \alpha^{122})(X - \alpha^{123})(X - \alpha^{124})(X - \alpha^{125})(X - \alpha^{126})(X - \alpha^{127})$$
$$\tag{8.5}$$

where α is a primitive element in GF (256). The encoding circuit is efficiently implemented via shift registers using arithmetic over GF (256).

8.4.1.2 Upstream Randomizer

At the R-S encoder, each packet is fed through the randomizing circuit to maintain the balance between the ones and zeros within the bit stream. The randomizer circuit uses a pseudorandom number (PN) generator, which employs a 13-bit shift register. Taking the exclusive-OR of the input bit stream with the PN sequence output generates the randomized bit stream. Figure 8.5 shows the upstream randomizer structure using linear shift registers. Taps are located at the output of stages 1,3,4, and 13 of the shift register. Stages 1 to 5 of the shift register are always initialized to zero for each packet. Stages 6–13 are initialized

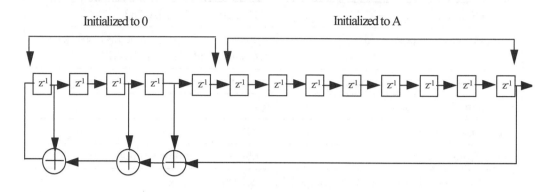

Figure 8.5 Randomizer structure for the upstream transmitter. After Ref. [12].

to a programmable value. The 13-bit default value for this initialization is 0xAA (0 0000 1010 1010), with the most significant bit shifted out first. The corresponding generating polynomial is defined as

$$g(X) = X^{13} + X^4 + X^3 + X + 1 \tag{8.6}$$

8.5 MPEG Video/Audio Demultiplexer and Decoder

The arriving MPEG-TS, which is sometimes called transport multiplex, to the STB consists of many different PESs, representing the different video and audio services. Each packet consists of 184 bytes of payload and 4 bytes of overhead. The transport header consists of a sync byte for packet delineation, a priority bit for congestion control, an ID for packet identification, scrambling control bits, an error indicator, and a continuity counter for packet-loss detection. The task of the MPEG processor is to demultiplex the MPEG-TS to the specific MPEG elementary stream according to the MPEG-2 service map, which identifies the program ID (PID), program type, and all other components that are associated with that service.

Each of the selected MPEG-2 video/audio and AC-3 audio programs is then decoded and processed. The MPEG processor also can utilize the main microprocessor in the STB for other functions such as decoding various MPEG formats. The MPEG-2 demultiplexer and decoder reconstruct the TV frames from the compressed MPEG-2 video, decompressing the "I" frame and using the "P" frame motion vectors to reconstruct the frame. The decoder/

Table 8.5 Baseband video performance parameters of OpenCable STB [2].

CHARACTERISTICS	SPECIFICATION
Video Standard	NTSC composite, EIA-563
Signal Level (composite video)	1.0 volt peak-to-peak, sync tip (−40 IRE) to reference white (100 IRE) ±10%
Long Time Distortion (Bounce)	± 1%, settle in less than 1 second
Field Time Distortion	± 4%
Line Time Distortion	± 2%
Short Time Distortion	± 6% (Rising and/or Falling)
Chrominance-to-Luminance Gain Inequality	≤ ±10 %
Chrominance to Luminance Delay	± 150 ns maximum
Frequency Response for Baseband Video Output	−2 to +2 dB, 0-kHz to 3.7 MHz
Terminal Contribution to Output Frequency Response for RF Output	−1 to +1 dB, 0 kHz to 3.75 MHz
Luminance Non-Linearity	5% p-p max.
Chrominance Nonlinear Phase Distortion	± 5°
Chrominance Nonlinear Gain Distortion	± 2%
Chrominance/Luminance Intermodulation	± 3%
Differential Gain (over 10% to 90% APL range)	10% p-p max. for RF modulated output; 5% p-p max. for baseband video output
Differential Phase (over 10% to 90% APL range)	10° p-p max. for RF modulated output 5° p-p max. for baseband video output
920 kHz Beat	−52 dBc
Video Signal-to-Noise Ratio (over the full input tuning range stated in "signal level" above)	53 dB for RF modulated video with a digital input signal and 48 dB with AM input signal at 0-dBmV. 57 dB for baseband video output with a digital input signal and 49 dB with AM input signal at 0-dBmV

CHARACTERISTICS	SPECIFICATION
Baseband Video Output Impedance	75 ohms ± 10%
Baseband Video Output Return Loss	≥ 16 dB across video bandwidth

decompressor predicts motion of the "B" frames and then reconstructs the "B" frames by applying the difference motion vectors that were developed by the MPEG-2 encoder at the video source. The decoder/decompressor receives 27-Mb/s (or 38-Mb/s for 256-QAM) multiplex by parsing the MPEG-2 transport stream syntax, and performing the video decompression process. It provides the digital video as described in Interface for Digital Component Video Signals in 525-line and 625-line Television Systems, CCIR 656 [17].

The MPEG-TS demultiplexer typically implements 32 PID filters to process transactions, messages, control and access communications, multicast data, video, audio, DOCSIS data, and other data traffic. Data traffic can be delivered to the filters via the OOB receiver or the RF tuner. The PID filters will have the ability to be addressed individually through the network address or be configured as a group address and respond to multicast address messages. The baseband video performance parameters of OpenCable STBs are summarized in Table 8.5.

An important and interesting application is the delivery of a 3D video program from the STB to the TV. The 3D video program can be derived and reconstructed at the STB from a 2D MPEG-2 video program stream enhanced by 3D information, which is carried as associated private data in the MPEG-2 data stream. This 3D service application can be viewed, for example, on a standard NTSC monitor, an HDTV monitor using IR synchronized eyeglasses, or in the future, direct 3D display systems.

8.5.1 VBI Retriever and Decoder

As we learned in Section 2.1.1, closed-caption information is carried on line 21, field 1 and/or field 2, of the vertical blanking interval (VBI) of an analog NTSC signal as mandated by the FCC. Closed captions, which are program-related data as defined in the EIA-608 standard, are given priority over other data, which may be carried on line 21[18]. The closed-caption format as defined in EIA-608 can be extended from lines 10 to 21 to support extended data services. The digital STB has a VBI data retriever to decode data from analog and digital video services. For digital video services, the VBI lines are constructed from the MPEG-2 transport streams using the defined extension to the picture user data syntax as defined in DVS-053 [19]. The STB receives and retrieves multicast IP data packets delivered over the VBI using the North American basic Teletext standard (NABTS) packet protocol, which is defined in EIA-516 [20]. For analog VBI pass-through and digital VBI reconstruction, the STB is reading or reconstructing data (captioning, text, HTML data, etc.) on all available lines (i.e., lines 10 through 20, plus line 21). In analog pass-through mode, all lines are passed simultaneously. EIA-516 specifies the transmission of 36 bytes NABTS packet (including sync bytes) per NTSC VBI horizontal line. Table 8.6 summarizes the effective VBI data rates for 1 and 11 VBI lines.

Table 8.6 Effective VBI data rates for one and eleven VBI lines.

Characteristic	Specification
Horizontal Scan Line Refresh Rate	1/60th second
Aggregate Bit Rate	36 bytes x 8 bits per byte x 60 lines per second = 17,280 b/s
Maximum Allowable Overhead	39.9 %
Minimum Effective Throughput Required	17,280 b/s x (1−0.399) = 10, 380 b/s per line
Minimum Total Throughput for 11 VBI Lines	10,380 b/s per line x 11 lines = 114.2 kb/s

8.6 Conditional Access and Control

Conditional access or security has a primary importance for existing and new STB applications [1]. There is an increasing need to protect the transmitted information in a multiuser cable TV network from unauthorized use. Stories of unauthorized users or "hackers" breaking into various restricted networks are continuously in the news. Millions of STBs currently operating in the U.S. running multiple applications further exacerbate this problem. To address this problem, a variety of security methods and systems can be implemented, depending on the value of the information that needs to be protected. For example, point-to-point applications over cable TV network such as video-on-demand (VOD) and IP telephony are highly personalized applications with limited resale value in the network. Thus, minimum-security level is typically required for point-to-point applications. In contrast, point-to-multipoint applications, which are sometimes called broadcast applications, such as broadcast sporting events, have a potentially lucrative opportunity due to their high demand. Thus, there is a need to implement a high-level security system in order to make financially impractical the theft of services. Before we can discuss the basic concepts of access control and authorization, let us briefly review the basic digital encryption and decryption methods that are currently used in digital communication systems.

8.6.1 Digital Encryption/Decryption Basics

Digital encryption is the primary method to obscure information under the control of one or more secret keys. Unencrypted information is called *clear* or *plaintext*, while encrypted information is called *ciphertext* [21, 22]. Block encryption always operates on a given size of block data such as 64, 128, 256, or more bits. Stream encryption, which is used to encrypt/decrypt high-speed and/or high-volume data, operates on the data a single bit at a time, typically through the generation of a keystream.

Let us denote plaintext message as M and a ciphertext message as C. K is called a key if it uniquely determines a one-to-one mapping, which is called the encryption function, from M to C. This definition can be written as

$$C = E_K(M) \tag{8.7}$$

The process of applying the encryption transformation is simply called encryption of message M. The reverse process is called decryption of ciphertext message C to obtain M. The term cryptosystem defines a complete system with encryption-decryption algorithms, and with all possible plaintexts, ciphertexts, and keys. There are two generic types of cryptosystem: symmetric-key (or secret-key), and public-key. In symmetric-key cryptography, the same secret key is used for both encryption and decryption. The most popular symmetric-key cryptosystem, which was developed by IBM in the middle 1970s and has been a federal standard since 1976, is called data encryption standard (DES).

In a public-key cryptosystem, each user has a public key, which becomes public, and a private key, which remains secret. All communication between the sender and the receiver is done using only public keys. Thus, it is no longer necessary to trust the security of the communication system. The transmitted encrypted message can be decrypted only with the secret key, which is known only to the recipient. The most popular public-key cryptosystem is called Rivest-Shamir-Adleman (RSA), which was named after its inventors [23].

The RSA key generation algorithm works as follows: generate two large prime numbers, each roughly the same size, say p and q, and compute their product $n = p \cdot q$ and $\phi = (p-1)(q-1)$. N is called the *modulus*. Choose a number e, $1 < e < \phi$, such that e is relatively prime to ϕ. Find another number d such that $(ed-1)$ is divisible by ϕ. The public key is the pair (n, e), while pair (n, d) is the secret key. The numbers p and q should be destroyed or kept secret. The security of the RSA algorithm is based on the assumption that it is difficult to obtain the private key d from the public key (n, e). To use the RSA public-key encryption, the sender creates a ciphertext C using the relation $C = M^e \bmod(n)$, where the message is M, and the recipient public key is (n, e). To decrypt, the decrypt the message from the sender uses the relation $M = C^d \bmod(n)$. Since only the recipient knows d, only the recipient can decrypt this message. Generally speaking, the larger the RSA modulus, the greater the security, but the longer it takes to decrypt. Thus, the size of the RSA modulus should fit one's security needs. Another valuable use of public-key encryption is called authentication or digital signatures. Using a hash function, the digital signature generates a message authentication code (MAC), which reflects the authenticity of the message. The MAC is encrypted using a secret key. This allows any public user to decrypt the message using a public key and confirm the authenticity of the message by confirming the MAC.

The DES algorithm is a Feistel cipher that operates on a plaintext block size of 64 bits to produce a ciphertext block of the same size [22]. The effective size of the secret key is 56 bits (eight bits can be used as parity bits, producing a 64-bit input key size). Thus, the number of possible keys is simply $256 \cong 7.2 \cdot 10^{16}$. Recently, a DES cracking machine was used to recover a DES key in 56 hours. Thus, with rapidly increasing computation power of modern day desktop PCs, 56-bit keys are becoming vulnerable to exhaustive search and thus may not be secure enough for some applications. To increase the security of the DES algorithm, the plaintext message can be encrypted multiple times, typically by cascading multiple block ciphers or stages, where each stage does not need to have independent keys. For example, triple encryption sometimes called triple DES can be written mathematically as

$$C = E_{K_3}\left\{E_{K_2}\left[E_{K_1}(M)\right]\right\}$$

(8.8)

where E_K denotes a block cipher with key K, and K_1, K_2, and K_3 are three secret keys, which may be independent. The case with $K_1 = K_3$ is often called two-key triple encryption.

8.6.2 Access Control Basics

As shown in Figure 8.1, the digital STB has an embedded security system. The task of the security system can be divided into two parts: access control and decryption. The term conditional access refers to providing authorization to subscribers to view only the digital STB services that they paid for. The conditional access system technology of General Instrument is called DigiCipher® II. The conditional access control is specified through the following components:

- The entitlement control message (ECM) defines particular program access requirements, and is primarily managed through a tiering structure with secret key delivery. The encrypted program keys are delivered in the ECM stream. The ECM is typically transmitted in the downstream in-band 64/256-QAM channels.

- The entitlement management message (EMM) defines the access or authorization rights for a given digital STB, which is processed in the security processor. The EMM is typically being transmitted in the OOB channel.

Both ECM and EMM are usually used with the previously discussed encryption methods, namely, either public-key or symmetric-key encryption, to control the distribution of the key used to encrypt the protected information. The access to a given encrypted service can be done in two ways: (A) direct delivery of the service key to decrypt the encrypted service, and (B) tiered structure coupled with encrypted key delivery.

As we discussed before, with millions of subscribers and with many services used by each user, the problem of direct-key delivery becomes very complicated. To achieve a significant reduction in the number of keys without loss of security, a tiered structure delivery is used. Improved security is achieved by changing the service keys many times per second. Specific bits are designated in the EMM for each service in order for tiered delivery to work. It should be pointed out that the entitlement specification in the EMM is very efficient compared with direct delivery, requiring a change by only one bit per tier.

When the STB is tuned to a particular program, it receives an ECM that is associated with that service and compares the tiers contained in the ECM with those stored in the STB from the previously received EMM. The previously delivered EMM has the list of all the entitlements purchased by or granted to the subscriber as well as a group key. A group key is a limited duration secret key that can be used to decrypt services from a given tier. If the EMM stored in the STB has one or more tier bits that are also in the received ECM, the STB is authorized to use the group key. If the ECM does not contain tier bits that are stored in the STB EMM, then the STB is not authorized and can't use the group key.

8.6.3 Renewable Security

The embedded security system in the digital STB can also be a disadvantage. The 1996 Telecom Act Order 98-116 mandated the availability of digital STB at retail outlets with a renewable and replaceable security system in order to "accelerate private sector deployment of STB." Such a removable security system is called a point-of-deployment (POD) module, and the digital STB is called the host. The advantage of the renewable security system is the availability of digital STB at retail electronic stores, which is portable across many different HFC networks and cable operators. Thus, when moving from one cable system to another, the consumer can replace only the POD module and not its digital STB. Without the POD module, the digital STB can only receive channels in the clear, namely, unencrypted services. The next-generation of digital STBs, which are named OpenCable set-top terminals, are required to be compliant with OpenCable specifications [2, 24-25].

Let us briefly review the system architecture of the OpenCable STB and POD module. Figure 8.6 shows a block diagram of the OpenCable front-end host with the POD module. When the input signal is an MPEG video stream, it is demodulated by the in-band 64/256-QAM receiver, and then fed to the conditional access processor in the POD module. The authorization rights information for the received MPEG stream, namely, the EMM, is demodulated by the OOB demodulator and processed in the POD module. The OOB processing unit in the POD module provides FEC coding for the OOB receiver as well as program-identification (PID) filtering for the selected stream. The host remains tuned to the OOB channel in order to continuously received in-band services. If the subscriber is authorized for the tuned service, then the POD module decrypts the requested MPEG stream, and it is demultiplexed and decoded in the host as shown in Figure 8.6. To protect the transmission of the MPEG stream across the interface between the POD module and the host, OpenCable copy protection (CP) is used [26]. If the requested MPEG stream isn't authorized or the POD module is removed, the output of the QAM receiver is directly routed to the MPEG transport stream demultiplexer. Then, a message may appear on the subscriber's TV monitor, saying that the host is not authorized for the requested service. Nonencrypted digital services are transmitted through the POD module to the MPEG demultiplexer. When the input signal is a clear analog channel, it is routed through the NTSC demodulator and the VBI decoder.

For upstream communications, the data packets are routed through the POD CPU to the OOB processing unit to the upstream burst transmitter as shown in Figure 8.6.

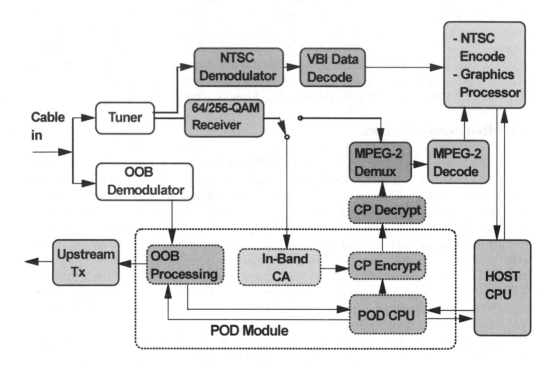

Figure 8.6 Block diagram of OpenCable Host with POD module for bidirectional cable plants. After Ref. [2].

8.7 Graphics Processor

There are various STB applications such as Web browsing and electronic program guide (EPG) that require the display of graphics content and video images on the TV monitor. The graphics processor, which sometimes is integrated in a single IC with the NTSC composite video encoder, is required to provide high-quality anti-aliased text with real-time two-dimensional (2D) and three-dimensional (3D) graphics effects, which can be incorporated with on-screen video images. Let us first review basic 2D/3D graphics concepts and methods before discussing the on-screen video and 2D/3D graphics rendering requirements in advanced digital STBs.

8.7.1 Basic 3D Graphics Concepts and Techniques

Before discussing the various techniques to map and display 3D graphics objects on a TV monitor, let us first explain the challenges in mapping a high-resolution color-separated graphics display (24-bits) to a 16-bit modulated TV channel. In particular, graphics content is digital with progressive scan format, has no TV synchronization signal, and can exist in various resolutions such as 600 x 400, etc. To properly display the graphics image on the TV monitor, the following functions are required:

- Video color conversion: The first step is to convert the digital red-green-blue (RGB) pixel data to digital video components, namely, Y, U, and V. Then, the color information has to be reduced to 16-bit in a 4:2:2 format.
- Interlacing: The progressive-scan graphics content must be interlaced before it can be displayed on a TV monitor.
- Anti-flicker filter: An artifact associated with the interlacing process is the flickering of the displayed image on the TV monitor. An adaptive 2-tap/3-tap anti-flicker filter can be used with other video enhancements to produce a high-quality display.
- Overscan compensation: Since the display area on a TV monitor is smaller than that on a graphics monitor, a portion of the displayed graphics may not be visible. The process in which the graphics content is contracted to accommodate the TV monitor's visible area is called overscan compensation.
- NTSC encoding: The encoding process consists of converting the flicker-filtered digital graphics data (YUV) to an analog composite NTSC waveform using a 9-bit A/D converter.

Let us now discuss the basics for 3D graphics [27]. The recent explosive growth in consumer appetite for 3D games and other virtual-reality applications has required the STB to deliver high-quality 3D texture-mapped images approaching photo-realism. Generally speaking, the implementation of 3D graphics can be divided into three distinct steps. The first step is called the scene manager, which may be integrated with a particular application, and is responsible for providing an internal presentation of the virtual scene to be imaged. The scene description is done using an internal database of virtual objects, what materials objects are made of, and how the scene is lit. The second step is the geometry processing including light modeling, which provides geometrical transformation of 3D objects onto 2D screen coordinates. The third process is called rendering or rasterization, which provides near photo-realistic scenic details to the 3D graphics objects, depending on its texture, light, viewing angle, and background [27]. Since rendering is a very pixel-intensive operation that

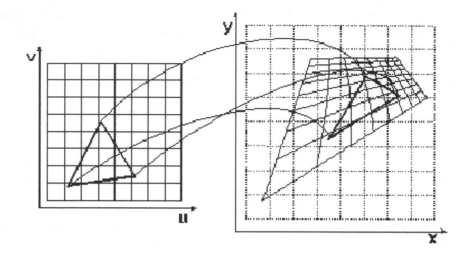

Figure 8.7 Triangle shape mapping from 2D geometry onto a 3D surface.

may be limited by the graphics hardware capabilities, let us briefly review its features. Texture mapping is by far the most important feature of a rendering engine because it provides realistic details to 3D objects without complicating the object's geometry. Without such texturing, mapping of complex 3D objects would be significantly more complicated since it would require the use of thousands of polygons to depict these objects. From a historical perspective, texture mapping has enabled the PC industry to embrace the use of 3D graphics.

Polygon mesh is the most common form to approximately represent a surface of an object. Texture on a polygon is generated by mapping a 2D texture pattern on each pixel of the polygon. Since polygons are rendered in perspective, it is important that texture be mapped in perspective as well. However, when an object is receded into the distance, texture is distorted, requiring texture perspective correction. Texture distortion can be minimized, for example, by reducing the size of the textured object. Perspective correction, which involves computational intensive operations of each pixel of the textured object, is necessary for photo-realistic 3D graphics.

There are several qualitative levels of texture mapping, depending on the complexity of the texture mapping logic. The simplest method is called point sampling, where the nearest texture element is mapped onto the corresponding pixel on the screen. Only one texture element is used for each pixel during the mapping. However, because of storage constraints, texture patterns are stored in a finite resolution, which may appear blocky on the screen from a close viewpoint. Another visual artifact of point-sampled images is that they scintillate or sparkle when the objects move. There are several anti-aliasing filtering techniques that can be applied to reduce this effect. In these techniques, multiple texture elements are filtered to derive a single texture element, which is mapped onto a pixel on the screen, resulting in a higher-quality image. Filtering really means weighted averaging of adjacent pixels (the more averaging the greater computing cost). For example, in the bilinear filtered texture method, the four nearest texture elements are bilinearly interpolated to determine the final texture element value, which is mapped onto a single pixel on the screen. This method smoothes out the blockiness caused by the point-sampled images, resulting in enhanced image quality. The drawback of this method is that it does not eliminate the "sparkling" artifact created by the point sampling method, thus requiring the use of more advanced texture techniques.

Another type of texture mapping visual artifact is depth aliasing, which occurs when objects that are further from the viewpoint appear smaller on the screen with less detail than closer objects. As the object moves further from the viewpoint, the texture image becomes more compressed when mapped onto a smaller object. To eliminate the depth-aliasing artifact, a *MIP-mapping* technique is used[2]. A MIP-map is a series of prefiltered, prescaled texture patterns. The original texture map is reduced by 1/4 (averaging every four pixels) for the next level of detail (LOD), and stored as the second MIP-map in memory. This MIP-map is further reduced by 1/4 to produce the third MIP-map for LOD. This process is repeated until the texture pattern is reduced to a 1x1 map. Then, the appropriate MIP-map is selected, depending on the distance of the object from the viewpoint. MIP-mapped textures reduce sparkling artifacts and appear to be more 3D realistic. However, the MIP-mapping introduces a visual artifact when the object is moving away or towards the viewer. The object jumps from one LOD to the next one whenever the MIP-map boundary is crossed. This artifact can be corrected using the *trilinear MIP-mapping* technique.

Retrieval of stored texture information in a dedicated graphics memory is one of the main factors that determine the performance of a 3D graphics engine. Typically, the application only stores the required texture for a particular scene in the dedicated graphics memory. Other unused texture information is generally stored in the STB memory such as SDRAM, and it is transferred over to the graphics memory using a dedicated 32-bit or 16-bit wide bus. To increase access to graphics memory, particularly for computationally intensive techniques such as the trilinear MIP-mapping, on-chip cache is used.

One of the important aspects of conveying depth perception to the viewer is the use of hidden surface removal. For example, in a scene where a house is covered by trees, the portion of the covered house has to be removed to convey realistic depth effects. The two common techniques for hidden surface removal are called *polygon sorting* and *Z-buffering* techniques. In the polygon sorting technique, the system CPU sorts the objects, either back-to-front or front-to-back, before being rendered into the frame buffer. The closer object overwrites the distant object if there is overlap between the two. The main drawback of this technique is its suitability for moving objects and accurately rendering objects that are interpenetrated. In the Z-buffering technique, the Z (depth) value of each pixel is stored, and the rendering engine compares its depth value with the previously rendered pixel for the same X-Y location. If the object is closer to the viewer, the graphics processor overwrites the

---

[2] MIP is derived from the Latin phrase *multum in parvo*, which means many in one.

previously rendered object in the same frame buffer. The advantage of the Z-buffering technique is relieving the CPU from sorting objects for hidden surface removal. In addition, it works well with moving objects, and renders accurately interpenetrating objects. The main drawback is the high system memory requirement. In practice, polygon sorting has been abandoned and Z-buffering is mainly used.

Another artifact, which degrades the quality of a 3D image, is called aliasing. Aliasing along the edges of an object, which results from high contrast between colors of adjacent pixels, causes visual anomalies such as stair stepping, scintillation, and break-up of solid lines to dotted lines. Polygon or edge anti-aliasing techniques remove such artifacts, resulting in enhanced image quality, and add realism to a scene. There are several anti-aliasing techniques that use the average area concept. In the *alpha blending* technique, which is a low-cost implementation of this concept, the percentage of a covered pixel by any polygon along an edge is calculated and used to blend its color with the already rendered pixels. Polygons are rendered from back to front. This technique improves the sharpness of jagged polygon edges and displayed text, is simple to implement, and does not require extra memory. The drawback of this technique is that the background color can bleed through the 3D object in front along the interior edges. Other techniques such as sub-pixel anti-aliasing with Z-buffering produce high quality, accurately rendered 3D objects.

An important feature of rendering objects such as water or glass is transparency or translucency. To create the effect of translucency, a 3D object is assigned an α value (0 to 1) that determines how much can be seen through it to whatever is behind it. A completely opaque object has $\alpha = 0$, while a translucent object has $\alpha = 1$. Translucency can also be used to gradually blend two adjacent LODs of an object, which is called LOD morphing. This feature is particularly useful for rendering moving objects in motion-driven scenes. In order to do alpha blending, the system needs an alpha channel. Each pixel consists of RGB color components, usually 24-bits, and the α component with an additional 8 bits, specifying how the pixel's colors should be merged with another pixel when the two are overlaid, one on top of the other.

Another interesting feature of rendered objects is called fogging. Fogging, which is a form of alpha blending, is used to create atmospheric effects such as low visibility conditions in flight-simulation games. With fog, distant objects can be rendered with fewer details, improving rendering speed. Fog is simulated by attenuating the color of an object with the fog color as a function of distance, where more distant objects appear less visible. Two techniques to implement fogging are called per-vertex (linear) fogging, and per-pixel (nonlinear) fogging. In the per-vertex technique, the fog value at the polygon vertices is interpolated to determine the fog factor at each pixel within the polygon. Realistic fogging is achieved as long as the polygons are small. Unnatural fogging is obtained with large polygons. However, the per-pixel fogging technique works well with large objects since fogging is a nonlinear function of the object's distance from the viewpoint.

8.7.2 On-Screen Video and 2D/3D Graphics Rendering Requirements

In an advanced digital STB, video and graphics rendering are required to be overlaid on TV screen backgrounds, which may consist of real-time video (analog or digital), graphics, or Internet web pages, blended with MPEG-2 decompressed displays, providing full-rate frame grabbing with scaling. The STB is required to render graphics-in-video, video-in-graphics, browser-in-video, video-in-browser, video-in-video, whether such video is still-frame or motion video.

As we learned in the previous section, a high-performance graphics processor, particularly for 3D display applications, requires the use of a low-latency, high-throughput data interface to its main memory. Accelerated graphics port (AGP), which was developed by Intel to be used with PCI bus, can provide real data throughput up to 1Gb/s.

Various applications in the future may require the delivery of 3D video to be displayed on a standard TV or HDTV monitor. One way to accomplish this task is to reconstruct the 3D-video program at the STB from a 2D MPEG-2 video program stream enhanced by 3D information, which is carried as associated private data in the MPEG-2 data stream. The various 3D programs can be viewed on a standard NTSC or HDTV set, using, for example, IR synchronized eyeglasses. It is expected that in the near future direct 3D display monitors will be available without the need for special eyeglasses.

The processing required for this application is described as follows, and is shown in Figure 8.7. The MPEG-2 stream with 3D information encoded in private user data is input to the MPEG-2 decoder, whereby such decoder outputs a 2D RGB image as a texture map to be input to the graphics processor; it is able to extract the private user data for input to the applications processor. The applications processor derives four outputs from the private user data input: (1) raw 3D data; (2) processed 3D data; (3) polygon data for input to the graphics processor; and (4) eyeglass shutter synchronization signals as input to the IR blaster for IR transmission to the IR synchronized eyeglasses. The first two (i.e., raw 3D data and processed 3D data) of these four data signals must be routed onto a 1394 link for use by external 3D receiver/display devices. Video/graphics is rendered in five planes:

- Background texture
- Media plane (MPEG-2 video)
- Graphics plane
- Cursor (minimum 16 x 16)
- Scaleable video-in-graphics plane

It should be pointed out that the display resolution rendering and digitizing modes must match and blend with the different resolutions to support animation; frame capture; transparent, opaque, and translucent partial or full-screen overlays on analog or digital video; per-pixel alpha blending; high-speed blitting; and display size scaling. Blitting means copying a large array of bits from one part of the STB's memory to another part, particularly when the memory is being used to determine what is shown on a display screen. The graphics processor in the STB should be able to scale graphics images as needed where the graphic images are designed for a different pixel aspect ratio.

The graphics plane can operate in various color modes such as

- 16-color indexed with color look-up table
- 256-color indexed with color look-up table
- 16-bit RGB/YUV (24-bit desired), 32-bit for α-blending
- Programmable graphics plane with respect to spatial size, screen position, and memory map location.

The important features of the video-in-graphics plane include its full scalability to any arbitrary size, and its capturing to memory. These features allow the analog, MPEG, or HD format, decoded digital video service to be scaled and captured to memory for display in a window within the graphics plane. The video-in-graphics plane will have the capability to be placed either above or below the graphics plane.

The alpha blending and chroma key modes for the graphics plane for each color mode can be supported, for example, as follows:

- 16-color indexed look-up table with reserved chroma key index
- 256-color indexed look-up table with reserved chroma key index
- 32,768-color RGB with chroma key bit (5:6:5:1)
- 4,096-color RGB with 4-bit alpha channel (4:4:4:4)
- 16-color indexed look-up table with 8-bit alpha

It should be pointed out that to get true color, one needs to use 24–32 bits per pixel.

The combination of pixel blitter with alpha blending allows one fast scrolling with graphical block moves, where all color modes can be supported with dedicated pixel color and pattern filling. Furthermore, pixel blitting supports nonrectangular shape instantiation[3], including the same alpha blending capabilities defined previously. Text blitter with anti-aliased font capabilities can be supported, for example, with dedicated text blitting hardware, and provide two-color (one bit per pixel) or four-color (two bits per pixel) font text instantiation.

The STB can also deliver a reconstructed ATV/HDTV signal in its original format with graphic additions, and ATV/HDTV signal with graphics, which has been converted to a NTSC formatted signal.   The required hardware and firmware implementation features include:

---

[3] Instantiation means producing a more defined version of some object by replacing variables with values (or other variables).

- The contents of the color map need to be modifiable under software control.
- The graphics display resolution should be configurable to match the MPEG-2 resolutions supported by the STB.
- The STB should have hardware acceleration for pixel blitter and text blitter transfers.
- Graphics rendering should be blended with the MPEG-2 decompressed display;
- Graphics rendering and display should be double-buffered for smooth animation.

8.8 CPU and Memory

8.8.1 Set-Top Box CPU

As we learned in the previous sections, the digital STB has various processors to perform the various functions such as MPEG-2 video decoding, AC-3 audio decoding, decryption and access control, and graphics. The main CPU in the digital STB, which is separated from the other processors, is responsible for handling the overall management of the STB and the various applications. This includes the real-time operating system (RTOS), application program interfaces (APIs), resident and special applications downloads and execution, arithmetic and logical operations, DSP algorithm acceleration, imaging, and 2D/3D graphics. In order for the main CPU to execute in a timely fashion the numerous instructions it receives from the RTOS and various applications, it must be high-performance with sufficient memory and have appropriate memory bus architecture. Such a CPU is, for example, a 64-bit superscalar million instructions per second (MIPS)-based microprocessor with reduced instruction set computer (RISC) architecture, operating at a clock frequency of 200-MHz or more [28]. Superscalar microprocessor refers to a microprocessor architecture that enables more than one instruction to be executed per clock cycle. The MIPS term refers only to the CPU speed, while real applications are generally limited by other factors, such as input/output (I/O) speed. Nearly all microprocessors, including the Pentium, PowerPC, Alpha, and SPARC microprocessors, are superscalar. Table 8.7 summarizes typical characteristic specifications of the main CPU processing capabilities.

The main CPU, which provides the flexibility to download new functions and applications, also supports a series of system hardware and software interrupts that can be dynamically prioritized, and provides a hardware timing function to supply programmable timing with a 1-ms resolution. Additional CPU features may include 5-volt tolerant I/O, per set instruction and data cache locking, power down wait instruction feature, and DSP multiply-add and three-operand multiply capability. Future advanced digital STB designs are migrating toward higher integration implementations, including more advanced microprocessor(s) utilizing 0.25-μm or smaller CMOS technology, five layer, with supply voltage of 3.3V or less. Furthermore, it is possible that a truly integrated advanced digital STB will be able to integrate the functionality of the various processors with the main CPU on a single chip.

Table 8.7 Processing capabilities for the Set-Top Box CPU microprocessor.

Characteristic	Specification
Instruction Set	MIPS IV instruction set
Address / Data Word Width	32, 64, or 128 bits as required to support features
CPU Clock Frequency	≥ 200 MHz
Superscalar Instruction Issue Rate	Integer + floating point
Integer/Floating Point	2.7/2.5
MIPS	233 minimum
Instruction/Data Cache	16-kB/16-kB
Set Associativity	Two-way
Write Policy	Write-back, write-through
Minimum System Memory Interface Bandwidth	Determined by STB manufacturer for sufficient support and render services
Upper Limit of Memory Frequency	> 87 MHz
Typical Power Consumption at 3.3 V	< 3 watts

8.8.2 Set-Top Box Memory

The digital STB uses four different types of memory modules for various purposes. The four types of memory modules are (A) random access memory (RAM), (B) read only memory (ROM), (C) synchronous dynamic RAM (SDRAM), and (D) flash memory. Let us briefly review the characteristics of each type of memory module. RAM, which is used in many electronic devices, is for temporary storage of data, which can be randomly accessed by the CPU and other processors for various tasks. When the power is disconnected, all data stored in RAM is lost. This is the reason why RAM is called volatile memory. In contrast, ROM is typically used to store specific STB codes (under particular conditions) such as self-diagnostics and navigational instructions.

Another type of volatile memory is SDRAM [29]. The use of SDRAM in digital STBs is designed to partially fulfill the increased demand by the main CPU for higher data rate transfers than conventional DRAM technology can provide. Synchronous DRAM technology combines recent industry advances in fast DRAM architecture with high-speed interface and miniature packaging technologies. The core of SDRAM memory is a standard DRAM with the important addition of synchronous control logic. By synchronizing all address, data, and control signals with a single system clock, SDRAM technology enhances performance, simplifies design, and provides faster data transfer. For example, one 1Mx16 SDRAM can replace four 256Kx16 page-mode DRAMs and provide the 135-200-Mbyte/sec data access required for MPEG-II television (NTSC and PAL) applications. SDRAM is typically used for graphics rendering and on-screen video and graphics display.

Flash memory modules are reprogrammable modules that retain their content after the power source is removed or disconnected. Thus, flash memory is called nonvolatile memory. Generally speaking, flash memory offers lower cost and higher performance than other competitive memory modules. Flash memory is a solid-state reliable module with no mechanical parts. Flash memory is based on a technology similar to electrically erasable programmable read only memory (EEPROM), but unlike EEPROM, flash is more scalable and cost effective. With production densities of up to 8 MB, flash memory modules are used in applications where moderate amounts of nonvolatile storage are desired such as STBs, PCs, digital cameras, and cellular phones. When the STB first powers up, the firmware typically is loaded from flash to DRAM, and it is executed in DRAM. Slower flash memory access times, compared to processors or other system components, have pushed set-top designers to this boot and download scheme.

Generally speaking, there are two system architectural methods to achieve the required memory bandwidth [30]. In the distributed method, which is also the more conventional one, the required memory module is typically attached to the processor performing a specific function such as 3D graphics. This approach works well if the memory can be easily partitioned and there are no problems with memory granularity. However, the increasing integration of the various functions in the digital STB has motivated the use of the unified memory architecture (UMA) method. In the UMA method, the same block memory can be effectively shared among the various processors. For example, the SDRAM memory is shared among MPEG-2 decompression, on-screen graphics, and other applications. However, the SDRAM memory sharing allocation process can provide priority to the MPEG video/audio decoding when competing against graphics or application uses, since it requires the most continuous use of SDRAM for buffering functions.

The different memory modules are needed in the STB to support the following software and/or firmware activities:

- MPEG-2 decompression
- Resident device drivers, utilities, RTOS kernel
- Graphics
- Resident applications
- Downloaded applications
- CM processing

Table 8.8 shows an example of the different memory module sizes in a digital STB that are needed for the various functions described. The STB design needs to support a hardware memory manager for memory protection of application processes and the operating kernel, and to provide a flat address space for user processes. The STB typically uses a secure, nonvolatile random access FLASH memory for essential information storage such as video channel maps, authorization tables, and OOB channel assignments. The contents of the Flash memory can be downloadable via the HFC network. A minimum of 4-kB nonvolatile memory is typically reserved for subscriber messaging.

Table 8.8 Different types of memory modules for advanced digital STB.

Characteristic	Specification	Description
RAM	8 MB	CPU, applications
ROM	2 MB	Boot strap and navigational diagnostics
SDRAM	8 MB	Shared with graphics and MPEG-2 decode, and on-screen display
FLASH	8 MB	Code download, booting

8.9 Advanced Set-Top Box with Built-in DOCSIS Cable Modem

The digital STB that we have discussed so far provides basic and interactive TV-based applications. However, it does not allow the subscriber to get a high-speed Internet access through the cable TV network. To get an Internet access, the subscriber needs a CM. An advanced digital STB combines the functionality and features of an interactive digital STB and a CM. Figure 8.8 shows a simplified block diagram of an advanced digital STB with a built-in DOCSIS CM [31]. Note that the advanced STB has three RF tuners as follows: (I) 54–860-MHz RF tuner for NTSC/MPEG video reception, (II) 54–860-MHz RF tuner for simultaneous Internet access through the DOCSIS CM, and (III) 70–130-MHz RF tuner for the OOB channel. The DOCSIS modem is essentially an RF QAM transceiver operating according to DOCSIS 1.0/1.1 protocols [14]. The DOCSIS protocol, which defines both the physical (PHY) layer and the media-access-control (MAC) layer, will be discussed in detail in Chapter 11. Currently, the basic transmission and MPEG video processing with limited graphics capabilities are integrated into a single chip, which is available today using 0.25-µm CMOS technology. With 0.18-µm CMOS technology, most of the building blocks of the advanced digital STB (excluding the analog-front-end and DRAM) shown in Figure 8.8 can be integrated into a single IC. The advanced digital STB represents a key technology platform for the cable TV operators to significantly increase their revenue base over the currently installed digital STBs. Consequently, the digital STB software architecture as well as the various TV-based and Internet-based interactive applications will be addressed in detail in Chapter 12.

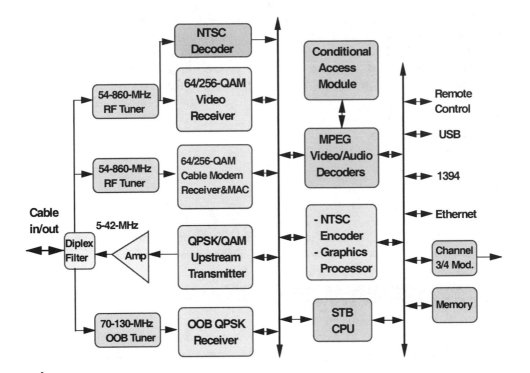

Figure 8.8 Block diagram of an advanced digital set-top box with a built-in DOCSIS modem. After Ref. [3] (© 1999 IEEE).

8.10 M-QAM Transmission Impairments in HFC Networks

The downstream transmission performance of a digital STB or CM in HFC networks can be divided into two class categories as (I) nonlinear distortions induced by lightwave impairment effects, and (II) electronic impairment effects. Chapters 9 and 10 discuss in detail the different types of lightwave effects, including single and multiple wavelength effects over cable TV networks without and with EDFAs, respectively. Generally speaking, the electronic impairment mechanisms, which are discussed in this section, also can be divided into three different groups: (A) electronic implementation impairments, (B) cable TV headend impairments, and (C) cable TV network transmission impairments. Before we discuss the impairment mechanisms caused by the cable TV network, let us briefly review type A and B impairments.

It is well known that the actual hardware implementation of the digital STB design can significantly affect not only the on-screen analog/digital and graphics display and audio output, but also the interaction of the subscriber with the different applications. Rather than discussing overall STB implementation issues, let us focus on the implementation losses of the front-end of the STB, including the QAM receiver, which are essential for robust operation. The implementation losses of RF and IF front-end of the STB are

- RF tuner noise figure (NF)
- RF tuner phase noise
- RF tuner nonlinear distortions
- A/B switch
- SAW filter
- QAM receiver

The generated 64/256-QAM channels at the cable TV headend also can be impaired before they are transmitted over HFC network. These impairments include

- QAM modulator modulation error ratio (MER)
- QAM channel up-converter phase noise
- Adjacent channel spurious noise.

Finally, the electronic impairments in cable TV networks can be divided as follows:

- Additive white Gaussian noise (AWGN)
- Multipath electrical echoes and group delay
- In-band QAM amplitude variations
- Phase noise
- Burst and impulse noise
- AM hum modulation
- Nonlinear distortions (CSO, CTB, XMOD)

The characteristics and impact of the nonlinear distortions, particularly their bursty behavior, on the downstream QAM receiver are discussed as part of the transmission impairments in lightwave systems in Section 9.2. The discussion and results that are presented in this section will focus on a downstream QAM receiver operating according to ITU-T J.83 Annex B specifications [32].

8.10.1 Set-Top Box Front-End Losses

Figure 8.9 shows a simplified block diagram for the RF input signal path through the front-end of the digital STB to the QAM receiver with the assumed signal gain or loss through each block. The RF tuner and IF amplifier are assumed, for example, to have a worst-case NF of 9.5 dB and 10 dB, respectively. In addition, the NF of the A/B switch and SAW filter is assumed to be equal to their loss. The overall worst-case NF of the STB is obtained using Equation (2.4). This yields an overall worst case NF of 13 dB for the STB. With 13-dB worst case NF, the thermal noise floor for 256-QAM channel (i.e., 5.36-MHz noise bandwidth) is −44.96-dBmV. As the RF input signal level is reduced below its nominal level, the 256-QAM implementation loss caused by the 13 dB NF of the STB is increased. Let us introduce the concept of an implementation loss (IL) for a receiver. The IL of a receiver caused by a given impairment can be defined as [33]

$$IL(dB) = SNR_{in} - SNR_{theory} \qquad (8.9)$$

where SNR_{theory} is the theoretical input SNR of an ideal receiver without the impairment at a given BER, and SNR_{in} is the resultant input SNR of the actual receiver with the impairment at the same BER, which can be calculated as follows:

$$SNR_{in}(dB) = -10 \cdot \log\left[10^{-SNR_{theory}/10} - 10^{-SNR_{imp}/10}\right] \qquad (8.10)$$

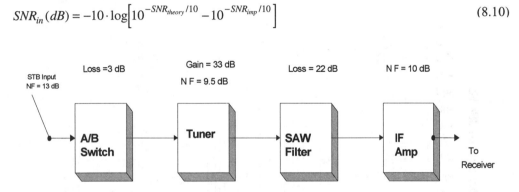

Figure 8.9 Simplified block diagram of an RF input signal path through the front-end of a digital STB.

Table 8.9 NF-related 256-QAM implementation loss at different RF input signal levels.

RF Input Signal Level	SNR due to NF	256-QAM Implementation Loss
−5 dBmV	40 dB	0.28 dB
−10 dBmV	35 dB	0.94 dB
−13 dBmV	32 dB	2.14 dB
−15 dBmV	30 dB	4.16 dB

where SNR_{imp} is the equivalent input SNR for Gaussian noise that would degrade the QAM receiver the same as the impairment, and it is typically measured by an RF spectrum analyzer. For example, the SNR caused by the NF of the STB is 30-dB at an RF input level of -15 dBmV (Table 8.9), and the theoretical SNR for 256-QAM at BER $= 1 \cdot 10^{-10}$ is 27.9 dB (Table 7.3). Substituting these values in Equation (8.10) yields $SNR_{in} = 32.06$ dB. When the NF impairment is added, the input AWGN must be reduced in order to maintain the same BER. Therefore, the implementation loss is 32.06 dB $-$ 27.9 dB $=$ 4.16 dB. The RF tuner also generates its own phase noise and nonlinear distortions. The worst-case single-sideband (SSB) phase noise of a typical dual-conversion tuner as a function of the frequency offset from the QAM carrier is shown in Figure 8.10. Notice that the phase noise behavior of the RF tuner drops approximately as $1/f^2$ up to about 1 kHz. From 1 kHz up to 100 kHz, the SSB phase noise drops as $1/f^3$. Furthermore, as the RF tuner warms up, its phase noise characteristic changes nonlinearly with frequency. Consequently, it is recommended to specify the phase noise requirement of the RF tuner at room temperature and

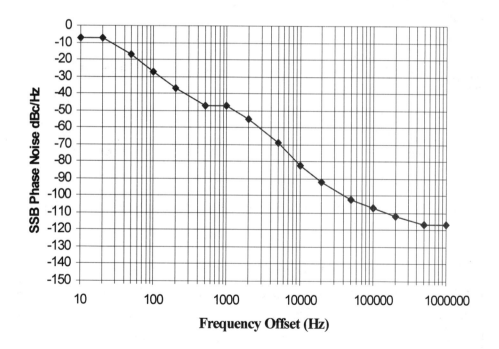

Figure 8.10 Measured single-sideband (SSB) phase noise versus the frequency offset from the carrier for a typical dual-conversion RF tuner.

Table 8.10 Typical characteristics of a dual-conversion cable TV tuner.

Characteristics	Typical Value
RF Input Level Range (AM video)	0 dBmV to +15 dBmV
Noise Figure (NF)	9 dB
Frequency Response Flatness (in 6-MHz)	0.5 dB
Conversion Gain	33 dB
Phase Noise at 10 kHz offset	−85 dBc/Hz
CSO Distortion	≤ −60 dBc
CTB Distortion	≤ −65 dBc
Cross-Modulation Distortion	≤ −60 dBc
Output Return Loss	≥ 10 dB
Image Rejection	≥ 64 dB
Input Return Loss	≥ 5 dB
Isolation	≥ 70 dB

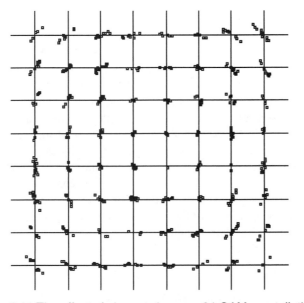

Figure 8.11 The effect of phase noise on a 64-QAM constellation.

at the maximum operating temperature at various offset frequencies. The recommended worst-case SSB phase noise specifications at room temperature of RF tuner are typically −78 dBc/Hz and −85 dBc/Hz at 10 kHz offset for 64-QAM and 256-QAM modulations, respectively. The rationale for these specifications can be explained as follows. The amount of phase noise that the QAM receiver "sees" is determined by the RF tuner phase noise, external phase noise, and its carrier-recovery loop. The carrier-recovery loop in the QAM receiver acts as a high-pass filter to the phase noise generated by the tuner. Thus, the total phase noise at the QAM receiver is obtained by the numerical integration of the SSB phase noise under the curve shown in Figure 8.10 and filtered by the receiver carrier recovery loop. This value can be converted into the total RMS phase noise jitter (in radians), which is given by [33, 34]:

$$\phi_{RMS} = \left[\frac{2\pi}{360}\right]\sqrt{P_t} \qquad\qquad (8.11)$$

where P_t is the total integrated power. Table 8.10 summarizes the typical characteristics of a dual-conversion tuner for a digital STB. The effect of phase noise on a 64-QAM constellation is shown in Figure 8.11. The outer constellation points are smeared the most, while the center constellation points remain almost unchanged. From a practical perspective, it is also important that the generated nonlinear distortions by the RF tuner are sufficiently small, and sufficient isolation ($\geq$ 70-dB) exists between downstream and upstream transmission.

8.10.2 QAM Transmitter Losses

In addition to front-end losses at the digital STB, the generated RF QAM signal at the cable TV headend is not an ideal QAM signal. In Section 7.9.2, we explained the concept of equalized MER for a QAM transmitter. The measured SNR of the QAM modulator needs to be corrected for the departure of the HP89441A instrument from an ideal receiver according to Equation (7.33). As we learned in Section 7.6.2, the SAW filter reduces spurious noise, which is out of the QAM channel. To translate the generated QAM channel at a 44-MHz IF to the desired RF channel frequency, an upconverter unit is typically used. Adjacent channel interference (ACI) is generated when multiple QAM channels are contiguously stacked one next to the other. Since the carrier level of the various QAM channels across the cable TV frequency plan can be different by as much as 10-dB or more, the measured QAM SNR will be further reduced because of the ACI effect. In addition, the upconverter unit has its own SSB phase noise, which will be transmitted downstream to the digital STB or CM receiver. If the upconverter is overdriven by the 44-MHz IF signal, it can result in compression of the digitally modulated carrier, resulting in a data/packet loss.

8.10.3 Additive White Gaussian Noise (AWGN)

Once the SNR of the QAM modulator has been measured, the performance of the downstream QAM receiver in the digital STB or CM needs to be measured first in a back-to-back configuration in the presence of a Gaussian noise. Figure 8.12 shows an example of the measured and calculated 256-QAM coded BER (ITU-T J.83 Annex B) versus E_s/N_o (in dB) for a Gaussian channel. The RF input signal level to the QAM receiver was 0 dBmV.

Pseudorandom bit sequence (PRBS) $2^{23} - 1$ data and the clock signals are typically used for this measurement. As we learned in Chapter 7, the ITU-T J.83 Annex B FEC consists of a 16-state trellis- coded modulation (TCM) at a rate of 19/20 concatenated with R-S T = 8 (128,122) code. The transmitted symbol rate is 5.36094 MBaud. The measured 256-QAM CNR (= E_s/N_0) has been corrected for the departure of the QAM modulator from an ideal modulator. For example, assuming 42-dB MER for the QAM modulator, the corrected 256-QAM CNR in a 5.36-MHz receiver noise bandwidth (= transmitted symbol rate) is 28.8-dB at BER = $1 \cdot 10^{-9}$. This results in about 1-dB implementation loss for the QAM receiver.

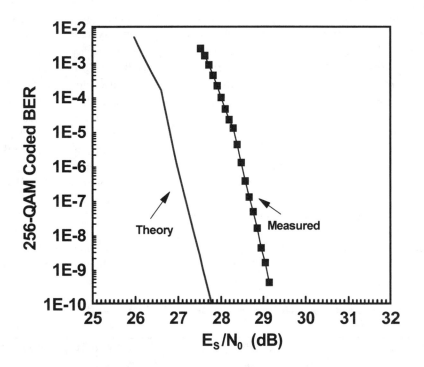

Figure 8.12 Measured and calculated 256-QAM BER as a function of the CNR, assuming only AWGN and RF input power level of 0-dBmV.

8.10.4 Multipath Reflections

Multipath microreflections in the coaxial portion of the HFC network can significantly degrade the CNR of the QAM channel. Static multipath reflections can be generated in a lab using, for example, an Agilent 11759A dynamic ghost simulator, which can generate single or multiple echoes with a specified magnitude, time delay, and phase relative to the transmitted signal. It is often more convenient and simple to analyze the adaptive equalizer (AE) behavior in the presence of a single dominant echo rather than multiple echoes. Figure 8.13 shows an example of the measured echo magnitude for a transmitted 256-QAM signal as a function of the echo time delay in microseconds. The data was obtained by setting the 256-QAM SNR to 30-dB for a single echo with 180° phase relative to the transmitted signal. Then, the echo magnitude was increased from −50-dBc until QAM carrier acquisition could not be achieved.

The QAM receiver used in this measurement features a 32-tap symbol-spaced equalizer (8 FFE taps and 24 DFE taps) with decision-directed adaptation. One of the FFE taps is typically used as the center tap, leaving 7 FFE taps for leading echoes and 24 DFE taps for lagging echoes. Since the modulation symbol duration is approximately 186.55 ns, the theoretical total range of the AE is −1.3 μs to +4.48-μs. For time delays up to 4.0 μ s, the ITU-T J.83B receiver can equalize lagging echoes with a relative magnitude up to −10 dBc. For delays up to 4.5 μs, lagging echoes with magnitude of −30 dBc can be equalized. Echoes with time-delays greater than 4.5 μs cannot be equalized, and thus behave like noise. This result demonstrates that the AE can "handle" any single echo with magnitude up to −10 dBc almost independently of the echo's time delay within its range. From a practical point of view, the dashed line (Fig. 8.13) can be used as an echo magnitude specification that can that can be reliably handled by the tested QAM receiver at various delays.

There are various models such as the IEEE 802.14 HFC channel model [35] and the DOCSIS 1.1 model [14], which specify the echo magnitude and time delay for an HFC network. Generally speaking, the multipath behavior of a cable TV network can vary widely from one area to another. These two multipath models may represent two extremes, and are shown here only for comparison. As we learned in Chapter 2, the DOCSIS 1.0 model assumes only a single dominant reflection with a specified maximum echo magnitude at each time delay interval. In contrast, the IEEE 802.14 model allows for one or more discrete

Table 8.11 Multipath echo profile according to the IEEE 802 HFC model.
After Ref. [35].

Echo Time Delay	Total Echo Magnitude (Downstream)	Total Echo Magnitude (Upstream)
0 to 0.2 μs	−11 dBc	−10 dBc
0.2 to 0.4 μs	−14 dBc	−12 dBc
0.4 to 0.8 μs	−17 dBc	−14 dBc
0.8 to 1.2 μs	−23-dBc	−18 dBc
1.2 to 2.5 μs	−32 dBc	−24 dBc
2.5 to 5 μs	−40 dBc	−30 dBc
5 to 15 μs	None	−35 dBc
> 15 μs	None	None

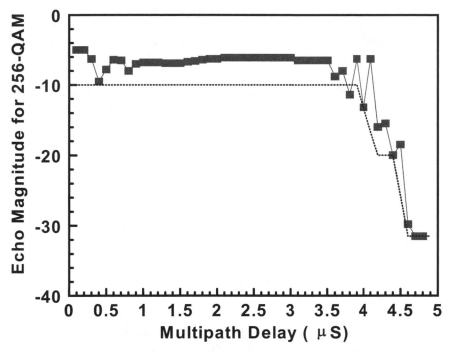

Figure 8.13 The measured echo magnitude relative to the transmitted 256-QAM signal (in dBc) as a function of the multipath time delay (μs).

reflections within each time interval, where the echo magnitude is randomly distributed within each time interval. Table 8.11 shows, for example, the maximum downstream and upstream echoes' magnitude for each time interval according to the IEEE 802 model. Specifying the echo magnitude at the maximum of the time-delay interval produces the largest GDV. But, it does not necessarily have the worse effect on the transmitted signal. It turns out that breaking the magnitude of a single echo into multiple echoes within the time-delay interval lowers the GDV, but actually has a worse effect on the transmitted signal because of a higher peak-to-RMS ratio for combining multiple echoes compared with a single echo.

The effect of multipath echoes on the transmitted QAM signal also can be analyzed using the DFE and FFE tap values within the time delay range of the AE. Figure 8.14 shows an example of a histogram of the tap values for a symbol-spaced AE with eight DFE taps and eight FFE taps in the presence of a −14-dBc echo with a time-delay of 1.2 μs relative to the received 64-QAM signal [36]. Since the ITU-T J.83 Annex B receiver was operating in the 64-QAM mode, the corresponding tap spacing is 0.1978 μs. One of the FFE taps is designated as the center tap (zero time-delay), which has been set to an arbitrary value of +10 dB.

The other seven FFE and eight DFE taps are available for echo cancellation. In order to cancel the effect of the multipath echo, the adaptation algorithm has set the sixth DFE tap weight to −14 dBc relative to the center tap. This indicates that there is an echo with a −14 dBc magnitude and time delay of 1.2 μs. The fifth tap, which is located at time delay of 1.0 μs, is also activated at about −25 dBc. However, the possibility of a second echo with a time delay of 1.0 μs is very low since the magnitude of the seventh tap is quite low. Thus, it probably means that the single echo is actually occurring between the fifth and sixth taps, or between time delays of 1.0 and 1.2 μs, but closer to 1.2 μs. Notice that most of the other taps are also partially activated, which could be due to the presence of very weak echoes, or there may be a slight tilt in the channel. In practice, the Agilent 89441A instrument provides a frequency response display that is calculated as the Fourier transform of the tap values (see Figure 7.16).

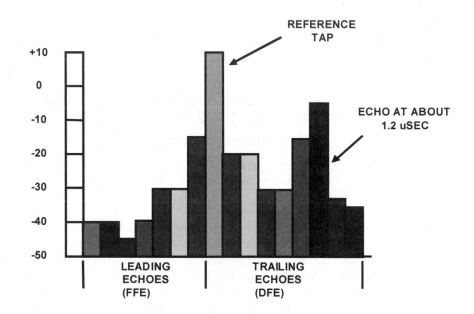

Figure 8.14 Tap weights for an adaptive equalizer in the presence of a −14-dBc echo at about 1.2-μs time-delay. After Ref. [36].

8.10.5 Amplitude and Group Delay Variations

The amplitude and the group delay (GD) of the transmitted downstream and upstream QAM channel can significantly vary across within its bandwidth and also across the cable TV frequency spectrum because of multipath reflections and poor frequency response. The in-band amplitude variation, which often is called amplitude ripple, is specified to be no more than 3-dB for downstream 64/256-QAM channels according to DOCSIS 1.1 [14]. However, there is no specification for the QAM channel amplitude variation across the cable TV frequency plan. Cable operators typically like to balance the QAM channel amplitude across the frequency plan and limit the amplitude slope to less than 10 dB. For upstream transmission, the maximum amplitude ripple is specified to be 0.5-dB per MHz according to the DOCSIS 1.1 specifications [14].

As we learned in Section 2.8, the group delay variation (GDV) is related to the magnitude and time delay of the multipath echoes. The downstream GDV, which is typically generated by the diplex filters in active coaxial components in the HFC network and an improper network alignment, is specified to be no more than 75-ns within the transmitted QAM channel. The upstream GDV is specified per 1-MHz of channel bandwidth because the channel bandwidth is not fixed, but is allowed to vary within specific steps. For example, the maximum upstream GDV is specified to be 200-ns/MHz across the upstream band (5–42 MHz) according to the DOCSIS 1.1 specifications [14].

8.10.6 Burst and Impulse Noise

As we learned in Chapter 2, burst and impulse noise, particularly on the return path portion of the HFC network, constitute another important degradation mechanism. The frequency of occurrence of burst noise longer than 1-μs can vary significantly from one cable TV network to another. Generally speaking, although such bursts occur only a handful of times in a 24-hour period, changing HFC network conditions can generate relatively long duration bursts (>10-μs) at a much higher rate. For example, the IEEE802.14 HFC model specifies the following assumptions for burst noise [35]:

- Burst events are Poisson distributed with a downstream rate of one burst per minute, and an upstream burst of one per ten seconds
- Each burst of the event is amplitude modulated with zero-mean Gaussian noise in 1 MHz RF bandwidth (0.5-MHz lowpass baseband noise) with standard deviation equal to 1/4 of carrier amplitude
- The carrier frequency of each event is selected from a uniform distribution between 5-MHz and 15-MHz
- Each burst event has the same amplitude, where the amplitude of each event is selected from a sawtooth distribution, which is zero at 0-Vrms and is peaked at 1-Vrms (= 60 dBmV across 75 ohms)
- With multiple bursts per event, the duration of each burst is selected from a uniform distribution between 1-μs and 30-μs. With a single burst, the burst length is exponentially distributed with a mean of 30-μs.
- The space between a burst and the following burst (if there is another in the event) is selected from a sawtooth distribution peaked at 10-μs and down to zero at 130-μs.

   To understand the downstream transmission degradation of QAM channels in the presence of burst noise, let us look at the following test result. The operating point of a 256-QAM receiver was set for a sufficiently high CNR (i.e., error-free transmission), say 35-dB, in the presence of a Gaussian noise. Then, a pulse generator was fed to a doubly balanced mixer, which was used to gate the Gaussian noise during the burst pulse, thus reducing the 256-QAM CNR. During the burst event, the QAM CNR was sufficiently reduced to corrupt the 256-QAM constellation, thus preventing the QAM receiver from reacquiring the QAM carrier. The gated burst noise signal was combined with the 256-QAM modulator output at IF, and upconverted to RF before it was fed to the QAM receiver. The measured burst noise duration was adjusted until uncorrectable errors were observed at BER $\leq 1 \cdot 10^{-6}$. This coded BER level is typically selected since it is approximately the threshold of visibility (TOV) for transmitted digital video and because it also allows one to collect a good error statistics for a reliable measurement. Figure 8.15 shows the measured duration of the gated burst noise at a coded BER $\leq 1 \cdot 10^{-6}$ as a function of the interleaver depth (J) at a 10 Hz burst noise repetition rate. The case of $I = 128$, $J = 1$ interleaver/deinterleaver combination, which is specified by DOCSIS 1.1, has a theoretical burst-noise protection value of 66-μs. There are two important issues, which are apparent in Figure 8.15.   First, the measured burst noise erasure

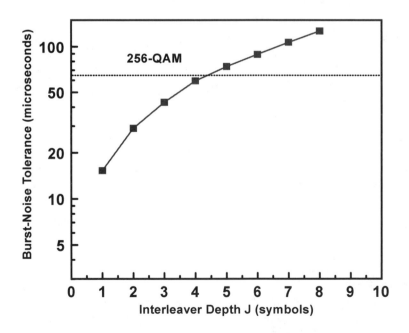

Figure 8.15 Burst-noise tolerance duration of 256-QAM receiver at BER $\leq 1 \cdot 10^{-6}$ versus the interleaver depth (symbols).

for $I = 128$, $J = 1$ interleaver mode is significantly below the theoretical value for 256-QAM transmission. This is because the analog front end of the QAM receiver, particularly the RF tuner, adds phase noise and nonlinear distortions, further degrading the QAM receiver performance. Second, a downstream receiver operating in 256-QAM mode is significantly more sensitive to the burst noise rate than in the 64-QAM mode as expected. Thus, in order to achieve the theoretical burst noise tolerance of 66-μs, the $I = 128$, $J = 4$ interleaver needs to be used. The drawback is the additional cost of SRAM memory and added latency in the QAM receiver for robust transmission.

8.10.7 AM Hum Modulation

AM hum modulation, as explained in Section 2.5.4, is an amplitude distortion of the transmitted signal caused by the modulation of the signal by the power line sources at 60-Hz and 120-Hz. According to the FCC rules in Section 76.605, cable TV operators are required to keep the AM hum modulation at or below 3% peak-to-peak (0.26 dB amplitude reduction) of the transmitted QAM signal [37]. The shape of the AM hum modulation signal is actually trapezoidal with rise and fall times of about 0.25-ms. It should be pointed out that the QAM channel degradation increases very rapidly as the AM hum modulation increases above 3% peak-to-peak. The AM hum degradation can be mitigated, for example, by implementing a fast AGC loop within the QAM receiver, allowing the equalizer to track the QAM signal variations. Lab measurements have shown that the 256-QAM CNR can be degraded by more than 1-dB at an AM hum modulation of 5% peak-to-peak.

8.10.8 64/256-QAM System Link Budget

To specify the required operating conditions for the digital STB, one needs first to construct a system link budget based on all the various impairment types. The system link budget takes into account the simultaneous presence of all the different types of impairments over the cable TV network. To develop the link budget for the STB, one needs to calculate the associated implementation loss for each type of impairment on the transmitted 64/256-QAM channels. Table 8.12 summarizes, for example, OpenCable STB specification for downstream transmission of both AM-VSB video and 64/256-QAM signals [31]. The nominal relative carrier power levels for both the analog and digital signals are given in Table 8.13. These values are needed to determine the adjacent channel interference between the analog and digital signals. Note that these relative power levels must be within the absolute RF input power range of −15 dBmV to +15 dBmV for the digital signals.

Using this procedure, the required 256-QAM C/(N+I) versus the RF input level (in dBmV) to a digital STB was calculated in the presence of white Gaussian noise and multiple simultaneous impairments. In this example, 79 AM-VSB channels (55.25–547.25 MHz) with a CNR of 45 dB at RF input level of 0-dBmV, and thirty-three 256-QAM channels (553.25–751.25) were transmitted to the digital STB. The RF power ratio of the AM-VSB

channels to the 256-QAM channels was fixed at 5 dB. As the RF input power level is reduced below 0-dBmV, the required C/(N+I) for error-free transmission steadily increases
from 34 dB in the presence of multiple simultaneous impairments. At an RF input level of
−12 dBmV, the required C/(N+I) is 39 dB. The operating margin of the digital STB is defined as the difference between the C/(N+I) for the cable TV plant and the STB in the presence of multiple simultaneous impairments. Thus, the operating margin represents a safety
margin for the cable TV operators to work with the required C/(N+I). The operating margin
in the presence of the actual impairments is likely to vary from 0 dB to 5 dB, depending
on the type of impairment and its magnitude. This means that the cable TV operator
must continuously monitor and balance the magnitude of all of the impairments over its
network. Failure to do so can easily reduce or even eliminate the STB operating margin,
resulting in interrupted analog and digital services.

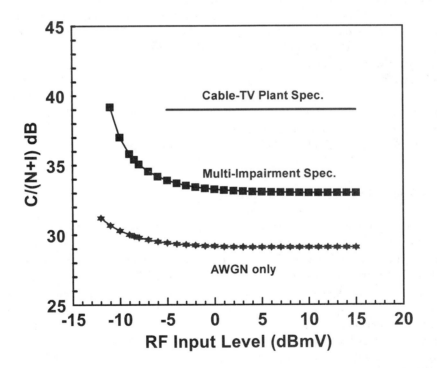

Figure 8.16 Calculated 256-QAM C/(N+I) in the presence of Gaussian noise and multiple impairments as a function of the RF input level to the STB. The solid
line represents the assumed cable TV network specification.

Table 8.12 OpenCable downstream STB specification for 64/256-QAM transmission (0°–40° C). After Ref. [31].

CHARACTERISTICS	OpenCable STB Analog and 64/256-QAM SPECIFICATIONS
RF Input Channel Bandwidth	6 MHz
RF Frequency Range	54 MHz-860 MHz (IRC/HRC/standard frequency plans)
RF Input Signal Level Range	−15 dBmV to +15 dBmV for 64-QAM −12 dBmV to +15 dBmV for 256-QAM 0 to +15 dBmV for Analog visual carrier −10 to −17 dBmV for Analog aural carrier
RF Input Return Loss	≥ 6 dB (over full tuning range)
RF Input Impedance	75 ohm unbalanced
RF Bypass Isolation	≥ 60 dB (over full tuning range)
$C/(N+I)^2$ @ BER=$1 \cdot 10^{-9}$	≥ 33 dB for 256-QAM[3] ≥ 27 dB for 64-QAM[3]
CSO Distortions	≤ −60 dBc[4]
CTB Distortions	≤ −63 dBc[4]
Cross-Modulation Distortions	≤ −57 dBc[4]
Spurious Emissions (within the output channel 3/4 bandwidth)	≤ −60 dBc[4]
Spurious Emissions (outside the output channel)	≤ −10 dBc[4]
Phase Noise Tolerance[5]	≤−88 dBc/Hz @ 10 kHz offset
In-Band Amplitude Ripple Tolerance	≤ 5 dB (peak-to-peak) for digital channels ≤ 4 dB (peak-to-peak) for analog channels
Burst Noise Duration	Not longer than 25 μsec at10 Hz repetition rate
CW Interference	Included in C/(N+I)

---

[2] C/(N+I) includes the simultaneous presence of all impairments in the 6-MHz channel bandwidth.

[3] Channel loading assumptions: at least 110 AM-VSB channels with CNR = 43 dB, and at least 20 QAM channels at RF input level of −10 dBmV.

[4] Channel loading assumptions: at least 110 AM-VSB video channels at RF input level of +15 dBmV, and at least 20 QAM channels at RF input level of +5 dBmV.

[5] The phase noise tolerance specification is for a phase noise generated by the cable TV network relative to the center of the QAM signal band, excluding the contribution of the tuner in the set-top box.

CHARACTERISTICS	OpenCable STB Analog and 64/256-QAM SPECIFICATIONS
Multipath Reflections (One dominant echo with maximum specified magnitude relative to the primary QAM channel)	−10 dBc at time-delay ≤ 0.5 μsec −15 dBc at time-delay ≤ 1 μsec −20 dBc at time-delay ≤ 1.5 μsec −30 dBc at time-delay < 5.0 μsec −35 dBc at time-delay ≥ 5 μsec
Group Delay Variation Tolerance	≤ 0.25 μsecond across the 6-MHz channel
AM Hum Modulation	Not greater than 3% (p-p)
Adjacent Channel Interference	≤ 10 dBc between QAM channels
Conversion isolation (RF input to channel 3/4 RF output)	≥ 65 dB

Table 8.13 Nominal and range of the relative carrier power levels for the transmitted analog and digital signals to the digital set-top box. After Ref. [31].

Channel Type	Nominal and Range of the Relative Carrier Power Level
Analog Channel	0 dBc (reference level)
Downstream OOB Channel	−8 ± 5 dBc
Downstream 64-QAM Channel	−10 ± 2 dBc
Downstream 256-QAM Channel	−5 ± 2 dBc

References

1. R. K. Jurgen, *Digital Consumer Electronics Handbook*, McGraw-Hill, New York, 1997.
2. CableLabs, *OpenCable Set-top Terminal CORE Functional Requirements for Bi-directional Cable*, CFR-OCS-BDC-INTC03-00818 (2000).
3. W. S. Ciciora, "Inside the Set-Top Box," *IEEE Spectrum Magazine*, April, 70–75 (1995).

4. H. Samueli, "Broadband Communications ICs: Enabling High-Bandwidth Connectivity in the Home and Office," *IEEE International Solid-State Circuits Conference Proceedings*, paper MA 1.3 (1999).

5. IEEE, IEEE Standard for High Performance Serial Bus, IEEE 1394–1995 (1996).

6. Information on Edwin H. Armstrong can be found in the Web site: http://www.world.std.com/~jlr/doom/armstrng.htm.

7. M. Waight, "Wireless Industry Opens Doors for Cable," *NCTA Technical Papers* (1996).

8. P. Putman, *Question of Balance: A Look at Voltage Current, Video, Cables and Where the Balun Fits In*, Intertec Publishing (1999).

9. D. Agahi, "Monolithic Ceramic Block Combline Bandpass Filters," *RF Design*, March, 35–41 (1989).

10. N. Scheinberg, "A GaAs Up Converter Integrated Circuit for a Double Conversion Cable TV "STB" Tuner," *IEEE Journal of Solid-State Circuits*, **29**, No. 6, June, 688–692 (1994).

11. DVS-167, *Digital Broadband Delivery System: Out-Of-Band Transport*, Society of Cable Telecommunications Engineers (1998).

12. DVS-178, *Cable System Out-Of-Band Specifications*, Society of Cable Telecommunications Engineers (1998).

13. S. Lin and D. J. Costello Jr., *Error Control Coding: Fundamentals and Applications*, Prentice-Hall, Englewood Cliffs, N.J. (1983).

14. CableLabs, *Data-Over-Cable Service Interface Specifications: Radio Frequency Interface Specification* SP-RFIv1.1-I01-990311 (1999).

15. The ATM Forum, *ATM User-Network Interface (UNI) Specification Version 3.1*, Prentice-Hall PTR (1995).

16. H. T. Nicholas III and H. Samueli, "A 150-MHz Direct Digital frequency Synthesizer in 1.25-μm CMOS with -90-dBc Spurious Performance," *IEEE Journal of Solid State Circuits* **26**, 1959–1969 (1991).

17. International Radio Consultative Committee (CCIR) Recommendation 656, *Interface For Digital Component Video Signals in 525-Line and 625-Line Television Systems* (1994).

18. EIA-608, Recommended Practice for Line 21 Data Service, Electronic Industries Association (1994).

19. DVS-053, *Standard for Carriage of NTSC VBI Data in Cable Digital Transport Streams*, Society of Cable Telecommunications Engineers (1998).

20. EIA-516, *Joint EIA/CVCC Recommended Practice for Teletext: North American Basic Teletext Specification (NABTS)*, Electronic Industries Association (1998).

21. D. R. Stinson, *Cryptography Theory and Practice*, CRC Press, Boca Raton, Florida (1995).

22. A. J. Menezes, P. C. Van Oorschot, and S. A. Vanstone, *Handbook of Applied Cryptography*, CRC Press (1997).

23. RSA Labs, *Frequently Asked Questions About Today Cryptography 4.0* (1998).

24. CableLabs, *OpenCable Host-POD Interface Specification*, IS-POD-131-WD01-991008 (1999).

25. CableLabs, *OpenCable Set-top Terminal CORE Functional Requirements for Unidirectional Cable*, CFR-OCS-UDC-WD04-991015 (1999).

26. CableLabs, *OpenCable POD Copy Protection System*, IS-POD-CP-WD02-991027 (1999).

27. See for example A. B. Tucker Jr., Editor-in-chief, *The Computer Science and Engineering Handbook*, CRC Press (1997).

28. For details on RISC microprocessors see http://developer.intel.com/design/strong/sapps.html.

29. For technical discussion on SDRAM see http://www.chips.ibm.com/products/memory/sdramart/sdramart.html.

30. J. Mitchell, "Minimizing Memory Cost in Set-top Boxes," *NCTA Technical Papers*, 57–64 (1996).

31. CableLabs, *OpenCable Set-top Terminal with DOCSIS Modem CORE Functional requirements for Bidirectional Cable* (1999).

32. ITU-T Recommendation J.83 Annex B, Digital Multi-program Systems for Television Sound and Data Services for Cable Distribution, April (1997).

33. M. Kolber, "Predict Phase-Noise Effects in Digital Communications Systems," *Microwaves and RF Magazine* **9**, 59–70 (1999).

34. D. Barker, "The Effects of Phase Noise on Higher-Order QAM Systems," *Communication Systems Design* **10**, 35–42 (1999).

35. T. Kolze, "Proposed HFC Channel Model," General Instrument IEEE802.14 contribution, IEEE 802.14-96/196, Enschede, Netherlands (1996).

36. M. Kolber and M. Ryba, "Measuring Multipath in the Wireless Cable Environment," *RF Design Magazine* **2**, 52–74 (1999).

37. FCC Code of Federal Regulations, Title 47, Part 76, Section 605, Subpart K, Technical Standards, Cable Television Service (1997).

CHAPTER 9

TRANSMISSION IMPAIRMENTS IN WDM MULTICHANNEL AM/QAM LIGHTWAVE SYSTEMS

In general, one can categorize the different subcarrier multiplexed (SCM) multichannel AM/QAM video lightwave transmission systems into four main groups as follows:

1. DM laser transmitter-based systems operating around 1310-nm
2. EM laser transmitter-based systems operating around 1319-nm
3. EM laser transmitter-based systems operating around 1550-nm
4. DM laser transmitter-based systems operating around 1550-nm

There are many technical and economic design considerations that determine which laser transmitter system one should select to address the access network needs. The purpose of this chapter is to explain the theoretical models for the different lightwave impairment mechanisms, and how the performance of such systems operating at both single and multiple wavelengths are degraded. The degradation mechanisms for EDFA-based WDM SCM multichannel lightwave systems are discussed in Chapter 10. The reasons for partition will become clear later when we discuss the evolution of the current HFC network architecture toward dense-wavelength-division-multiplexed (DWDM) architecture. The main lightwave degradation mechanisms are:

(A) Clipping-induced impulse noise
(B) Bursty nonlinear distortions
(C) Multiple optical reflections
(D) Dispersion-induced distortions
(E) Stimulated Brillouin scattering (SBS)
(F) Self-phase modulation (SPM)
(G) Stimulated Raman scattering (SRS)
(H) Cross-phase modulation (XPM)
(I) Polarization-dependent distortions
Each of these impairment mechanisms is discussed in detail in the following sections.

9.1   Clipping-Induced Nonlinear Distortions

Recently, there have been many studies on QAM channel degradation due to the loading of the AM-VSB video channels at the laser transmitter [1–5]. It has been shown experimentally by Maeda, et al., [1] that the bit error rate (BER) of a transmitted 16-QAM channel is significantly degraded at the optical receiver when transmitting multiple (e.g., >40) AM-VSB video channels. A BER floor is formed, caused by the "clipping" behavior of the laser transmitter. Figure 9.1 shows the clipped output modulated waveform from the ideal laser L-I curve. This phenomenon occurs when the instantaneous voltages of each of the loaded AM-VSB signals add up to temporarily drive the laser diode below its threshold, thus clipping the transmitted modulated signal [5, 6]. The clipping noise can be observed on electrical spectrum analyzer by setting the center frequency to the QAM channel, resolution and video bandwidth each to their maximum (3-MHz), and frequency span to zero.   In this

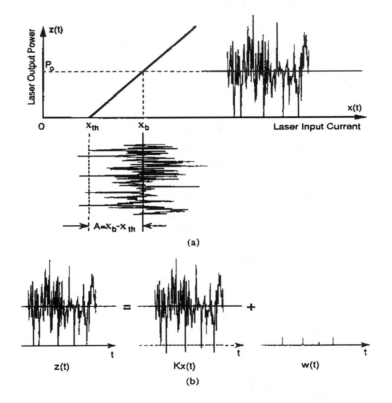

(a)

(b)

Figure 9.1 (a) Laser clipping noise at the biased input point (X_b) of the laser L-I curve; (b) laser output decomposed as a sum of the attenuated input signal and noise. After Ref. [6] (© 1995 IEEE).

way, the spectrum analyzer is used as a bandpass filter, and the observation of the clipping impulse events in a given time period is controlled by the sweep time of the spectrum analyzer [3].

9.1.1 Asymptotic Statistical Properties of Clipping Noise

In order to understand the model for the BER of M-ary QAM channels, which will be discussed in the Section 9.1.2, let us briefly review the asymptotic statistical properties of clipping-induced impulse noise. We note that the laser transmitter is essentially a memoryless device over the cable TV frequency range. Thus, the laser clipping problem shown in Fig. 9.1(a) is essentially equivalent to the A-level crossing problem discussed by Mazo [6]. Let us consider the signal x(t) to be a stationary zero-mean Gaussian process that models the composite input signal of multiple subcarrier signals. Using Bussgang's theorem, the laser output signal z(t) can be written as   (9.1)

$$z(t) = K \cdot x(t) + w(t)$$

where w(t) is the clipping noise and K is chosen so that x(t) and z(t) are uncorrelated. Thus, multiplying both sides of Equation (9.1) by x(t) and taking the expectation value, K is given by

$$K = -\frac{E[x \cdot w]}{E[x^2]} = -\frac{1}{\sigma A \sqrt{\pi}} \int_A^\infty x(x+w) \exp[-x^2 / 2\sigma^2] dx \qquad (9.2)$$

Based on these assumptions, we can write the following three asymptotic statistical properties of clipping noise.

1. The sequence of upcrossing epochs of a high-level A by input signal x(t) asymptotically becomes a Poisson process. Based on a paper by Rice [7], the expected rate of upward crossing of level $A > 0$ by a stationary zero-mean Gaussian process x(t) is given by

$$\lambda_A = \frac{1}{2\pi} \frac{\dot{\sigma}}{\sigma} \exp[-A^2 / 2\sigma^2] \qquad (9.3)$$

where $A = x_b - x_{th}$, as shown in Figure 9.1(a),

$$\sigma^2 = E[x^2(t)] \qquad (9.4)$$

is the power variance of Gaussian process x(t), and $d[\sigma^2]/dt$ is the variance of the first derivative of x(t). Furthermore, if the spectrum of x(t) is flat over a band of frequencies $[f_1, f_2]$, then according to Rice [7] we have

$$\left[\frac{d\sigma / dt}{\sigma}\right]^2 = (2\pi)^2 \frac{f_2^3 - f_1^3}{3(f_2 - f_1)} \qquad (9.5)$$

and

$$\lambda_A = \left[\frac{f_2^3 - f_1^3}{f_2 - f_1} \right]^{1/2} \exp[-A^2 / 2\sigma^2] \tag{9.6}$$

2. The probability density function (pdf) of the time intervals between an upcrossing and a downcrossing (e.g., clipping intervals) asymptotically becomes a Rayleigh distribution. Denoting the time interval length as τ, the pdf is given by [6]

$$p(\tau) = \frac{\pi}{2} \frac{\tau}{\bar{\tau}^2} \exp[-\frac{\pi}{4}(\frac{\tau}{\bar{\tau}})^2] \tag{9.7}$$

for $\tau \geq 0$, where $\bar{\tau}$ is the average clipping time interval. The clipping rate is related to average clipping interval by

$$\lambda_A \bar{\tau} = \mathrm{Pr}\,ob\{x(t) > A\} = \frac{1}{\sigma\sqrt{2\pi}} \int_A^\infty dx \exp[-x^2 / 2\sigma^2] = \frac{1}{2} erfc(\frac{A}{\sigma\sqrt{2}}) \tag{9.8}$$

where erfc($\bullet$) is the complementary error function.

3. The shape of the clipping process between any upcrossing and downcrossing, that is, the shape of w(t) over the clipping interval τ, asymptotically becomes a parabolic arc,

$$f(t) = \frac{1}{2} \ddot{x}(t)[t^2 - \frac{1}{4}\tau^2] \tag{9.9}$$

for $|t| < \tau/2$, and $f(t) = 0$ otherwise.
Based on these asymptotically statistical properties, the clipping-noise waveform can be modeled as a random Poisson pulse train

$$w(t) = \sum_{k=-\infty}^{k=\infty} p(t - t_k; \tau_k) \tag{9.10}$$

with $p(t)$ denoting the parabolic arc given by Equation (9.9), t_k denoting the Poisson clipping times, and τ_k denoting the Rayleigh distributed-clipping intervals.
    Another useful relationship, which we will use in the next section, is the area under $f(t)$ which is given by

$$A_\tau = \left[\frac{\dot{\sigma}}{\sigma} \right]^2 \frac{A\tau^3}{12} \tag{9.11}$$

where dσ/dt is given by Equation (9.5).

9.1.2 BER of M-ary QAM Channels Due to Clipping Noise

Now we are ready to discuss the BER performance of M-ary QAM channels in the presence of asymptotic clipping impulse noise and additive white Gaussian noise. The analytical model for M-ary QAM transmission system is shown in Figure 9.2.

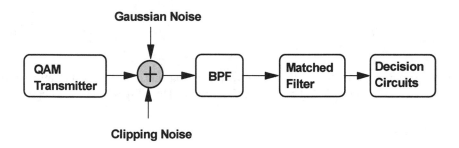

Figure 9.2 Analytical model of an M-ary QAM system in the presence of clipping-induced impulse noise and Gaussian noise (BPF = bandpass filter). After Ref. [6] (© 1995 IEEE).

According to work by Middleton [7], the first-order characteristic function of the clipping noise can be written as:

$$\Psi_I(\xi;t) = \exp(\lambda_A \int_0^\infty \int_0^\infty g(\tau, A_\tau) dA_\tau d\tau \int_{-\infty}^\infty d\alpha \{\exp[j\xi y(t-\alpha;\tau)] - 1\}) \qquad (9.12)$$

where $g(\tau, A_\tau)$ is the joint probability function of the random variables τ and A_τ, and $y(t; \tau)$ is the impulse noise output of the QAM receiver, which is given by the convolution of the following terms:

$$y(t;\tau) = p(t;\tau) \otimes h_{BPF}(t) \otimes h_M(t) \qquad (9.13)$$

where h_{BPF} and h_M are impulse responses of the bandpass filter (BPF) and matched filter, respectively. Using the Fourier transform on Equation (9.13), the QAM receiver output power is then given by

$$\left\langle y^2 \right\rangle = \int_{-\infty}^{\infty} S_p(f) \left| H_{BPF}(f) \right|^2 \left| H_M(f) \right|^2 df$$

(9.14)

where $S_p(f)$ is the power spectral density of the clipping noise, and H_{BPF} and H_M are the Fourier transform functions for the BPF and matched filter, respectively.

To simplify Equation (9.12), let us make the following assumptions:

(A) The average clipping duration interval τ is much smaller than the symbol duration T_s (= 1/B) of M-ary QAM signal in the frequency band of interest. The clipping pulse train can be approximated by

$$w(t) \cong \sum_{k=-\infty}^{k=\infty} A_\tau \delta(t - t_k)$$

(9.15)

(B) We assume that the BPF at the QAM receiver is an ideal BPF and the impulse noise power spectral density is flat across the filter band. Thus, Equation (9.13) becomes

$$y(t; \tau) = A_\tau h_M(t)$$

(9.16)

With these assumptions, Equation (9.12) simplifies to

(9.17)

$$\Psi_I(\xi) = \exp\left(\lambda_A \int_0^{\infty} g(A_\tau) dA_\tau \int_{-\infty}^{\infty} \{ \exp[jA_\tau \xi h_M(t - \tau)] - 1 \} d\alpha \right)$$

For M-ary QAM channel, the impulse response of the matched filter $h_M(t)$ is either $(2/T_s)^{1/2} \cos(\omega_c t)$ or $(2/T_s)^{1/2} \sin(\omega_c t)$ for $0 \leq t \leq T_s$ and 0 otherwise for either the I or the Q channel. Substituting $h_M(t)$ in Equation (9.17) and letting $t = T_s$, we obtain the following relation:

$$\Psi_I(\xi) = \exp\{ \gamma \int_0^{\infty} [J_0(\xi A_\tau \sqrt{2/T_s}) - 1] g(A_\tau) dA_\tau \} = e^{-\gamma} \exp\{ \gamma \langle J_0[\xi A_\tau \sqrt{2/T_s}] \rangle \}$$

(9.18)

where $\gamma = \lambda_A t_s$ is sometimes referred to as the clipping or impulsive index, that is, the clipping noise during one symbol interval, $J_0(\bullet)$ is the zero-order first-kind Bessel function, and $\langle \bullet \rangle$ is the expectation value of the function inside the brackets.

Expanding J_0 in power series of γ, and assuming a small probability for clipping ($\gamma \ll 1$), Equation (9.18) can be further simplified to

$$\Psi_I(\xi) \cong 1 - \gamma + \gamma \cdot [1 + \frac{\gamma C_4 \xi^4}{4^3} \cdot \exp(-\xi^2 \langle A_\tau^2 \rangle / 2T_s) - \frac{\gamma^2 C_6 \xi^6}{2 \cdot 4^3 (3!)^2} \exp(-\xi^2 \langle A_\tau^2 \rangle / 2T_s)] \cdot \exp(-\xi^2 \langle A_\tau^2 \rangle / 2T_s)$$

(9.19)

where terms with ξ^8 or higher order have been neglected, and C_4 and C_6 are given by the following equations:

$$C_4 = 72\langle A_\tau^2 \rangle^2 / T_s^2$$
$$C_6 = 7 \cdot 12^3 \langle A_\tau^2 \rangle^3 / T_s^3$$

(9.20)

The $1-\gamma$ represents the probability of no clipping during one symbol interval, while the last term represents the probability that clipping occurred in the same symbol interval.

Since the additive Gaussian noise and the clipping noise are statistically independent, the total characteristic function can be written as

$$\Psi_{I+G}(\xi) = \Psi_I(\xi)\Psi_G(\xi) = \left[1 - \gamma + \gamma \left\{ 1 + \frac{\gamma C_4 \xi^4}{4^3} e^{-\frac{\xi^2 \sigma_I^2}{2\gamma}} - \frac{\gamma^2 C_6 \xi^6}{2 \cdot 4^3 (3!)^2} e^{-\frac{\xi^2 \sigma_I^2}{2\gamma}} \right\} e^{-\frac{\xi^2 \sigma_I^2}{2\gamma}} \right] e^{-\frac{\xi^2 \sigma_G^2}{2}}$$

(9.21)

where the Gaussian noise variance is $\sigma^2{}_G$, and is given by

$$\sigma_G^2 = B \cdot \left[N_{th}^2 + 2qI_r + RIN \cdot I_r^2 \right]$$

(9.22)

where N_{th} is the optical receiver noise equivalent current, q is the electronic charge, I_r is the receiver photocurrent, and RIN is the laser transmitter relative-intensity noise.

The clipping noise variance is given by

$$\sigma_I^2 = \langle y^2 \rangle = \lambda_A \langle A_\tau^2 \rangle = \gamma \langle A_\tau^2 \rangle / T_s$$

(9.23)

The pdf of the total noise (clipping + Gaussian) at the QAM receiver output can be obtained by taking the Fourier transform of Equation (9.21):

$$P_{I+G}(n) = \frac{1-\gamma}{\sqrt{2\pi}\sigma_G} e^{-\frac{n^2}{2\sigma_G^2}} + \gamma \left[\frac{\phi^{(0)}(n/\sigma_{\gamma 1})}{\sigma_{\gamma 1}} + \frac{\gamma C_4 \phi^{(4)}(n/\sigma_{\gamma 2})}{4^3 \sigma_{\gamma 2}^5} + \frac{\gamma^2 C_6 \phi^{(6)}(n/\sigma_{\gamma 3})}{4^3 (3!)^2 \sigma_{\gamma 3}^7} \right]$$

where

(9.24)

$$\sigma_{\gamma i}^2 = \sigma_G^2 + i\sigma_I^2 / \gamma = \sigma_G^2 [1 + i/\gamma G]$$

(9.25)

i = 1,2,3, G is the Gaussian noise to clipping noise variance ratio, and

(9.26)

$$\phi^{(k)}(z) = (-1)^k H_k(z)\frac{e^{-z^2/2}}{\sqrt{2\pi}}$$

where $H_k(z)$ are the Hermite polynomials.

We now can use Equation (9.24) to obtain the probability of error of M-ary QAM channels in the presence of impulsive clipping noise. Assuming independent I and Q channels and Gray coding, the average bit-error probability (BER) at the output of a coherent receiver is given by

$$P_e = \frac{2}{\log_2 M}(1 - M^{-1/2})\int_{d_{min}/2}^{\infty} P_{I+G}(n)dn \tag{9.27}$$

where d_{min} is the minimum distance between adjacent states of M-ary QAM constellation and is given by

$$d_{min} = 2\sqrt{\frac{3T_s P_{av}}{(M-1)}} \tag{9.28}$$

where P_{av} is the average power per symbol. Using Equation (9.24) in Equation (9.27), carrying out the integration, and applying Equation (9.28), the BER is given by

$$P_e^{I+G} = \frac{2(1-\gamma)}{\log_2 M}\left[1 - M^{-1/2}\right]erfc\left(\sqrt{\frac{3SNR_G}{2(M-1)}}\right) + \frac{2\gamma(1-M^{-1/2})}{\log_2 M}\left[erfc(\frac{\Delta_1}{\sqrt{2}}) - \frac{9\gamma\phi^{(3)}(\Delta_2)}{4(2+\gamma G)^2} - \frac{21\gamma^2\phi^{(5)}(\Delta_3)}{4(3+\gamma G)^3}\right] \tag{9.29}$$

where

$$\Delta_i = \frac{d_{min}}{2\sigma_{\gamma i}} = \sqrt{\frac{3SNR_G}{(M-1)(1+i/\gamma G)}} \tag{9.30}$$

i = 1,2,3, and SNR_G is the ratio of average energy per information symbol to the Gaussian noise variance of M-ary QAM signal. The first term in Equation (9.29) represents the BER of the M-ary QAM channel due to Gaussian noise, while the second term represents the BER due to clipping noise. Notice from this equation that the BER behavior is mainly affected by three independent parameters: SNR_G, G, and γ. We will see some examples in the next section for these parameters.

The BER model of the transmitted M-ary QAM channels, which was discussed so far, assumes a single-level clipping for directly modulated laser transmitters. However, this model underestimates the clipping noise in externally modulated laser-transmitter-based transmission systems. The transfer function of the externally modulated laser transmitter can be expanded to a two-level clipping model, in which single-level clipping is a special case [9]. The generalized two-level clipping model for the BER is similar to the model discussed here. It turns out that the first-order approximation to the BER of the transmitted M-ary QAM channels in externally modulated-based systems is twice that of directly modulated-based systems under the same operating conditions. Furthermore, if one clipping level is more than 10% larger than the other clipping level, the two-level clipping model can be closely approximated by the single-level clipping model.

9.1.3 Comparison to Experimental Results

In recent years, many experiments were done to explain the effect of clipping-induced impulse noise on the BER performance of M-ary QAM channels. Maeda, et al., [1] was the first to measure the effect of BER degradation of the 16-QAM channel due to the clipping noise from 42 simulated analog cable TV channels using directly modulated DFB laser transmitter operating at 1310-nm. The AM modulation index per channel was 5% and 6% at

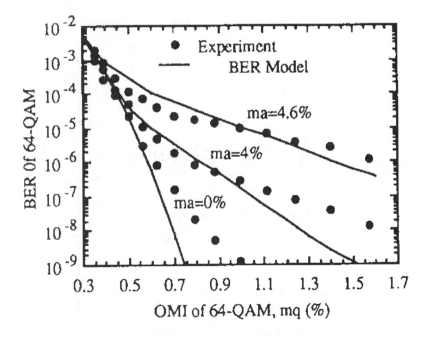

Figure 9.3 Measured and calculated 64-QAM BER as a function of the 64-QAM modulation index at AM modulation indices of 0%, 4%, and 4.6% for a directly modulated lightwave system. After Ref. [8] (© 1995 IEEE).

an optical received power of −1-dBm, and the 16-QAM channel was centered at 601.25-MHz. The calculated BER of the 16-QAM channel was based on Middleton's class A noise model, assuming that the impulsive index is equivalent to the probability of clipping. This assumption it turns out significantly underestimates the impulsive index, resulting in lower BER curves than the measured data. Thus, the validity of the calculated results was questioned. Other authors have obtained good agreement with Middleton's class A model [8–10]. Pham [10], for example, used a directly modulated 1550-nm DFB laser transmitter, which operated at a very large AM modulation index (5–6%) per channel, and thus probably contributed an additional amount of nonlinear distortion in the QAM channel band.

Using the Section 9.1.2 model of the BER of the QAM channel due to clipping-induced impulse noise, Q. Shi [8] calculated the BER of the 64-QAM channel based on the Lu, et al.[3] experimental results. Figure 9.3 shows the calculated and measured 64-QAM BER versus the 64-QAM optical modulation index (OMI) at AM modulation indices of 0%, 4%, and 4.6%. Sixty simulated AM channels (55.25 – 493.25-MHz), according to the US NTSC frequency plan, were transmitted with a single 64-QAM channel centered at 601.25-MHz using a directly modulated DFB laser transmitter. The AM CNR was 52-dB with CSO and CTB distortions less than −60-dBc. Figure 9.3 indicates that a good agreement between the

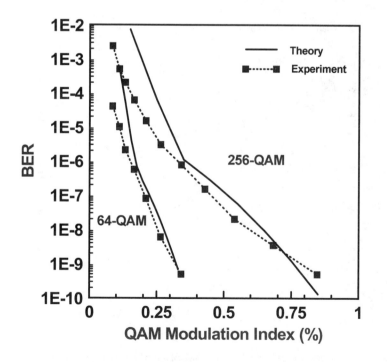

Figure 9.4 Measured and calculated uncoded 64-QAM and 256-QAM BER versus the QAM modulation index for an externally modulated 1550-nm lightwave system with 79 AM channels. After Ref. [4] (© 1995 IEEE).

Table 9.1 The parameters used to calculate the 64/256-QAM BER result in Figure 9.4

Parameter Name	Value
AM CNR	53.66 dB
AM Modulation Index	3 %
QAM Symbol Rate	5 Mbaud
Clipping Index (γ)	$5.3 \cdot 10^{-5}$
G (clipping-to-Gaussian-noise variance ratio)	32.6 dB
Laser Relative-Intensity Noise (RIN_L)	-164 dBc/Hz
Optical Receiver Responsivity	0.85 A/W
Receiver Noise Equivalent Current	$7\text{-pA/(Hz)}^{0.5}$
Received Optical Power	0 dBm
QAM Channel Bandwidth	6 MHz

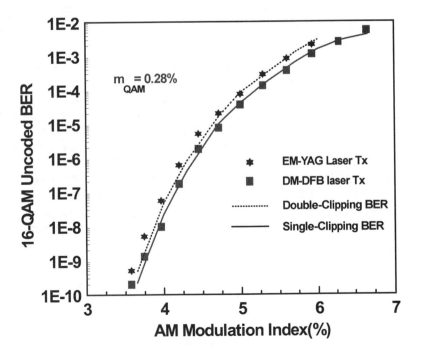

Figure 9.5 Measured (symbols) and calculated (lines) 16-QAM BER versus the AM modulation index for a hybrid 61AM/16-QAM transmission system using a 1310-nm directly modulated DFB laser transmitter and a 1319-nm externally modulated YAG laser transmitter. After Ref. [9] (© 1998 IEEE).

experimental results and the BER model presented in Section 9.1.2 was obtained. Furthermore, the BER of the QAM channel is significantly degraded as the AM modulation index is increased.

An additional comparison between the measured and calculated uncoded 64-QAM and 256-QAM BER versus the QAM modulation index for an externally modulated 1550-nm lightwave transmission system is shown in Figure 9.4. In this experiment, 79 simulated AM channels (55.25–547.25-MHz) along with four QAM channels, which were centered above the AM channels band, were transmitted over 50-km of SMF. The AM CNR was 53.6-dB with CSO and CTB distortions less than −65-dBc. The experimental results are also in good agreement with the calculated 64/256-QAM BER. Table 9.1 lists the parameters used to calculate the uncoded 64/256-QAM BER result in Figure 9.4. The average 64-QAM signal power level relative to the AM signal can be also found from Figure 9.4. Consider, for example, BER = 10^{-8} and AM modulation index (m_A) of 4%. Then, the calculated 64-QAM modulation index (m_{QAM}) is 1.25%. This corresponds to 20·log (1.25/4) $\cong$ −10 dBc, whereas the experimental result is − 8 dBc (m_{QAM} = 1.53%).

To further understand the difference between the single-level clipping and the double-level clipping BER models, consider the next result. Shi and Ovadia [9] have compared the calculated BER of 16-QAM channel operating at QAM modulation index of 0.28% in the presence of 61 simulated AM channels using the single-level clipping and the double-level clipping models based on the Ovadia, et al., [2] experimental results. The 16-QAM channel was operating at 5-Mbaud and was centered at 539.25-MHz. Figure 9.5 shows the measured and calculated uncoded 16-QAM BER versus the AM modulation index for a 1310-nm directly modulated DFB laser-transmitter-based system and a 1319-nm externally modulated YAG laser-transmitter-based system. The measured BER results are in excellent agreement with the calculated results. As expected, the 16-QAM BER increases dramatically with the AM modulation index. This result also confirms the assumption that the 16-QAM BER in an externally modulated transmission system is approximately twice that of a directly modulated transmission system operating at the same AM and QAM modulation indices.

It should be pointed out that symmetric double-clipping events have been taken into account. It has been shown [9] that the BER of QAM signals in the presence of double-clipping events, to the first-order approximation, is twice the BER from single-level clipping events under the same conditions.

9.1.4 BER of QAM Channel Due to "Dynamic Clipping" Noise

So far, we have discussed the BER of a QAM channel, which was "statically" clipping by the nonlinear transfer function of a laser diode or an optical modulator. In Section 9.1.2, we assumed that the laser diode can be modeled as an ideal memoryless device; that is, the laser output voltage responds instantaneously to changes in its input current. Recent measurements have shown that this is not the case. Figure 9.6 shows, for example, the normalized time dependence of the laser output (solid line) signal after being pulse modulated [11]. The dashed line shows the input pulse. Notice the significant turn-on delay, which is the difference between the output and input signals and typically varies from 0.3- to 0.5-ns, in addition to the frequency oscillations. This experimental result supports the explanation that dynamic clipping generates frequency-dependent nonlinear distortions that are different from those generated by static clipping. The observed frequency oscillations correspond to the laser resonance frequency. Anderson and Crosby [11] have calculated the uncoded 64-QAM BER as a function of the QAM channel center frequency and compared it with their experimental results. The clipping distortion is modeled as the difference between the clipped output and input signals. This difference can be approximated by a Poisson sequence of random parabolic pulses with a Rayleigh distribution for the pulse width τ. The laser diode is assumed to have a turn-on time delay $\Delta\tau$ independent of τ, after which the laser diode turns on immediately. Neglecting the frequency oscillations, the clipping impulse is approximated as a truncated parabola. Figure 9.7 shows the measured and calculated (solid lines) uncoded 64-QAM BER as a function of the QAM channel center fre-

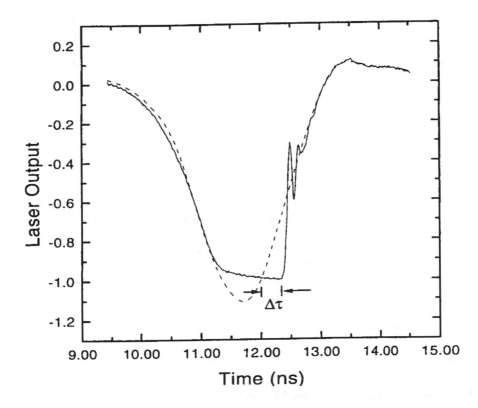

Figure 9.6 Normalized time dependence of the laser output (solid line) after being pulse modulated, showing the turn-on delay. The dashed line shows the input pulse. After Ref. [14] (© 1996 IEEE).

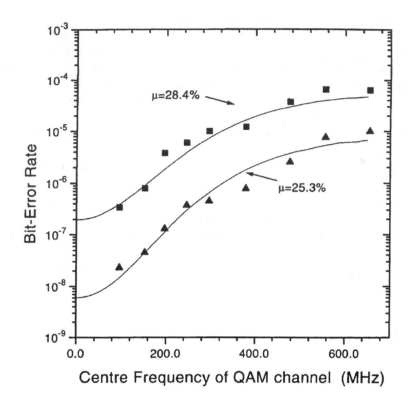

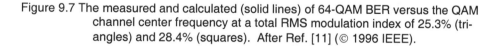

Figure 9.7 The measured and calculated (solid lines) of 64-QAM BER versus the QAM
channel center frequency at a total RMS modulation index of 25.3% (triangles) and 28.4% (squares). After Ref. [11] (© 1996 IEEE).

quency for total RMS modulation indices of 28.4% (squares) and 25.3% (triangles), assuming 0.4-ns turn-on delay. Sixty-three simulated AM channels (55.25–555.25 MHz) with a
single 64-QAM channel were used to directly modulate a DFB laser transmitter. The 64-
QAM channel to the AM channel power ratio was −10 dBc. The 64-QAM channel was operating at 6.25-Mbaud. The BER was calculated based on the stated assumptions, and it is
in good agreement with the experimental results. Notice that the observed uncoded BER at
600-MHz is degraded by approximately three orders of magnitude compared with the BER
at 50-MHz.

9.1.5 Clipping Noise Reduction Methods

There are various methods, which have been proposed and demonstrated recently, to improve the transmission performance of SCM AM/QAM systems. Several recent works have

introduced the concept of a preclipping circuit to compensate for the statically clipped laser diode [12–15]. In this method [12], the multiplexed AM-VSB signal is preclipped by a limiter in order to avoid below-threshold operation of the laser diode. Then, a bandpass filter is applied to the preclipped signals, since they contain various distortions over a wide frequency band. The filtered and preclipped signals are multiplexed with the digital signals before they are used to drive the laser diode. Since the QAM modulation index is much lower than that of the AM modulation index, the multiplexed preclipped signals do not momentarily drive the laser diode below the threshold. Thus, the QAM channels are transmitted without clipping distortion-induced degradation. This method provides an AM CNR improvement of 1.3-dB.

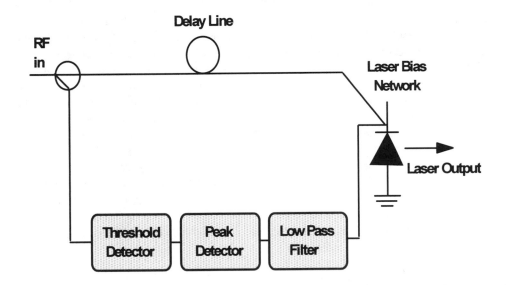

Figure 9.8 Block diagram of the clipping-reduction circuit. After Ref. [12]
(© 1995 IEEE).

Another preclipping method [13] uses a bandstop filter instead of a low-pass filter. The bandwidth of the stopband filter spans all the transmitted QAM channels, such that the filter will cut off the preclipping noise in this band. Thus, the preclipped waveform is altered much less than for the low-pass filter case since the bandwidth of the stopband filter is typically much larger. For a stopband of 200-MHz with the preclipping method, the authors calculated that the 64-QAM BER is reduced by more than three orders of magnitude at a QAM modulation index of 0.5% for a directly modulated transmission system with 65 AM channels and one 64-QAM channel centered at 600-MHz. Furthermore, this approach also applies to other frequency plans, where for example, the QAM channels are located between the AM channels. In this case, the low-pass filter in the preclipping method cannot be used.

Another intrinsically different approach from the preclipping methods to reduce the clipping distortions before the laser diode is shown in Figure 9.8 [14]. In this approach, part of the RF input signal is split and delayed. The clipping-reduction circuit allows the clipping events to be detected, amplified, shaped, and then fed forward to the DFB laser. The laser bias current is then increased momentarily by this current pulse in coincidence with the clipping event. Thus, this scheme permits an increase in the total RMS modulation index in the SCM lightwave system without increasing the magnitude of the clipping distortions. Using this approach, the authors were able to increase the AM CNR by more than 1.5-dB without further degradation of the CSO and CTB distortions for their 80 AM channel transmission system operating at an AM modulation index of 4% per channel. The clipping reduction circuit reduces the required the 64-QAM SNR at BER of 10^{-8} by about 14-dB.

Another approach conceptually similar to the one shown in Figure 9.8 to reduce the clipping distortions is called a dissymmetrization scheme [15]. The difference between the dissymmetrization and the clipping reduction circuits is that the peak detector in Figure 9.8 is replaced by a short-time hold circuit, and an inverting amplifier is added after the low-pass filter before the split signal is recombined with the delayed portion of the RF input signal to drive the DFB laser. The dissymmetrization scheme allows the clipping waveform of the multiplexed AM signal below a given threshold to be detected, low-pass filtered, amplified, and inverted. This signal, which is called the dissymmetrization signal, is then added to the delayed SCM signal to obtain the dissymmetrized SCM signal to drive the laser. The purpose of the short-time hold circuit is to reduce the pulse width variations before the low-pass filtering such that the amplitude of the dissymmetrized signal does not depend on the width of the clipped waveform. The author was able to improve the RMS modulation index by up to 5-dB for an SCM lightwave system operating in the 200–600-MHz band using a 1550-nm directly modulated DFB laser transmitter.

9.2   Bursty Nonlinear Distortions

As we learned in Section 2.5, when AM-VSB video channels are transmitted through a nonlinear device such as a laser transmitter, second-order distortions, namely, $f_1 \pm f_2$ type where f_1 and f_2 are two AM channel frequencies, and the third-order distortions, namely, $f_1 \pm f_2 \pm f_3$, are generated. Due to the temporal properties of modulated AM-VSB signals, the CSO/CTB distortions are typically time-varying distortions, which generate burst errors in a hybrid SCM AM/QAM lightwave system. As expected, the burst errors degrade the BER performance of the transmitted QAM channels in the lightwave link [16].

To understand the effect of the bursty behavior of the nonlinear distortions, the following experiment was constructed. Seventy-nine modulated AM-VSB channels were combined with two 256-QAM channels, operating at 40.5-Mb/s at RF frequencies of 571.25-MHz and 643.5-MHz to directly modulate a 1310-nm DFB laser transmitter. A broadband AWGN source was used in the QAM link for BER-SNR measurements. After transmission through 10.6-km of an SMF, the combined signal was detected at an optical power of 0-dBm at the cable-TV receiver. The AM CNR was 51.8-dB with an average CSO distortion of –56.8-dBc, and a CTB distortion of –60-dBc in the QAM channel band centered at 571.25-MHz. It turns out that the CSO distortion located at 572.5-MHz is the dominant nonlinear distortion in the QAM band centered at 571.25-MHz. The DFB laser transmitter was operating at an AM modulation index of 3.5% per channel with a clipping index $\gamma \approx 4{\cdot}10^{-4}$. As we will see later, CTB distortion is typically the dominant distortion in the QAM band for EM laser transmitter-based lightwave systems. Figure 9.9 shows typical 100-μs time traces of the average (trace A) and peak (trace B) CSO distortion at 572.5-MHz on a spectrum analyzer (SA) in a zero-span mode [16]. In this mode, the SA acts as a sampling oscilloscope. The SA trigger was set to be 30-dB below the 256-QAM carrier and to start the sweep for any event of the peak CSO distortion, which is above the solid line. The 30-

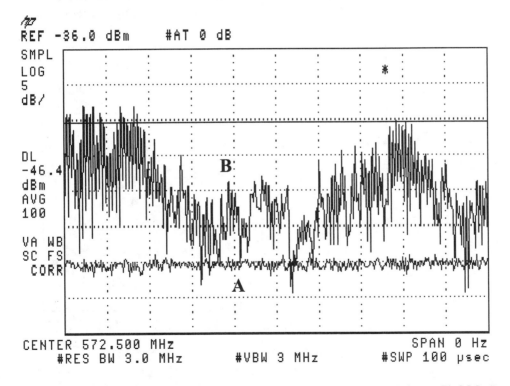

Figure 9.9 Typical 100-μs traces of the average (trace A) and peak (trace B) CSO distortions at 572.5-MHz on a spectrum analyzer in zero span mode. The solid horizontal line is the single sweep trigger for the peak CSO distortion. After Ref. [16] (© 1998 IEEE).

dB 256-QAM CNR corresponds to a coded BER = 1.5 x 10^{-9} in the presence of AWGN. Thus, the trigger line represents the threshold level for impulses with higher amplitudes that would degrade the coded BER above 10^{-9}.

The observed bursty behavior of the peak CSO distortion, which has non-Gaussian statistics, can be explained as follows. Unlike CW carriers from a multitone generator, the peak envelope power of the modulated AM-VSB video signals can vary by as much as 18-dB, depending on the picture content [17]. Furthermore, the synchronization pulses of the AM-VSB channels may temporarily align with each other, causing the corresponding video carriers to be at their maximum power at the same time, resulting in increased CSO/CTB distortions. Thus, the use of CW carriers to simulate modulated video carriers does not accurately represent the time dependence of the peak CSO/CTB distortions.

The use of CW carriers instead of modulated video carriers also affects the laser clipping distortion. In particular, the automatic-gain control (AGC) circuitry in the DM-DFB laser transmitter will maintain the average RF power at a fixed level. Under these conditions, the AGC circuitry reduces the AM modulation indices of the CW carriers by a factor of $\sqrt{\delta}$ ($\delta \approx$

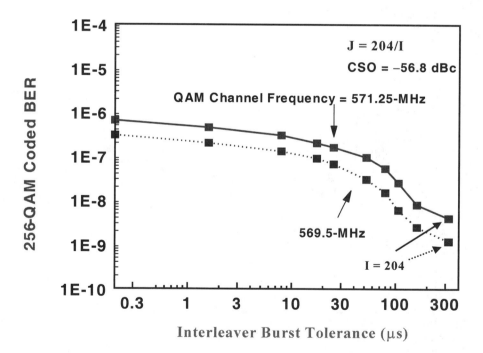

Figure 9.10 Measured 256-QAM coded BER versus the interleaver burst length tolerance with and without QAM channel frequency offset at QAM SNR of 30-dB and CSO distortion of −56.8-dBc. After Ref. [16] (© 1998 IEEE).

0.46), and the corresponding CNR is lowered by a factor of δ (about 3.4-dB) compared with modulated video carriers, resulting in a reduced nonlinear distortion [18].

In order to combat the generated burst errors, a variable convolutional interleaver in the QAM modem, with a depth of up to I = 204 symbols and with J = 204/I symbols, was used with the R-S code [19]. Interleaving the R-S symbols before transmission and deinterleaving them after reception evenly disperses the burst errors in time, and thus the QAM receiver can handle them as if they were random errors. It can be shown that the maximum burst noise tolerance, which is provided by the interleaver and deinterleaver combination, is given by [16]

$$\tau = \frac{(J \cdot I + 1) \cdot T}{R_{RS}} \left[\frac{I}{N} \right] \tag{9.31}$$

where I and J are the interleaver parameters, N = 204 symbols is the R-S block size, R_{RS} is the transmitted R-S symbol rate, and T = 8 symbols. Figure 9.10 shows the measured

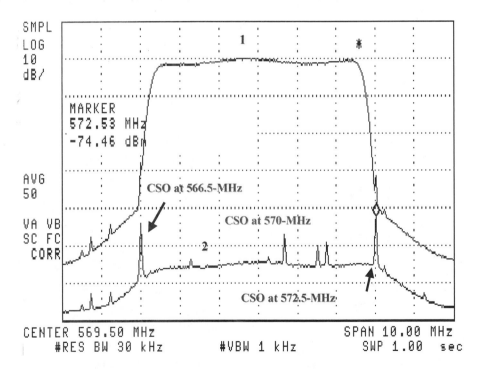

Figure 9.11 The RF spectrum with (trace 1) and without (trace 2) the transmitted 256-QAM channel centered at 569.5-MHz. After Ref. [16] (© 1998 IEEE).

256-QAM coded BER versus the interleaver burst tolerance $\tau(\mu s)$ for QAM channel center frequencies of 571.25-MHz and 569.5-MHz. The 256-QAM SNR was set to 30-dB, corresponding to a coded BER= 1.5×10^{-9} with no AM channel loading. Since the interleaver with I x J = 204 symbols was used, increasing I according to Equation (9.31) results in increasing the burst tolerance of the QAM receiver as shown in Figure 9.10. For the I = 204, J = 1 interleaver mode, the nearly four times reduction in the 256-QAM BER is obtained by the QAM channel frequency offset alone. For a burst duration of 30-μs or less, the 256-QAM coded BER slowly decreases as the interleaver depth is increased, since the average observed burst length is still larger than the maximum burst tolerance of the interleaver/de-interleaver combination. However, as the interleaver burst tolerance becomes significantly larger ($\geq$ 3X) than a burst duration of about 30-μs, most of the generated burst errors are effectively corrected by the interleaver, resulting in a steeper reduction in the coded BER. In fact, the 256-QAM coded BER is reduced by more than two orders of magnitude compared with the no-interleaver case.

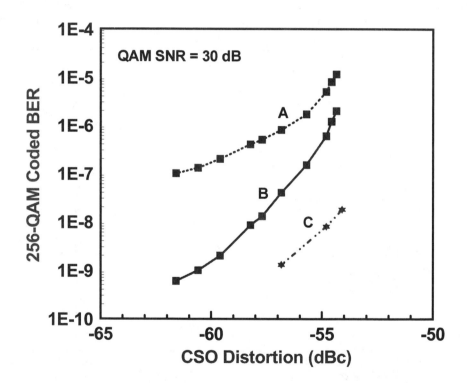

Figure 9.12 The 256-QAM coded BER versus the CSO distortion at 572.5-MHz for (A) no interleaver, (B) I = 68, J = 3 interleaver, and (C) downshifted QAM channel centered at 569.5-MHz and I = 204, J =1 interleaver. After Ref. [16] (© 1998 IEEE).

Examination of the RF spectrum of the transmitted QAM channel centered at 569.5-MHz with the CSO and CTB distortions provides further insight. Shifting the QAM channel frequency down to 569.5-MHz, the dominant CSO distortions at 572.5-MHz and at 566.5-MHz are now located outside the downshifted QAM channel band as shown in Figure 9.11 (trace 2). The other nonlinear distortions in the downshifted QAM band, namely, the CSO distortion at 570-MHz and the CTB distortion at 571.25-MHz, have smaller magnitudes relative to the QAM carrier. If the CTB distortion becomes the dominant distortion in the QAM band, then, a 3-MHz downshift of the QAM channel frequency to 568.25-MHz is required to improve the coded BER. Thus, the 256-QAM coded BER is reduced by more than 500 times to nearly error free using the $I = 204$, $J = 1$ interleaver with QAM channel frequency offset compared with the no-interleaver and QAM channel frequency offset case. The fixed QAM frequency offset method can easily be applied for multiple contiguous QAM channels, where the dominant CSO and CTB distortions will fall in the small gaps between the QAM channels.

Does the interleaver in the QAM receiver work as well at other CSO distortion levels?

To answer this question, the AM modulation index per channel at the laser transmitter was changed while keeping the power ratio between the QAM and the AM-VSB channels constant. Figure 9.12 shows the 256-QAM coded BER as a function of the relative magnitude of the CSO distortion located at a frequency of 572.5-MHz for the following three cases: (A) no interleaver, (B) the $I = 68$, $J = 3$ interleaver, and (C) a downshifted QAM channel to 569.5-MHz with a 30-dB SNR and $I = 204$, $J = 1$ interleaver. Notice that the $I = 68$, $J = 3$ interleaver works best at CSO distortion levels less than –60-dBc. At higher CSO distortion levels, the interleaver is being overwhelmed with burst errors, resulting in BER performance that approaches the no-interleaver case. Consequently, using the $I = 204$, $J = 1$ interleaver with a QAM channel frequency offset can provide robust transmission even in the presence of large CSO distortions (> –55-dBc).

9.3 Multiple Optical Reflections

It is known that multiple optical reflections in optical fiber link degrade the performance of SCM multichannel AM-VSB/QAM video transmission systems due to two or more discontinuities in the index of refraction and a Rayleigh backscattering effect in long fibers [20–25]. The laser transmitter's phase noise fluctuations are converted to intensity noise by interferometric phase-to-intensity noise conversion from multiple reflection points along the fiber link. The effect of multiple discrete optical reflections on the transmission performance of lightwave systems can be minimized using angled polished connectors (APC), fusion splices, and high-return loss fiber-optic components.

Rayleigh scattering is due to microscopic inhomogeneities in the fiber index of refraction along the optical transmission path, which presents a fundamental limitation that cannot be ignored even in a relatively short optical fiber length ($\approx$ 10-km).

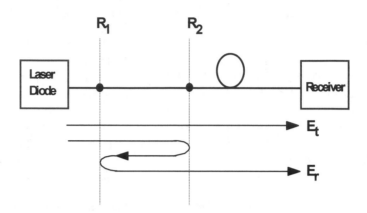

Figure 9.13 Interference intensity noise generated from the interference between the
double-Rayleigh backscattered light and the directly propagating light.

Double-Rayleigh backscattering (DRB) is due to the interference between the main propagating optical beam and the doubly reflected optical beam between two adjacent microreflectors (R_1 and R_2) along the fiber as shown in Figure 9.13 [20]. The system impact of
the interferometric phase-to-intensity conversion due to DRB and laser RIN will be
analyzed first [21]. Then, the effect on the transmitted AM and QAM channels in an SCM
multichannel AM/QAM lightwave system will be discussed.

9.3.1 Interferometric Noise Due to DRB

The interference noise between two or more discrete optical reflections is sometimes called
interference intensity noise (IIN). For two discrete reflectors, the IIN is proportional to the
equivalent reflectance,

$$R_{eq}^2 = P_{12}\eta_{12}R_1R_2 \tag{9.32}$$

where P_{12} is the polarization coupling between reflectors 1 and 2 with reflectance R_1 and R_2,
respectively; η_{12} is the fiber loss.

To model the effect of DRB, let us assume that a single fiber with length L is made up of
N ($N \to \infty$) equal discrete microreflectors ($L = N \cdot \Delta x$), which reflect light incoherently [20].

Each microreflector has a loss of $\eta = \exp(-\alpha\Delta x)$ where Δx is the differential fiber length, and a reflectivity $R = S\alpha_s\Delta x$, where S is the fraction of the scattered light that is captured by the fiber, and α_s the proportion of the signal scattered per unit length. The N microreflectors form a total of $N(N-1)/2$ Fabry-Perot etalons with a total equivalent reflectance given by

$$R_{eq}^2(N) = \sum_{i=1}^{N-1}\sum_{j=i+1}^{N} P_{ij}R_iR_j\eta_{ij} = PR^2 \cdot \frac{\eta}{(1-\eta)^2}\left[N(1-\eta)-(1-\eta^N)\right] \qquad (9.33)$$

where P_{ij} is the polarization coupling between the microreflectors i and j. To get the right side result of Equation (9.33), it was assumed that all of the microreflectors have the same reflectivity R, polarization coupling P, and equal spacing between adjacent microreflectors. To show the interaction between the fiber and discrete reflections, Equation (9.33) can be written in the following recursive form:

$$R_{eq}^2(N+1) = R_{eq}^2(N) + R_{N+1}R_{in}(N) \qquad (9.34)$$

where the last term represents the increased noise due to the interaction between the added reflector and the original group, whose total reflectance is given by

$$R_{in}^2(N) = PR\left[\frac{1-\eta^N}{1-\eta}\right] \qquad (9.35)$$

where, as before, equal spacing between adjacent microreflectors was assumed, and the same reflectivity R and polarization coupling.

Equations (9.34) and (9.35) can be further reduced if one assumes $P = 0.5$ since Rayleigh scattering is randomly polarized. The insertion loss of the micro-reflector is given by:

$$R_{in}^2(L) = \frac{S\alpha_s}{4\alpha}\left(1-e^{-\alpha L}\right) \qquad (9.36)$$

$$R_{eq}^2(L) = \frac{1}{2}\left[\frac{S\alpha_s}{\alpha}\right]^2\left(\alpha L - 1 + e^{-\alpha L}\right) \qquad (9.37)$$

where η was approximated as,

$$\eta \approx 1 - \alpha \cdot \Delta x \qquad (9.38)$$

For a typical 1310-nm single-mode fiber, $S = 0.0015$ and $\alpha = \alpha_s = 0.081$ nepers/km.

9.3.2 Laser RIN Due to Multiple Discrete Optical Reflections

Consider the case of a laser diode operating under CW conditions, where the emitted optical field can be written as

$$E_i(t) = A \exp(i\omega_0 t + \phi(t)) \tag{9.39}$$

where ω_0 is the laser center frequency and $\phi(t)$ is the laser random phase noise. The light intensity according to Figure 9.13, in which the directly propagating field $(\alpha_1\alpha_2)^{1/2}E_i(t)$ and the doubly reflected light with the same polarization $\alpha_1(\alpha_1\alpha_2 R_1 R_2)^{1/2}E_i(t-\tau)$ interfere at the optical receiver, is given by

$$I(t) = \left|\sqrt{\alpha}E(t) + R\sqrt{\alpha}E(t-\tau)\right|^2 \cong \alpha|A|^2[1+\rho(t,\tau)] \tag{9.40}$$

where R is the effective reflection coefficient given by $R = \alpha_1(R_1 R_2)^{1/2}$, R_1 and R_2 are the intensity reflection coefficients of the two fiber-optic connectors, $\alpha = \alpha_1\alpha_2\alpha_1$, α_1 and α_2 are optical attenuation coefficients between the connectors and between the second connector and the receiver, respectively, τ is the time delay between the direct and doubly reflected light, and $\rho(t,\tau)$ is given by

$$\rho(t,\tau) = 2R \cdot \cos[\omega_0\tau + \phi(t) - \phi(t-\tau)] \tag{9.41}$$

Thus, ρ contains the interference component of the light intensity $I(t)$. The autocorrelation function of the power spectral density of $\rho(t,\tau)$ is given by

$$S_\rho(t,\tau) = \langle \rho(t',\tau)\rho(t'+t,\tau)\rangle = 4R^2\langle\cos[\omega_0\tau + \theta(t',\tau)]\cdot\cos[\omega_0\tau + \theta(t+t',\tau)]\rangle \tag{9.42}$$

The random variable $\theta(t,\tau) = \phi(t) - \phi(t-\tau)$ is normally distributed with $\sigma^2 = 2\pi\tau\cdot\Delta v$, where Δv is the 3-dB Lorentzian linewidth of the laser diode.

It can be shown that the autocorrelation function is given by

$$S_\rho(t,\tau) = 2R^2 \begin{cases} e^{-2\pi\tau\Delta v}\cdot[1+\cos(2\omega_0\tau)], & |t| > \tau \\ e^{-2\pi\Delta v\cdot|t|}\cdot[1+\cos(2\omega_0\tau)e^{-4\pi\Delta v\cdot(\tau-|t|)}], |t| < \tau \end{cases} \tag{9.43}$$

The single-sided noise power spectral density is obtained by taking the Fourier transform of Equation (9.43) and subtracting the DC term,

$$RIN(f) = 4R^2 \left[1 + \left[\frac{-\sin(\gamma f)}{\pi f} \right] e^{-\gamma \Delta v} \cos(2\omega_0 \tau) \right] +$$

$$\frac{4R^2}{\pi} \left[\frac{\Delta v}{f^2 + (\Delta v)^2} \right] \cdot \left\{ 1 - e^{-\gamma \Delta v} \left[\cos(\gamma f) - \frac{f}{\Delta v}\sin(\gamma f) \right] - \cos(2\omega_0 \tau) \left[e^{-2\gamma \Delta v} - e^{-\gamma \Delta v} [\cos(\gamma f) + f\sin(\gamma f)/\Delta v] \right] \right\}$$

(9.44)

where $\gamma = 2\pi\tau$. From Equation (9.44) we can see that the maximum conversion of phase noise to intensity noise occurs when the main propagating light beam interferes in quadrature with the doubly reflected light beam, that is, $\omega_0 \tau = (n \pm 1/2)\pi$. This result is given by [22]

$$RIN(f) = \frac{4R_1 R_2}{\pi} \left[\frac{\Delta v}{f^2 + \Delta v^2} \right] \cdot \left\{ 1 + e^{-4\pi\tau\Delta v} - 2e^{-2\pi\tau\Delta v} \cos(2\pi f\tau) \right\}$$

(9.45)

where R_1 and R_2 are the effective reflectivities of two fiber connectors, Δv is the 3-dB Lorentzian linewidth of the laser transmitter, f is the frequency, and τ is the time delay between the direct and the doubly reflected light. When $2\pi\Delta v\tau \ll 1$, Equation (9.45) becomes $16\pi R^2 \Delta v\tau \cdot 2\text{sinc}^2(f\tau)$, which is the limit of small phase fluctuations, and the maximum RIN is proportional to the laser linewidth. In the incoherent interference limit, where $2\pi\Delta v\tau \gg 1$, Equation (9.45) becomes $(4R^2/\pi)[\Delta v/(\Delta v^2 + f^2)]$. When the laser linewidth is very narrow such as for YAG lasers, the effect of this interferometric RIN becomes negligible for both of these limits. In contrast, the modulated linewidth of the DM-DFB laser transmitter is broad ($\Delta v > 1$GHz), and the interferometric RIN is significantly increased in the presence of high multiple optical reflections ($R > -40$-dB) due to bad optical connectors. Sections 9.3.3 and 9.3.4 discuss the effect of multiple optical reflections on both the AM-VSB and QAM channels.

9.3.3 Multiple Reflections Effect on AM-VSB Channels

In a typical transmission link, the sources of optical reflections are fiber-optic splices, connectors, and couplers. Figure 9.14 shows a typical multichannel SCM hybrid AM/QAM lightwave transmission system with a frequency-allocation plan [23]. Sixty-one simulated AM channels are combined with a single 16-QAM signal centered at 539.25-MHz to directly modulate either a 1310-nm DFB laser transmitter or a 1319-nm EM-YAG laser transmitter. An additive white Gaussian noise (WGN) source is used for the measurement of BER versus SNR for the QAM channel. Multiple discrete optical reflections are generated by placing two variable back reflectors (VBRs) with a double-loop polarization controller in a 4.4-km of lightwave link as shown in Figure 9.14. The polarization controller is typically used to maximize the phase-to-intensity conversion [21]. The optical isolation between the VBRs and each of the laser transmitters is better than 50 dB. In this measurement, FC/APC SMF connectors (return loss < −60 dB) were used to minimize an additional interferometric noise.

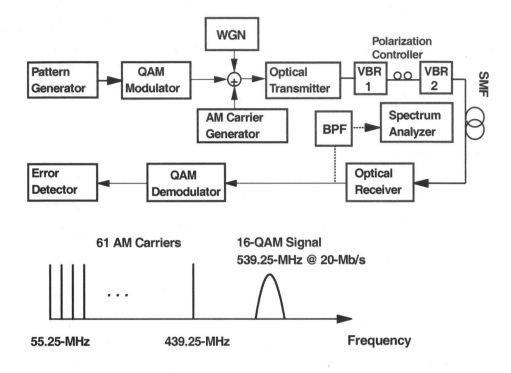

Figure 9.14 Typical experimental setup and frequency-allocation plan. After Ref. [23] (©1994 IEEE).

To provide a fair comparison between the 1310-nm-based laser transmitters, a two-part experiment was conducted [23]. First, the AM modulation index of the DFB laser transmitter was reduced from 4.2% per channel (manufacturer specified optimum value) to 3.3% in order to minimize the clipping-induced impulse noise effect on the BER of the 16-QAM channel. Figure 9.15 shows the AM CNR as a function of the reflection level at channel frequencies of 55.25-MHz, 205-MHz, and 439.25-MHz in lightwave systems using a 1310-nm DM-DFB laser transmitter and a 1319-nm EM-YAG laser transmitter operating at an AM modulation index of 3.3% per channel. Notice that the 53-dB AM CNR drops rapidly for reflection levels greater than −50-dB. At a reflection level of −14-dB, the resultant CNR is only about 20-dB as shown in Figure 9.15. In contrast, the 55-dB AM CNR is reflection independent for the same AM modulation index using the EM-YAG laser transmitter.

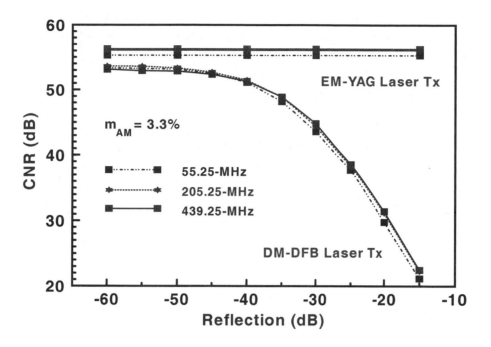

Figure 9.15 CNR versus the reflection level for a directly modulated 1310-nm DFB
laser transmitter and a 1319-nm externally modulated YAG laser transmitter operating at an AM modulation index of 3.3% per channel. After
Ref. [23] (© 1994 IEEE).

The AM CNR degradation due to optical reflections for the DM-DFB laser transmitter-
based systems can be explained as follows: When the AM modulation index per channel is
reduced to 3.3%, the interferometric intensity noise (IIN) becomes the dominant noise
source. The effective backscatter coefficient of the 4.4-km long SMF is −35.9-dB, compared with the theoretical value of −30 dB for an infinite-length fiber. The IIN is generated
from two sources, namely, multiple discrete reflections between the two VBRs and from the
fiber's Rayleigh backscatter interacting with those discrete reflections. Thus, an infinite
number of "optical cavities" are formed that degrade the lightwave system's performance in
the same manner as do cavities formed by discrete reflections. The CNR degradation due to
multiple optical reflections in a 1550-nm EM-DFB laser transmitter-based system is different than both previous cases. Figure 9.16 shows, for example, the AM CNR as a function of
the reflection level for a 79 AM channel SCM lightwave system. All of the channels were
transmitted over a 10.8-km of SMF using a 1550-nm EM-DFB laser transmitter operating at
an AM modulation index of 3.2% per channel [24]. This result demonstrates two important
new features. First, the CNR of the AM channels exhibits a strong frequency-dependent
degradation. Channel 2 (= 55.25 MHz) degrades the most, while channel 79 (= 547.25
MHz) degrades the least. Second, the magnitude of the AM CNR degradation at a given

AM frequency is larger than that of a 1319-nm externally modulated YAG laser transmitter-based SCM system, but less than that of a 1310-nm directly modulated DFB laser-transmitter-based SCM system.

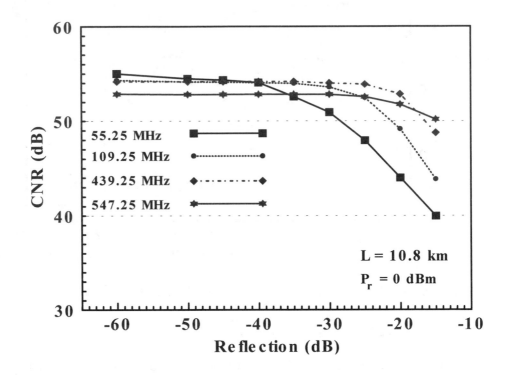

Figure 9.16 CNR versus the reflection level for a 79 AM channel SCM lightwave system using a 1550-nm externally modulated laser. After Ref. [24] (© 1995 IEE).

To understand the AM CNR degradation results, recall from Chapter 5 that the CNR at the optical receiver for an AM-VSB system can be written as

$$CNR = \frac{(m_a I_r)^2}{2B\left[2qI_r + N_r^2 + RIN \cdot I_r^2\right]} \tag{9.46}$$

where m_a is the AM modulation index per channel, I_r is the receiver photocurrent, B is the noise video bandwidth (= 4 MHz), q is the electronic charge, N_r is the receiver noise equivalent current, and RIN is the total relative-intensity-noise in the link. In addition, we recall that the worst-case RIN caused by multiple reflections is given by Equation (9.45). Let us examine the following two interesting cases. For a very narrow linewidth laser such as the EM-YAG laser transmitter ($\Delta v < 10$ kHz), Equation (9.45) is reduced to $(16/\pi)R_1 R_2 \Delta v \cdot \tau^2$, which does not have a frequency dependence. The very low RIN as a result of the narrow linewidth does not degrade the observed AM CNR. In contrast, Equation (9.45) is reduced to $(4/\pi)\, R_1 R_2/\Delta v$ for a broad modulated linewidth ($\Delta v > 1$ GHz) as is the case for a DM-DFB laser transmitter. The high RIN (RIN = −130 dBc/Hz at −15 dB reflection) reduces significantly the observed CNR without frequency dependence. In the intermediate case such as for the EM-DFB laser transmitter ($\Delta v = 1$ MHz), the RIN generated from multiple reflections has strong frequency dependence, particularly at the lower cable-TV frequencies based on Equation (9.45).

9.3.4 Multiple Reflections Effect on the QAM Channels

The effect of multiple optical reflections on the QAM channel is even more dramatic [23].

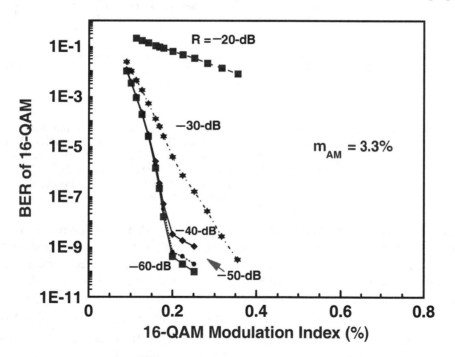

Figure 9.17 16-QAM BER versus the QAM modulation index at reflection levels of −20-dB, −30 dB, −40 dB, −50 dB, and −60 dB for a directly modulated 1310-nm DFB laser transmitter. After Ref. [23] (© 1994 IEEE).

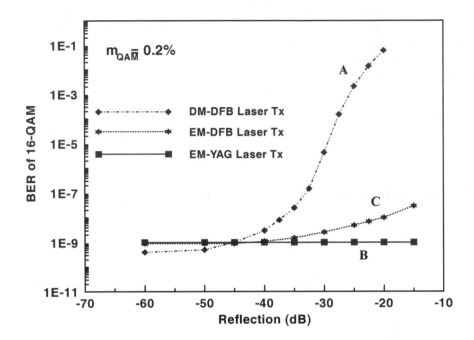

Figure 9.18 16-QAM BER versus the reflection level at 0.2% QAM modulation index for (A) a 1310-nm directly modulated DFB laser transmitter with AM modulation index of 3.3% per channel, (B) a 1319-nm externally modulated YAG laser transmitter at the same AM modulation index, and (C) a 1550-nm externally modulated DFB laser transmitter with AM modulation index of 3.2% per channel. After Ref. [24] (© 1995 IEE).

Figure 9.17 shows, for example, the 16-QAM uncoded BER as a function of the QAM modulation index with an AM modulation index of 3.3% at reflection levels from −60 dB to −20 dB (on both VBRs) for a DM-DFB laser transmitter-based lightwave system. At a −60 dB reflection level, the 16-QAM BER shows only a small impairment due to the clipping-induced impulse noise. As the reflection level is increased to −40-dB, the 16-QAM BER is increased from 10^{-10} to 10^{-9} for a 0.22% QAM modulation index. The 16-QAM BER is significantly increased to 10^{-6} at a reflection level of −30-dB. In fact, one needs twice the QAM modulation index at a reflection level of −30-dB compared with the −60-dB reflection level case to achieve the same 16-QAM BER of 10^{-9}. At a reflection level of −20-dB or higher, the 16-QAM BER is unacceptably high.

In the case of the EM-YAG laser transmitter-based system, the 16-QAM BER is independent of the reflection level, even for high reflections (> −15-dB). It has been argued in a previous study that the AM video signal and the interference noise are highly correlated [25]. Under standard CNR measurement, when the video carrier is turned off, the noise is also effectively turned off, resulting in no measurable degradation. However, the QAM results support the interpretation that the EM-YAG laser transmitter has a negligible amount of phase noise to be converted to intensity noise, resulting in immunity to even high reflection levels [25]. Only a small 16-QAM SNR degradation (≈ 0.7 dB) is observed for the 1550-nm EM-DFB laser transmitter-based system. The comparison of the QAM BER degradation caused by multiple optical reflections among the different laser transmitters is illustrated in Figure 9.18. The 256-QAM coded BER versus the reflection levels at a QAM modulation index of 0.8% is shown for the following three cases:

(A) DM-DFB laser transmitter with an AM modulation index of 3.3%,
(B) EM-YAG laser transmitter with an AM modulation index of 3.6%, and
(C) 1550-nm EM-DFB laser transmitter with an AM modulation index of 3.2%.

Consequently, the 1319-nm EM-YAG laser-transmitter-based lightwave system has the best immunity to multiple optical reflections for both the AM and the QAM channels.

9.4 Dispersion-Induced Nonlinear Distortions

When a DM-DFB laser transmitter operating at 1550-nm is used in an AM/QAM video light wave link, the laser frequency chirp interacts with the fiber chromatic dispersion to produce nonlinear distortions [26–28]. The fiber chromatic dispersion, which is caused by the wavelength dependence of the index of refraction, is about 17-ps/mn-km at 1550-nm for a standard SMF. As we learned in Section 3.3.2, the gain saturation effect in the DFB laser diode is responsible for its optical frequency chirping. When a DFB laser is amplitude modulated, dynamic wavelength or frequency changes are produced. Because of the fiber dispersion, different wavelengths propagate with different group velocities through the fiber.

A closed-form analytical expression can be obtained by solving the rate equations for the DFB laser using frequency-domain analysis [26, 28]. However, a simpler and more intuitive time-domain analysis will be presented here [27]. Assuming a directly modulated DFB laser by only two RF subcarriers, the injected optical power to the fiber can be written as

$$P_{in}(t) = P_0 + P_0 m\left[\sin(\Omega_1 t + \phi_1) + \sin(\Omega_2 t + \phi_2)\right] \tag{9.47}$$

where P_0 is the average transmitted optical power, m is the modulation index of each subcarrier, and Ω_i and ϕ_i (i = 1, 2) are the frequencies and phases of the two subcarriers, respectively. The transmitted optical power through the fiber can be written as

$$P_t(t) = \alpha \cdot \frac{d(t - \Delta\tau_g)}{dt} P_{in}\left[t - \Delta\tau_g(I)\right] \tag{9.48}$$

where α is the fiber attenuation and $\Delta\tau_g(I)$ is the group delay variation due to the fiber dispersion, which depends on the driving current of the DFB laser. Consequently, $\Delta\tau_g(I)$ is given by the following equation:

$$\Delta\tau_g(I) = D \cdot L \cdot \Delta\lambda(t) \cong D \cdot L \cdot \frac{d\lambda}{df}\frac{df}{dI}I(t) = -\frac{\lambda^2}{c}D \cdot L \cdot \frac{df}{dI}I(t) \tag{9.49}$$

where D is the fiber dispersion, L is the fiber span, and df/dI represents the frequency chirp of the DFB laser. The DFB laser driving current can be written as

$$I(t) = I_{max}\left[\sin(\Omega_1 t + \varphi_1) + \sin(\Omega_2 t + \varphi_2)\right] \tag{9.50}$$

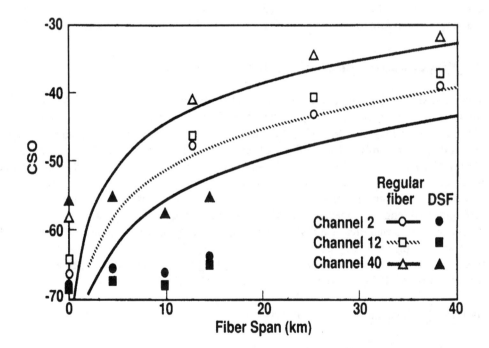

Figure 9.19 Measured (symbols) and calculated (lines) CSO distortion at channel 2, 12, and 40 versus the fiber span for a standard SMF and DSF using a DM-DFB laser transmitter with a chirp of 720-MHz/mA. After Ref. [27] (© 1991 IEEE).

The second term in Equation (9.48) represents the power-level compression or expansion after transmission through the fiber. The last term in Equation (9.48) can be expanded as

$$P_{in}(t - \Delta\tau_g) = P_{in}(t) - \Delta\tau_g \cdot \frac{dP_{in}}{dt} \tag{9.51}$$

where higher orders have been neglected. Substituting Equations (9.49), (9.50), and (9.51) into Equation (9.48), the second-order intermodulation distortion products relative to the subcarrier are given by

$$IM_2 / C = \left[D \cdot L \cdot I_{max} \left(\frac{\lambda^2}{c} \right) \frac{df}{dI} \right]^2 \cdot (\Omega_1 + \Omega_2)^2 \tag{9.52}$$

For a multichannel video transmission system, the CSO distortions are obtained by multiplying Equation (9.52) by the CSO product counts, N_{CSO}, and replacing $\Omega_1 + \Omega_2$ by the CSO distortion frequency Ω_d as follows:

$$CSO = N_{CSO} \left[D \cdot L \cdot I_{max} \left(\frac{\lambda^2}{c} \right) \frac{df}{dI} \right]^2 \cdot \Omega_d^2 \tag{9.53}$$

Notice that the CSO distortion is proportional to the square of the distortion frequency, the frequency chirp, the laser driving current, and the fiber dispersion.

Figure 9.19 shows the measured and calculated CSO distortion at channels 2, 12, and 40 versus the fiber span for a 42 AM video channel system using a DM-DFB laser transmitter with a frequency chirp of 720-MHz/mA in both a standard SMF and a dispersion-shifted fiber[1] (DSF) [26]. Notice that an acceptable level of CSO distortion (< –62-dBc) limits the fiber length to only a few kilometers in a standard SMF. This is the primary reason why DM-DFB laser transmitters operating at 1550-nm cannot be used for long-distance transmission over a standard SMF.

There are two main options to overcome this limitation. The first option is to use an EM-DFB laser transmitter operating at 1550-nm, which has a negligible residual frequency chirp from the LiNbO$_3$ MZ modulator. The other option is to use dispersion-compensating fibers or chirped-fiber gratings to compensate for the fiber dispersion and reduce the nonlinear distortions to acceptable levels.

---

[1] For a dispersion-shifted fiber, the zero-dispersion point is shifted from about 1310-nm to 1550-nm.

9.5 Optical Fiber Nonlinear Effects

In order to achieve longer transmission distances or to provide more fiber splits, higher optical power needs to be launched into the optical fiber. This has been accomplished by the recent development of high-power 1550-nm laser transmitters and high-power in-line EDFAs with output powers as high as +27-dBm. However, the various nonlinear effects in an optical fiber limit the upper bound of the injected optical power in HFC or DWDM access networks. The nonlinear fiber-optic effects can be divided into single-wavelength effects and multiple-wavelength effects. The single-wavelength fiber nonlinearities include the stimulated Brillouin scattering (SBS) effect and self- and external-phase modulation (SPM and EPM) effects, while the multiwavelength nonlinearities include the stimulated Raman scattering (SRS) and cross phase-modulation (XPM) effects.

9.5.1 Stimulated Brillouin Scattering Effect

The SBS effect converts the transmitted optical signal in the fiber to a backward-scattered signal, and thus limits the maximum optical power that can be launched into the SMF. Consider a refractive index grating in the fiber, which is formed by acoustic (sound) waves that are traveling at a velocity of approximately 6-km/s. For a Bragg grating, the grating period is $\Lambda = \lambda/2n = 1.55\text{-}\mu\text{m}/2\text{x}1.5 \cong 0.516\text{-}\mu\text{m}$. The grating scattering is stimulated by the injected optical power in the fiber. In order to conserve energy and momentum, an acoustic phonon is generated at a frequency of 6-km/s/0.516-μm = 11.6-GHz lower than the 1.55-μm incident light for the directly reflected beam. This is known as the Brillouin-Stokes frequency shift. Therefore, the SBS process generates a backward-propagating light beam at a lower frequency than the 1.55-μm incident beam. At low incident optical power levels, the scattering cross section is sufficiently small so that one can neglect the optical power losses. However, above the so-called SBS power threshold, the optical power of the backward-scattered signal increases exponentially when the incident power is increased, leading to significant optical power losses. Based on Smith's condition [29], the SBS power threshold is defined when the backward-scattered power is equal to the injected power, which for a uniform fiber is given by [30]

$$P_{th} = \frac{21\alpha \cdot A_{eff}}{(1-e^{\alpha L})g_0}\left[\frac{\Delta v_B \otimes \Delta v_L}{\Delta v_B}\right] \tag{9.54}$$

where A_{eff} is the effective fiber core area, α is the fiber loss coefficient, L is the total fiber length, and g_0 is the peak Brillouin gain and is given by

$$g_0 = \frac{2\pi K n^7 \cdot p_{12}^2}{c\lambda^2 \cdot \rho_0 V_a \cdot \Delta v_B} \tag{9.55}$$

where n is the fiber refractive index, ρ_o is the material density, V_a is the acoustic velocity, p_{12} is the elasto-optic coefficient, and $0.5 < K < 1$ accounts for the optical field polarization. The operator $\otimes$ denotes the convolution of Δv_L, which is the laser linewidth, and Δv_B, which is the spontaneous Brillouin linewidth of the fiber in which the Brillouin gain can be accumulated along the fiber length.   For Gaussian profiles, the convolved linewidth is $(\Delta v_L^2 + \Delta v_B^2)^{1/2}$, while for Lorentzian profiles, the convolved linewidth is $\Delta v_L + \Delta v_B$. For silica- based fibers, the Brillouin gain bandwidth is about 20-MHz at 1550-nm. Equations (9.54) and (9.55) indicate that the SBS threshold is directly proportional to Δv_B. For a standard SMF, $g_0 = 2 \times 10^{-9}$ cm/W, and the SBS threshold without modulation is approximately 6-dBm (4-mW) at 1550-nm and 9-dBm (8-mW) at 1310-nm. In addition to the dependence of the SBS threshold on the fiber type and uniformity along its length, the SBS threshold depends on the specific optical system requirements such as modulation format, symbol rate, and modulation type (i.e., direct or external). The effects of SBS on the CNR performance of SCM lightwave system can be demonstrated through the following experimental result shown in Figure 9.20 [31]. The AM and the scattered optical power were measured as a function of the injected optical power in the fiber using a 1550-nm EM-DFB laser transmitter. Forty-two AM-VSB channels with an optical modulation index of 4% per channel were transmitted over 13-km of dispersion-shifted fiber.   As injected optical power

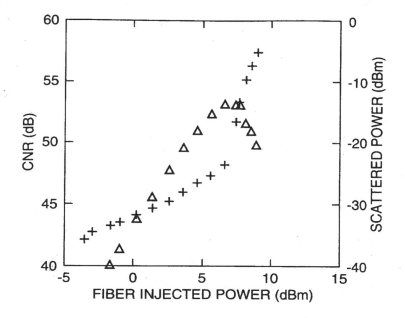

Figure 9.20 Measured CNR (Δ) and the scattered optical power (+) versus the fiber injected power (in dBm). After Ref. [31] (© 1992 IEEE).

is increased above +6-dBm, the SBS scattered power dramatically picks up and the AM CNR rapidly degrades. The SBS also induces CSO and CTB distortion degradations in externally modulated 1550-nm laser transmitters, which are typically used due to the chromatic dispersion degradation (see next section). Since the Δv_B, which is about 20-MHz at 1550-nm for silica fibers, is lower than the NTSC channel 2 frequency, most of the reflected power is frequency independent when the injected power is above the SBS threshold. The CSO and CTB distortions can be approximated by [32]

$$CSO(dB) = 10 \cdot \log \left\{ N_{CSO} \left[\frac{(1-\sigma)m}{4\sigma} \right]^2 \right\}$$

(9.56)

$$CTB(dB) = 10 \cdot \log \left\{ N_{CTB} \left[\frac{3(1-\sigma)m^2}{16\sigma} \right]^2 \right\}$$

(9.57)

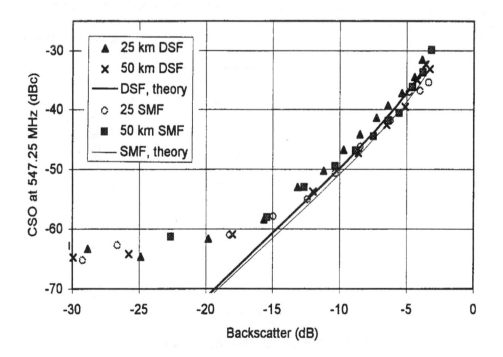

Figure 9.21 Measured (symbols) and predicted (lines) CSO distortion at 547.25-MHz caused by SBS in 25-km and 50-km of both a DSF and SMF. After Ref. [32] (© 1997 Optical Society of America).

where $\sigma = (1 - R_{BS}/0.85)/2$ is the fractional transmission coefficient for the SMF that was empirically estimated from the ratio of the backward-scattered power to the injected power (R_{BS}), m is the modulation index, and N_{CSO} and N_{CTB} are the CSO and CTB product counts, respectively. Figure 9.21 shows, for example, the measured and predicted CSO distortion at 547.25-MHz caused by SBS in 25-km and 50-km of a DSF and a standard SMF versus the backward-scattered power. To maintain the CSO distortion below −62-dBc, the backward-scattered power should be less than −20-dB, corresponding to a SBS threshold power of about 14-dBm, to achieve the necessary phase-modulation index. For a single-tone phase modulation, the increase in the SBS threshold is given by [33, 34]

$$\Delta P_{SBS}^{th}(\beta) = -10 \cdot \log \left[\max_{k \in \{0,1,..\}} \left\{ J_k^2(\beta) \right\} \right] \tag{9.58}$$

The effect of SBS can be countered by two methods. In method A, a single tone with a phase modulation at a frequency above about twice the maximum channel frequency (≈ 1.8 GHz) is used to avoid additional distortions. The phase-modulation method is more effective in reducing the interferometric noise, but requires relatively high electrical powers

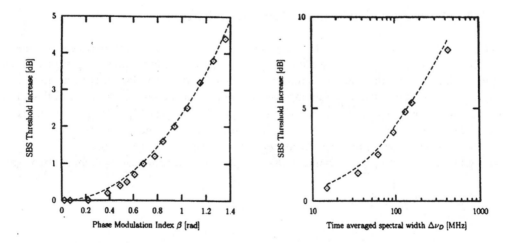

Figure 9.22 Measured (symbols) and calculated (dash lines) SBS threshold increase (dB) versus (A) the phase modulation index (β) using method 1, and (B) the time-averaged effective laser linewidth (Δv_D). After Ref. [33] (© 1994 IEEE).

Figure 9.22(A) shows the measured and calculated SBS threshold increase versus the phase modulation index (β). Using this method, the SBS threshold can be increased by about 10-dB, corresponding to optical injected power of +16-dBm (40-mW) at 1550-nm.

In method B, a low frequency dithering of the optical frequency of the laser transmitter is applied by modulating the laser bias current, resulting in a broader effective laser linewidth. Assuming a Lorentzian linewidth, the SBS threshold power increase can be obtained from Equation (9.47) and is approximated by

$$\Delta P_{SBS}^{th}(\Delta v_D) = 10 \cdot \log\left[1 + \frac{\Delta v_D}{\Delta v_B}\right] \tag{9.59}$$

Figure 9.22(B) shows the measured and calculated SBS threshold increase versus the time-averaged effective laser linewidth. This method can easily provide an SBS threshold above +20-dBm by allowing a frequency excursion above 1 GHz, and requires only a simple low-power laser current modulation. However, there are two possible limitations to this method. First, the package parasitics of high-power DFB lasers limit the maximum frequency of the modulating signal. Second, broadening the laser linewidth beyond 50-MHz may result in the CNR degradation of the transmitted AM channels.

9.5.2 Self- and External Phase-Modulation Effects

The self-phase-modulation (SPM) effect occurs due to the interaction between the fiber's chromatic dispersion and the modulated optical spectrum of the propagating signal in the fiber. Second- and third-order nonlinear distortions of either a DM-DFB or an EM laser transmitter are generated when high-power (below the Brillouin threshold) optical signals are transmitted through a dispersive nonlinear optical fiber [35–37]. The SPM effect is so named since the lightwave signal propagating through the fiber modulates its own phase.

An analytical solution to the SPM induced by CSO and CTB distortions can be obtained using the wave envelope equation in a lossy, dispersive, and nonlinear medium as follows:

$$\frac{\partial A}{\partial z} + \beta_1 \frac{\partial A}{\partial t} - \frac{i}{2}\beta_2 \frac{\partial^2 A}{\partial t^2} + \alpha A = -ikn_2|A|^2 A \tag{9.60}$$

where α is the fiber loss coefficient, β is the propagation constant, $\beta_1 = d\beta/d\omega = 1/v_g^1$ is the first-order dispersion coefficient, $\beta_2 = d^2\beta/d\omega^2$ is the second-order dispersion coefficient, n_2 is the nonlinear refractive index of the fiber; $n = n_o + n_2|A|^2$, where n_o is the fiber refractive index at low optical power, and $A(z, t)$ is the envelope of the propagating electric field represented by

$$A(z,t) = \sqrt{x(z,t)}\, \exp\left[iy(z,t)\right] \tag{9.61}$$

where the electric field is propagating along the z-axis. Substituting Equation (9.61) into Equation (9.60), and requiring that both the intensity modulation information $x(z, t)$ and the phase modulation information $y(z, t)$ are real quantities, the following equations for the intensity and phase are obtained:

$$\frac{\partial x}{\partial z} + \beta_1 \frac{\partial x}{\partial t} = -\beta_2 \left[\frac{\partial x}{\partial t}\frac{\partial y}{\partial t} + x\frac{\partial^2 y}{\partial t^2} \right] - 2\alpha x \tag{9.62}$$

$$\frac{\partial y}{\partial z} + \beta_1 \frac{\partial y}{\partial t} = \frac{\beta_2}{8x^2} \left[-\left(\frac{\partial x}{\partial t}\right)^2 + 2x\frac{\partial^2 x}{\partial t^2} - 4x^2\left(\frac{\partial y}{\partial t}\right)^2 \right] - kn_2 x \tag{9.63}$$

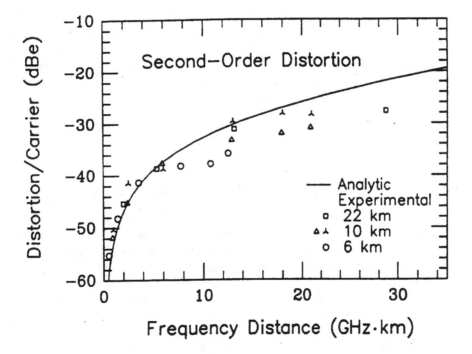

Figure 9.23 Measured second-order distortion due to SPM for a directly modulated DFB laser transmitter with 20-GHz chirp versus the frequency distance product. The solid line is the analytical solution for dual-tone input. After Ref. [36] (© 1991 IEEE).

At the fiber input ($z = 0$), the light intensity and phase are given by

$$x(0,t) = P_0\left[1 + m\sum_{i=1}^{N}\cos(\Omega_i t + \phi_i)\right] \tag{9.64}$$

$$y(0,t) = \gamma m\int_0^t dt'\sum_{i=1}^{N}\cos(\Omega_i t' + \phi_i) \tag{9.65}$$

$$y(0,t) = \beta\sum_{i=1}^{N}\cos(\Omega_i t' + \phi_i) \tag{9.66}$$

where P_0 is the average optical power of the optical signal, m is the modulation index per channel, Ω_i is the subcarrier frequency, and N is the number of cable TV channels. Equation (9.65) applies to a DM-DFB laser transmitter, where $\gamma = G(I_b - I_{th})$ is the laser-chirp parameter (GHz/mA), and I_b and I_{th} are the bias and threshold currents of the DFB laser, respectively. Equation (9.66) applies to an EM-DFB laser transmitter, where β is the phase-modulation index. Substituting Equations (9.64) and (9.65) into the intensity and phase equations (9.62 and 9.63), and after lengthy manipulations, the SPM-induced CSO distortion (CTB distortion is much smaller) is given by [36]

$$CSO_{SPM}(\Omega) = N_{CSO}\left[\frac{mP_0}{2A_{eff}}\left(\frac{L - L_{eff}}{\alpha}\right)\beta_2 k n_2 \Omega^2\right]^2 \tag{9.67}$$

where $\beta_2 = -D\lambda^2/2\pi c$, D is the dispersion coefficient of the fiber (17-ps/nm·km at 1550-nm), A_{eff} is the effective fiber core area, α is the fiber loss coefficient, $L_{eff} = (1 - e^{-\alpha L})/\alpha$ is the effective interaction length, and N_{CSO} is the CSO product count. Figure 9.23 shows the measured and calculated second-order distortion due to SPM for a DM-DFB laser transmitter with 20-GHz frequency chirp versus the frequency distance product for a standard SMF. Since the resulting CTB distortion caused by SPM was more than 30-dB smaller than the CSO distortion, it is neglected here. Figure 9.23 result clearly demonstrates that CSO distortions produced by the SPM effect using a DM-DFB laser transmitter operating at 1550-nm will limit the transmission distance to less than 10-km.

There are three main solutions to the SPM-induced CSO distortions, and they are as follows: (A) using an EM laser transmitter, which has a very low frequency chirp; (B) using a DSF; and (C) compensating for the fiber optic dispersion. Method A is the most widely used in the U.S. since it allows the use of a standard SMF. The other methods are usually more expensive to implement since they require either the installation of a new type of fiber-optic cables (instead of the standard installed cables) or using complex fiber optic components to compensate for the fiber dispersion.

The specific magnitude of the resultant CSO distortion is determined by the complex interplay of the SPM, the fiber length dispersion, and the sign of the phase-modulation index caused by the external modulator's residual chirp. Desem [37] has shown that the SPM contribution to the CSO distortion could be equal but opposite in sign to the CSO due to its initial phase modulation. Consequently, a total cancellation of the CSO distortion at a particular distance can be achieved. For fiber-optic links around 100-km, the phase modulation caused by the optical modulator's residual chirp is typically not a problem.

A related effect to the SPM is the external phase modulation (EPM) [38]. As previously explained, a high SBS threshold is achieved by applying a high phase-modulation index (β) in the integrated MZI and phase modulator for an EM laser transmitter. The EPM effect is the conversion of this external phase modulation to intensity modulation that interacts with the nonlinear dispersive optical fiber. The EPM effect generates CSO distortions in addition to the SPM effect, and thus needs to be carefully analyzed to correctly predict the resulting CSO distortions, particularly at long distances. For further discussions on EPM and SPM effects in optical links with cascaded EDFAs, see Section 10.1.5.

9.5.3 Stimulated Raman Scattering Effect

As we learned in Chapter 1, wavelength-division multiplexers and demultiplexers are used in HFC networks in order to deliver a mixture of broadcast and narrowcast services to the subscriber's home. The downstream and upstream DWDM architecture for cable TV networks will be discussed in Chapter 10. The analysis in this section concentrates on the crosstalk between two or more high-power WDM channels or wavelengths due to the stimulated Raman scattering (SRS) effect. In SCM WDM lightwave systems, each WDM wavelength is modulated by its own RF subcarrier frequencies.

Let us first consider the simple case of two WDM channels, where each wavelength is modulated by M RF subcarriers [39]. Then, the launch optical power to the fiber is given by

$$P_j(t)|_{z=0} = P_0\left[1 + \sum_{i=1}^{M} m\cos(\Omega_i t + \theta_{ji})\right] \qquad (9.68)$$

where $j = p, s$ represents the pump (shorter λ) and signal (longer λ) wavelengths, respectively. It is assumed that both wavelengths have the same average optical power P_0, and are modulated with the same set of RF subcarrier frequencies Ω_i, which have the same optical modulation index, m, but random phases θ_{pi} and θ_{si}.

A given SRS interaction is proportional to $g_{ps}P_pP_s$, where g_{ps} is the Raman efficiency between the pump wavelength with power P_p and signal wavelength with power P_s. Substituting P_s and P_p from Equation (9.68), the total SRS crosstalk from RF subcarrier i at the pump wavelength is composed of the following three terms:

1. $mg_{ps}P_0^2\cos(\Omega_i t + \theta_{pi})$ – This term is due to the SRS interaction between RF subcarrier i at the pump wavelength and the signal wavelength; this is just the optical power loss, $mg_{ps}P_0^2$, of RF subcarrier i.

2. $mg_{ps}P_0^2\cdot\cos(\Omega_i t + \theta_{si})$ – This term is due to the SRS interaction between the pump wavelength and the RF subcarrier i in the signal wavelength. This is the most important SRS crosstalk term in SCM WDM lightwave systems.

3. $(M/8)^{1/2}m^2\cdot g_{ps}P_0^2$ – This term is due to the SRS interaction between the RF subcarriers in the pump and signal wavelengths. This term is about one order of magnitude smaller than the first two terms, and thus can be neglected.

It should be pointed out that the SRS interactions between the pump and signal wavelengths cause only DC optical power loss or gain, and thus do not contribute to the SRS crosstalk at the subcarrier frequencies.

An analytical solution to the SRS crosstalk can be obtained by solving the following equations for the pump and signal wavelength in an optical fiber [40]

$$\frac{\partial P_s}{\partial z} + \frac{1}{v_{gs}}\frac{\partial P_s}{\partial t} = (g_{ps}P_p - \alpha)P_s \tag{9.69a}$$

$$\frac{\partial P_p}{\partial z} + \frac{1}{v_{gp}}\frac{\partial P_p}{\partial t} = (-g_{ps}P_s - \alpha)P_p \tag{9.69b}$$

where V_{gs} and V_{gp} are the group velocities of the transmitted light at the signal and pump wavelengths, respectively, and it is assumed that both channels have the same fiber-loss coefficient, α. An analytical solution to Equations (9.69a, 9.69b) is obtained by assuming that the SRS interaction is in the linear regime. The electrical crosstalk level suffered by subcarrier i in the signal or pump wavelength due to SRS is given by

$$XT_i^{sp} = \frac{\left(2g_{ps}P_0\right)^2\left[\left(\frac{\alpha L_{eff}}{2}\right)^2 + e^{-\alpha L}\sin^2\left(d_{ps}\Omega_i L/2\right)\right]}{\left[\alpha^2 + (d_{ps}\Omega_i)^2\right]\left[1 \pm 2g_{ps}P_0 L_{eff}\right]^2} \tag{9.70}$$

where $d_{ps} = |1/V_{gp} - 1/V_{gs}|$ is the group velocity mismatch between the pump and the signal wavelengths, L is the fiber length, L_{eff} is the effective fiber length. The right-hand side term in the denominator (Equation 9.70), where + is for the signal channel and – is for the pump channel, is attributed to the subcarrier loss or gain due to the SRS interaction. With $d_{ps} = 0$, the SRS crosstalk is independent of the subcarrier frequency. Figure 9.24 shows the calculated electrical SRS crosstalk level for two WDM channels with 20-km of a standard SMF at 1300-nm, and at increasing channel spacings from 2- to 16-nm. An effective fiber core area of 50-μm^2 and an optical fiber loss of 0.4-dB/km were assumed. This result demonstrates that crosstalk levels can easily exceed –62 dBc as required for robust AM-VSB transmission even for modest input optical power levels per WDM channel (< +10-dBm). The SRS crosstalk is significantly increased with the number of WDM channels and the optical input power per channel. For example, SRS crosstalk level of –60 dBc is obtained for 4 WDM channels with 4-nm spacing, 20-km of standard SMF, and with optical input power of 0-dBm.

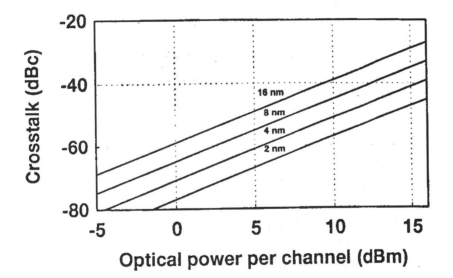

Figure 9.24 Calculated SRS crosstalk level (electrical) versus the input optical power per WDM channel at 1300-nm with two WDM channels and 20-km of standard SMF. The crosstalk dependency for channel spacings of 2-nm, 4-nm, 8-nm, and 16-nm are shown. After Ref. [36] (© 1995 IEEE).

For WDM SCM lightwave systems operating at 1550-nm, the SRS crosstalk is generally lower than in the zero-dispersion case, but it depends on the RF subcarrier frequency. In particular, the SRS crosstalk is lower at higher subcarrier frequencies (≥ 400-MHz) because the fiber dispersion is significant and can cause walkoff between RF subcarriers at different optical wavelengths.

9.5.4 Cross Phase-Modulation Effect

In a WDM SCM system, the fiber nonlinearities lead to crosstalk between subcarriers on different wavelengths. Due to the fiber dispersion at 1550-nm, the phase of the transmitted signals is converted to intensity modulation leading to the so-called cross-phase modulation (XPM) induced crosstalk [41, 42]. Let us assume two optical waves with the same polarization propagating in a SMF, and let $A_1(z, t)$ and $A_2(z, t)$ be the slowly varying field envelope of each wave, with $\lambda_1 > \lambda_2$. Assuming the slowly varying envelope approximation, the XPM is described in terms of the coupled equations as follows [43]:

$$
\frac{\partial A_1}{\partial z} + \frac{1}{v_{g1}} \frac{\partial A_1}{\partial t} = \left[-2j\gamma P_2 - \alpha/2 \right] A_1
$$

$$
\frac{\partial A_2}{\partial z} + \frac{1}{v_{g2}} \frac{\partial A_2}{\partial t} = \left[-2j\gamma P_1 - \alpha/2 \right] A_2
$$

(9.71)

where v_{gi} (i = 1, 2) are the group velocities for wave 1 and 2, γ is the nonlinearity coefficient, and α is the absorption coefficient. The normalized XPM crosstalk at λ_1 is given by [43]

$$
XT_{XPM_1} = \frac{-2\ddot{\beta}_1 \Omega^2 \gamma P_c}{\left(\alpha - j\Omega d_{12} \right)^2} \left\{ \left[e^{-\alpha L} \cos(\Omega d_{12} L) - 1 + \alpha L \right] + j \left[e^{-\alpha L} \sin(\Omega d_{12} L) - \Omega d_{12} L \right] \right\}
$$

(9.72)

where β_1 is the phase constant of wave 1, Ω is the RF modulation frequency, P_c is the average optical power, $d_{12} = (1/v_{g1} - 1/v_{g2})$ is the walkoff parameter between wave 1 and 2, and L is the fiber length. A similar expression for the XPM crosstalk at λ_2 can be easily obtained by changing $d_{12} \rightarrow d_{21}$, $\beta_1 \rightarrow \beta_2$. Notice that the XPM-induced crosstalk at both λ_1 and λ_2 have the same sign. In contrast, the SRS-induced crosstalk at λ_1 is 180° out of phase with the SRS-induced crosstalk at λ_2. This is because shorter wavelengths provide gain for the longer wavelengths through the SRS interaction. Thus, subcarriers at λ_1 acquire SRS crosstalk through gain, while subcarriers at λ_2 acquire SRS crosstalk through depletion. The total observed crosstalk is the combination of the SRS-induced and the XPM-induced intensity modulations. The SRS-induced crosstalk and the XPM-induced crosstalk are in phase for λ_1, but out of phase for λ_2. Figure 9.25 shows the measured and calculated total crosstalk versus the modulation frequency at λ_1 = 1546-nm, where λ_2 = 1542-nm, P_c = 17-dBm, and L = 25-km [43]. The crosstalk is dominated by the XPM effect. The solid line represents the calculated total crosstalk, while the dashed lines represent the SRS and XPM crosstalk contributions. At low modulation frequencies or large $\Delta\lambda = |\lambda_1 - \lambda_2|$, the crosstalk is dominated by the SRS effect, and at high modulation frequencies or small $\Delta\lambda$, the crosstalk is dominated by the XPM effect. At λ_1 and at modulation frequency of about 600-MHz, the SRS-induced crosstalk is added in phase with the XPM-induced crosstalk, increasing the total crosstalk (XPM + SRS) by about 20-dB compared with their out-of-phase crosstalk contributions.

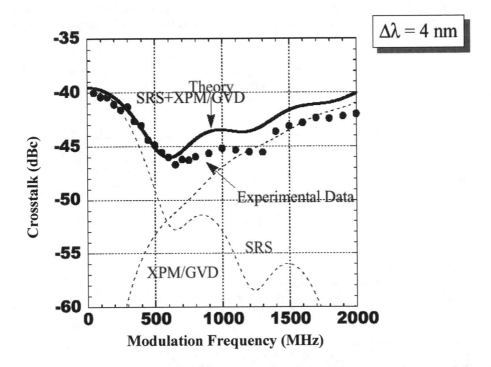

Figure 9.25 Measured (solid circles) and calculated (lines) crosstalk level versus the modulation frequency at λ_1 for $\Delta\lambda$ = 4-nm, P_c = 17-dBm, and L = 25-km. The solid line represents the total crosstalk level from both SRS and XPM. After Ref. [43] (© 2000, IEEE).

9.6 Polarization-Dependent Distortion Effects

In the polarization-dependent distortion effects, which were first reported in 1992 [44–46], the magnitude of the CSO or CTB distortion levels depends on either the polarization of the input light beam to the fiber, or to the optical isolator, or to cascaded in-line EDFAs. The polarization-induced nonlinear distortions in cascaded EDFAs will be explained in Section 10.1.5, when the performance of AM video trunking links is discussed.

Optical isolators, as was explained in Section 3.1 on the DFB laser module, are passive, nonreciprocal optical devices based on the Faraday effect. Birefringent optical crystals with a thickness of about 5-mm are typically used. Optical isolators can be divided into two groups: isolators with an optical isolation that are either sensitive or insensitive to the input polarization state. The type-A isolator separates the input port from the backward-reflected light. In a type-B isolator, the effective numerical aperture of the reflected light is much higher than that of the input port.   The mechanism of the polarization-induced distortion in

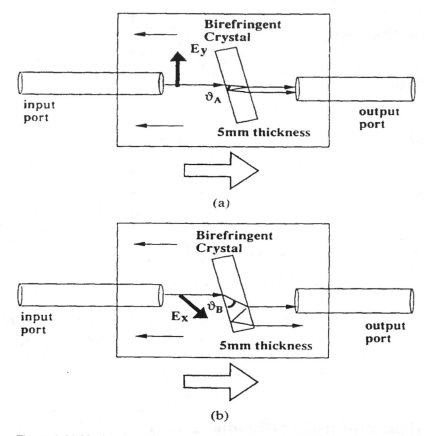

(a)

(b)

Figure 9.26 Mechanism of the polarization-dependent distortion in optical isolators when (a) multiple optical reflections are injected into the output port and (b) multiple optical reflections are not injected to the output port. After Ref. [45] (© 1994 IEEE).

optical isolators is illustrated in Figure 9.26. The birefringent crystal is typically rotated by a small angle relative to the input light beam to avoid specular reflection at the front facet of the crystal. Some of the multiple optical reflections are transmitted through the back facet of the birefringent crystal, depending on the orientation of the input beam's electric field. A Birefringent crystal has two optical axes, which are typically referred to as the "slow" and "fast" axes with corresponding refractive indices. θ_A and θ_B denote the refraction angles at the front facet of the birefringent crystal for the two input light polarization states, respectively. For a vertically polarized light (smaller refraction angle), the main propagating beam as well as some of the reflected beams are coupled into the output port of the fiber as shown in Figure 9.25(a). The doubly reflected beam (higher order reflections are neglected) will be relatively delayed compared to the main propagating beam, and thus generates interferometric interference, as we learned in Section 9.1.2. For DM-DFB laser transmitters, the interaction between the laser frequency chirp with polarization-induced multiple optical reflections

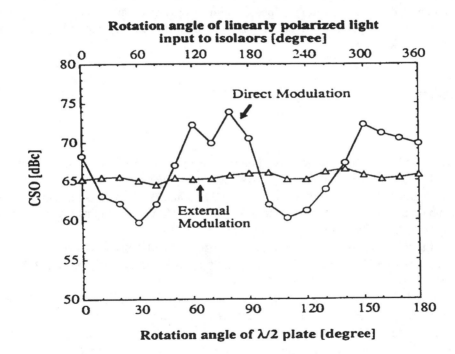

Figure 9.27 measured CSO distortions versus the rotation angle of λ/2 plate for type-A isolator using a direct-modulation scheme (circles) and an external-modulation scheme (triangles). After Ref. [45] (© 1994 IEEE).

will generate nonlinear distortions, mainly CSO distortions. In contrast, for the horizontally polarized light (larger refraction angle), only the main propagating beam is coupled into the output port of the fiber as shown in Figure 9.26(b). This type of polarization-induced nonlinear distortion can be avoided by using EM laser transmitters.

To understand the performance differences between using a direct modulation and external modulation of a laser transmitter, let us examine the following result. Figure 9.27 shows the measured CSO distortions versus the rotation angle of linearly polarized light into a type-A isolator under both a direct-modulation scheme (circles) and an external- modulation scheme (triangles). Notice that while the CSO distortion changes by only a few decibels, the CSO distortion varies by almost 15-dB. Kikushima, et al. [45] have also reported that the CSO distortion levels induced by SBS effect can vary by about 5–7 dB, depending on the polarization state of the input light in a fiber optic link. This is attributed to a factor of two higher Brillouin gain when linearly polarized light is input compared with circularly polarized light [47]. The degradation issues in the transmission of AM/QAM video signals due to polarization-induced nonlinear distortions in cascaded EDFAs, which typically use many optical isolators, are discussed in more detail in Chapter 10 on AM video lightwave trunking links.

References

1. K. Maeda, H. Nakata, and K. Fujito, "Analysis of BER of 16QAM Signal in AM/16QAM Optical Transmission System," *Electronics Letters* **29**, 640–641 (1993).

2. S. Ovadia, L. Eskildsen, J. T. Sciarabba, C. Lin, and W. T. Anderson, "Bit-Error-Rate Impairment in Hybrid Multichannel AM-VSB/16-QAM Video Lightwave Transmission System," *IEEE Photonics Technology Letters* **7**, 869–871 (1994).

3. X. Lu, G. E. Bodeep, and T. E. Darcie, "Broadband AM-VSB/64-QAM Cable TV System over Hybrid Fiber/Coax Network," *IEEE Photonics Technology Letters* **7**, 330–332 (1994).

4. S. Ovadia, H. Dai, C. Lin, and W. T. Anderson, "Performance of Hybrid Multichannel AM/256-QAM Video Lightwave Transmission Systems," *IEEE Photonics Technology Letters* **7**, 1351–1353 (1995).

5. Q. Shi, "Asymptotic Clipping Noise Distribution and Its Impact on M-ary QAM Transmission over Optical Fiber," *IEEE Transaction on Communications* **43**, 2077–2084 (1995).

6. J. E. Mazo, "Asymptotic Distortion Spectrum of Clipped, DC-Biased Gaussian Noise," *IEEE Transactions on Communications* **40**, 1339–1344 (1992).

7. S. O. Rice, "Distribution of the Duration of Fades in Radio Transmission," *Bell System Technical Journal*, vol.37, 581–635 (1958).

8. Q. Shi, R. S. Burroughs, "Hybrid Multichannel Analog/Digital CATV Transmission Via Fiber Optic Link: Performance Limits and Trade Offs," *NCTA Technical Papers*, 182–193 (1994).

9. Q. Shi and S. Ovadia, "Effects of Clipping-Induced Impulse Noise in Externally Modulated Multichannel AM/M-QAM Video Transmission Systems," *IEEE Transactions on Communications* **46**, 1448–1450 (1998).

10. K. Pham, J. Conradi, G. Cormack, B. Thomas, and C. W. Anderson, "Impact of Noise and Nonlinear Distortion due to Clipping on the BER Performance of a 64-QAM Signal in Hybrid AM-VSB/QAM Optical Transmission System," *Journal of Lightwave Technology* **13**, 2197–2205 (1995).

11. T. Anderson and D. Crosby, "The Frequency Dependence of Clipping Induced Bit-Error rates in Subcarrier Multiplexed Systems," *IEEE Photonics Technology Letters* **8**, 1076–1078 (1996).

12. A. Kanazawa, M. Shibutani, and K. Emura, "Preclipping Method to Reduce Clipping-Induced Degradation in Hybrid Analog/Digital Subcarrier-Multiplexed Optical Transmission Systems," *IEEE Photonics Technology Letters* **7**, 1069–1071 (1995).

13. Q. Pan and R. J. Green, "Pre-Clipping AM/QAM Hybrid Lightwave Systems with Bandstop Filtering," *IEEE Photonics Technology Letters* **8**, 1079–1081 (1996).

14. C. Y. Kuo and S. Mukherjee, "Clipping Reduction for Improvement of Analog and Digital Performance Beyond Clipping Limit in Lightwave CATV Systems," *OFC Post-deadline*, paper PD18-1 (1996).

15. L. Pophillat, "Optical Modulation Depth Improvement in SCM Lightwave Systems Using a Dissymmetrization Scheme," *IEEE Photonics Technology Letters* **6**, 750–753 (1994).

16. S. Ovadia, "The Effect of Interleaver Depth and QAM Channel Frequency Offset on the Performance of Multichannel AM-VSB/256-QAM Video Lightwave Transmission Systems," *IEEE Photonics Technology Letters* **10**, 1174–1776 (1998).

17. G. Hutson, P. J. Shepherd, and W. S. J. Brice, *Colour Television Systems Principles, Engineering Practice and Applied Technology*, McGraw-Hill International (1990).

18. S. S. Wagner, T. E. Chapuran, and R. C. Menendez, "The Effect of Analog Video Modulation on Laser Clipping Noise in Video-Distribution Networks," *IEEE Photonics Technology Letters* **8**, 275–277 (1996).

19. B. Sklar, *Digital Communications Fundamentals and Applications*, PTR Prentice Hall (Englewood Cliffs, New Jersey, 1988).

20. A. F. Judy, "Generation of Interference Intensity Noise from Fiber Rayleigh Backscatter and Discrete Reflections," *OFC Technical Digest*, 1991, paper WL4, p. 110.

21. J. L. Gimlett and N. K. Cheung, "Effects of Phase-to-Intensity Noise Conversion by Multiple Reflections on Gigabit-per-Second DFB Laser Transmission Systems," *Journal of Lightwave Technology* **7**, 888 (1989).

22. W. I. Way et al., "Multiple-Reflection-Induced Intensity Noise Studies in a Lightwave System for Multichannel AM-VSB Television Signal Distribution," *IEEE Photonics Technology Letters* **2**, 360–362 (1990).

23. S. Ovadia, L. Eskildsen, C. Lin, and W. T. Anderson, "Multiple Optical Reflections in Hybrid AM/16-QAM Multichannel Video Lightwave Transmission System," *IEEE Photonics Technology Letters* **6**, 1261–1264 (1994).

24. S. Ovadia, L. Eskildsen, C. Lin, and W. T. Anderson, "Reflection-Dependence of CNR and 16-QAM BER in AM/QAM Video Lightwave Systems Using Two Different Externally-Modulated Laser Transmitters," *IEE Electronic Letters* **31**, 1277–1278 (1995).

25. T. E. Darcie, G. E. Bodeep, and A. M. Saleh, "Fiber-Reflection-Induced Impairments in Lightwave AM-VSB CATV Systems," *Journal of Lightwave Technology* **9**, 991 (1991).

26. E. E. Bergmann, C. Y. Kuo, and S. Y. Haung, "Dispersion-Induced Composite Second-Order Distortion at 1.5-μm," *IEEE Photonics Technology Letters* **3**, 59–61 (1991).

27. D. B. Crosby and G. J. Lampard, "Dispersion-Induced Limit on the Range of Octave Confined Optical SCM transmission System," *IEEE Photonics Technology Letters* **6**, 1043–1045 (1994).

28. C. Y. Kuo, "Fundamental Second-Order Nonlinear Distortions in Analog AM CATV Transport Systems Based on Single Frequency Semiconductor Lasers," *IEEE Journal of Lightwave Technology* **10**, 235–243 (1992).

29. R. G. Smith, "Optical Power Handling Capacity of Low-Loss Optical Fibers as Determined by Stimulated Raman and Brillouin Scattering," *Applied Optics* **11**, 2489–2494 (1972).

30. X. Mao, R. W. Tkach, A. R. Chraplyvy, R. M. Jopson, and R. M. Derosier, "Stimulated Brillouin Threshold Dependence on Fiber Type and Uniformity," *IEEE Photonics Technology Letters* **4**, 66–69 (1992).

31. X. P. Mao, G. E. Bodeep, R. W. Tkach, A. R. Chraplyvy, T. E. Darcie, and R. M. Derosier, "Brillouin Scattering in Externally-Modulated Lightwave AM-VSB CATV Transmission Systems," *IEEE Photonics Technology Letters* **4**, 287–289 (1992).

32. M. R. Philips and K. L. Sweeney, "Distortion by Stimulated Brillouin Scattering Effect in Analog Video Lightwave Systems," *OSA TOPS vol.12 System Technologies*, 182–185 (1997).

33. F. W. Willems, W. Muys, and J. S. Leong, "Simultaneous Suppression of Stimulated Brillouin Scattering and Interferometric Noise in Externally Modulated Lightwave AM-SCM Systems," *IEEE Photonics Technology Letters* **6**, 1476–1478 (1994).

34. F. W. Willems, J. C. van der Plaats, and W. Muys, "Harmonic Distortion Caused by Stimulated Brillouin Scattering Suppression in Externally Modulated Lightwave AM-CATV Systems," *IEE Electronics Letters* **30**, 343–345 (1994).

35. M. R. Phillips, T. E. Darcie, D. Marcuse, G. E. Bodeep, and N. J. Frigo, "Nonlinear Distortion Generated by Dispersive Transmission of Chirped Intensity-Modulated Signals," *IEEE Photonics Technology Letters* **3**, 481–483 (1991).

36. D. A. Atlas, "Fiber Induced Distortion and Phase Noise to Intensity Noise Conversion in Externally Modulated CATV Systems," *NCTA Technical Papers*, 289–293 (1996).

37. C. Desem, "Composite Second Order Distortion Due to Self-Phase Modulation in Externally Modulated Optical AM-SCM Systems Operating at 1550-nm," *IEE Electronics Letters* **30**, 2055–2056 (1994).

38. M. C. Wu, C. H. Wang, and W. I. Way, "CSO Distortions Due to the Combined Effects of Self- and External-Phase Modulations in Long Distance 1550-nm AM-CATV Systems," *IEEE Photonics Technology Letters* **11**, 718–720 (1999).

39. Z. Wang, A. Li, C. J. Mahon, G. Jacobsen, and E. Bodtker, "Performance Limitations Imposed by Stimulated Raman Scattering in Optical WDM SCM Video Distribution Systems," *IEEE Photonics Technology Letters* **7**, 1492–1494 (1995).

40. G. P. Agrawal, *Nonlinear Fiber Optics*, Chapter 8, Academic Press (New York, 1995).

41. T. K. Chiang, N. Kagi, M. E. Marhic, and L. Kazovsky, "Cross-Phase Modulation in Fiber Links with Multiple Optical Amplifiers and Dispersion Compensators," *Journal of Lightwave Technology* **14**, 249–260 (1996).

42. R. Hui, Y. Wang, K. Demarest, and C. Allen, "Frequency Response of Cross-Phase Modulation in Multispan WDM Optical Fiber Systems," *IEEE Photonics Technology Letters* **10**, 1271–1273 (1998).

43. F. S. Yang, M. E. Marhic, and L. G. Kazovsky, "Nonlinear Crosstalk and Two Countermeasures in SCM-WDM Optical Communication Systems," *Journal of Lightwave Technology* **18**, 512–520 (2000).

44. T. E. Darcie and C. D. Poole, "Polarization-Induced Performance Variables," *Communication Engineering and Design Magazine*, 50–58 (1992).

45. K. Kikushima, "Polarization Dependent Distortion Caused by Isolators in AM-SCM Video Transmission Systems," *IEEE Photonics Technology Letters* **5**, 578–580 (1993).

46. K. Kikushima, Ko-ichi Suto, H. Yoshinaga, and E. Yoneda, "Polarization Dependent Distortion in AM-SCM Video Transmission Systems," *Journal of Lightwave Technology* **12**, 650–657 (1994).

47. R. H. Stolen, "Polarization Effects in Fiber Raman and Brillouin Lasers," *IEEE Journal of Quantum Electronics* **QE-15**, 1157–1160 (1979).

CHAPTER 10

EDFA-BASED WDM MULTICHANNEL AM/QAM VIDEO LIGHTWAVE ACCESS NETWORKS

Chapter 9 dealt with the transmission impairments in WDM SCM lightwave systems operating over relatively short links (< 30-km). Consequently, these links do not require the use of in-line EDFAs. However, typical HFC access networks, particularly in densely populated urban areas, employ relatively long optical links (> 50-km) or require a large number of optical splits. Cable TV operators are responding to the increasingly high optical power demands by deploying high-power in-line optical amplifiers, particularly EDFAs, throughout their networks. To address these issues and to build on the lightwave transmission impairments discussion in Chapter 9, Section 10.1 starts with the current architecture in WDM multichannel AM/QAM video lightwave access networks using cascaded EDFAs. The transmission performance results in the presence of various lightwave degradation mechanisms in these networks with cascaded EDFAs such as nonlinear distortions, SPM, EPM, polarization effects, and gain tilt are reviewed for both AM-VSB and QAM channels. In Section 10.2 the problem with the current HFC networks is explained. This problem leads the current evolution of the HFC architecture toward two-way DWDM access networks. The architecture of a DWDM downstream network carrying analog/digital video channels, high-speed data, as well as IP telephony is reviewed in Section 10.3. The architecture of return-path DWDM network is also reviewed in Section 10.4, including both frequency-stacking and digitized return-path schemes.

10.1 Architecture and Performance of Multichannel AM-VSB/QAM Video Lightwave Networks

As we discussed in Chapter 9, in order to maintain an acceptable AM CNR, the optical power budget of a video lightwave system based on a 1310-nm DM-DFB laser transmitter is typically limited to 10-dB. This moderate power margin imposes an upper limit on the length of the optical link and the number of optical splits. As pointed out earlier, many applications such as long-distance super-trunking and broadband networks based on passive optical network architectures require a much higher power link budget than can be provided using a DM-DFB laser transmitter. Figure 10.1 shows, for example, an EDFA-based multi-wavelength multichannel hybrid AM-QAM video lightwave transmission network, in which the master cable TV headend is typically connected by primary self-healing fiber-optics rings to the primary headends or the hubs in a large metropolitan area [1]. A self-healing fiber-optic ring has at least one redundant fiber ring with automatic-protection switching, which switches the traffic when equipment or point-to-point failures occur along the ring. Ultra-high-capacity video trunking can be achieved by using high-density WDM multiplexers and demultiplexers with high-power cascaded EDFAs [2]. In the primary fiber ring, multiple wavelengths in the 1550-nm band are transmitted bidirectionally with each wavelength carrying a mixture of AM and digital video signals at both RF passbands, using 64/256-QAM modulation formats, and at baseband at various rates such as OC-48 and OC-192. The secondary rings are connecting the primary hubs to the secondary hubs or headends. Notice that in a secondary fiber ring, only a few wavelengths are demultiplexed and transmitted using 1550-nm laser transmitters and in-line EDFAs. At the secondary hubs, the 1550-nm based broadcast traffic can be switched to a 1310-nm based traffic for both narrowcasting and broadcasting services.

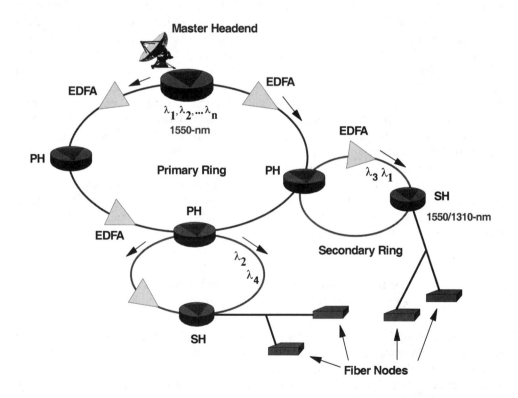

Figure 10.1 Multiwavelength and multichannel AM/QAM video lightwave trunking network using 1550-nm laser transmitters with cascaded EDFAs. PH = primary hub, SH = secondary hub. After Ref. [1] (© 1998 IEEE).

10.1.1 Multichannel AM/QAM Video Lightwave Trunking Systems

Recently, there have been many studies on the transport of only AM-VSB video signals in optically amplified 1550-nm lightwave systems [3–7]. The lightwave link performance has to overcome the transmission limitations imposed by SBS, SPM, clipping-induced impulse noise, and optical reflections. 1550-nm transmission systems using a DM-DFB laser transmitter, which were studied by Bulow, et al. [4], and Clesca, et al. [5], had an impressive optical power budget, partly due to the high laser-modulation index, with about 35 AM channels transmitted using a dispersion-shifted fiber (DSF). Muys, et al. [6] reported on 50 AM channel transmission using a 1550-nm EM-DFB laser transmitter with three cascaded 980-nm pumped EDFAs and a standard SMF. The advantage of such a system over the DM system is that the intermodulation distortions are essentially independent of the number of EDFAs used, since the EM laser transmitter generates virtually no frequency chirp, thereby preventing interaction with the fiber dispersion. Kuo, et al. [7] demonstrated the transmission of 80 AM channels with low CSO and CTB distortions over 100-km of a standard SMF. Using a differential detection scheme, which consisted of a pair of cable TV receivers followed by a broadband out-of-phase combiner, the authors showed that the AM CNR can be increased by 3-dB to at least 52.5-dB while maintaining a CSO distortion below –65-dBc over the AM-frequency band. A 120-km transmission of 79 simulated AM channels and four 64/256-QAM channels over a standard SMF with two cascaded EDFAs has been demonstrated by the authors [8].

Before reviewing the performance characteristics of a multichannel AM/QAM video lightwave trunking system, let us discuss the experimental setup, which is similar to the one shown in Figure 9.14 [8]. The lightwave link consisted of 120-km of a standard SMF with two in-line EDFAs as shown in Figure 10.2. The EM-DFB laser transmitter has built-in mechanisms for linearization and suppression of the SBS effect. The optical output from the laser transmitter was 15-dBm, and was 13-dBm from each in-line EDFA when operating in deep saturation. At

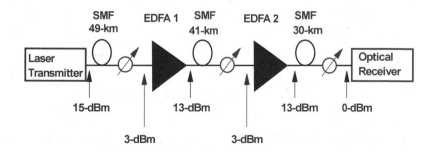

Figure 10.2 Block diagram of a 120-km lightwave transmission system experiment. After Ref. [8] (© 1996 IEEE).

the cable TV receiver, the composite video signal was detected and separated into analog and QAM channels using high/low-pass filters.   The signal corresponding to each of the four QAM channels was downconverted to the IF frequency, demodulated, and then fed to an error detector for BER analysis.

10.1.2 Multichannel AM-VSB Video Lightwave Trunking Results

Performance of the multichannel AM-QAM video lightwave trunking system is first analyzed for SMF lengths up to 50-km without using the in-line EDFAs. The received optical power is set at 0-dBm using a variable optical attenuator. Figures 10.3 and 10.4 show the measured AM CNR with the CSO and CTB distortions (without the EDFAs) at an AM frequency of 295.25-MHz (channel 36) as a function of the AM modulation index per channel. At AM modulation index of 3.1% per channel, the worst-case measured CNR (channel 78) was 53.7-dB with CTB distortions less than –65-dBc and CSO distortions less than –70-dBc.

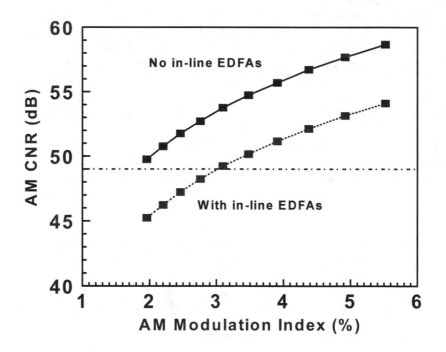

Figure 10.3 Measured AM CNR at frequency 295.25-MHz (channel 36) as a function of the AM modulation index with and without the in-line EDFAs. After Ref. [8] (© 1996 IEEE).

In order to select the proper operating point of the AM-QAM video lightwave trunking system with the cascaded EDFAs, the optical input power to each in-line EDFA needs to be determined. It is desirable to operate the EDFA in saturation to minimize the CNR penalty due to the amplifier's beat noise. With an optical power of 0-dBm at the receiver, the optimum optical input power of 3-dBm to each in-line EDFA is obtained by balancing the maximum possible AM CNR with the overall system link budget requirement. The AM CNR is increased by only 1-dB when the optical input power to the EDFA is increased from 3-dBm to 6-dBm, for example. Of course, the 1-dB addition to the AM CNR reduces the optical link budget by 3-dB. In contrast, some high-power in-line EDFAs, which have an NF greater than 5-dB, may require optical input power levels greater than 3-dBm [9]. This is due to the requirement from AM-QAM video lightwave trunking links to be able to transport at least 70 AM channels, each with a CNR greater than 50-dB, at a transmission distance of 100-km or more.

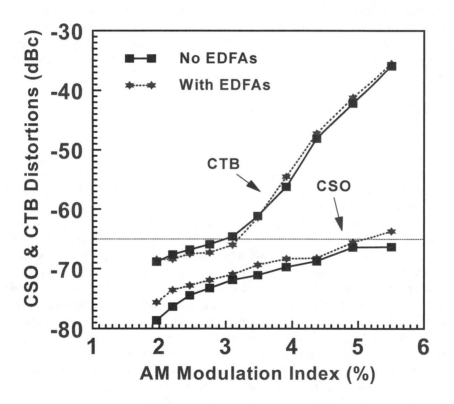

Figure 10.4 Measured CSO and CTB distortions versus the AM modulation index with and without the in-line EDFAs. After Ref. [8] (© 1996 IEEE).

The measured AM CNR with the CSO and CTB distortions at the AM frequency 547.25-MHz (channel 78) as a function of the AM modulation index per channel for an AM video trunking system with the cascaded EDFAs are shown in Figures 10.3 and 10.4. The optical input power to each in-line EDFA is set at 3 dBm, which is well above the input power for reaching the saturation point of these amplifiers. Notice that the measured AM CNR is decreased by about 4.5-dB caused by the addition of two in-line EDFAs at the optical input power level of 3-dBm. The transmitted AM CNR can be easily calculated using Equation (6.35). The measured CNR degradation is in good agreement with the calculated CNR penalty. Figure 10.4 shows that the use of in-line EDFAs has little effect on the CSO and CTB distortions for a single-wavelength operation. Furthermore, it suggests that degradation in the CSO and CTB distortions due to the residual chirp from the optical modulator in the 1550-nm EM-DFB laser transmitter is also negligible. The actual operating point of the transmitter is chosen by maximizing the AM CNR while maintaining distortion below acceptable levels. For example, according to Figures 10.3 and 10.4, the worst-case AM CNR at the AM frequency 547.25-MHz is 49.3-dB with CSO/CTB distortions less than −65-dBc at an AM modulation index of 3.1% per channel. Correspondingly, the optical power budget for this 120-km AM/256-QAM video lightwave link is 35-dB.

10.1.3 Multichannel AM-VSB/QAM Video Lightwave Trunking Results

The performance of 64/256-QAM digital channels is typically analyzed in terms of their uncoded BER. Figure 10.5 shows, for example, the measured BER of 64-QAM and 256-QAM channels as a function of the QAM SNR. The RF power of the 256-QAM channels was 12-dB below that of the AM channels. The solid lines in Figure 10.5 represent the theoretically predicted 64-QAM and 256-QAM BER (uncoded) versus the QAM SNR in a back-to-back configuration (no laser transmitter) with the assumptions of additive white Gaussian noise and Gray coding, included for comparison. While nearly error-free transmission was obtained for the 64-QAM channels, it is observed that a BER floor starts to emerge at around 35-dB SNR for 256-QAM channels, which were transmitted with the 79 AM channels at an AM modulation index of 3.1% per channel. This is likely due to the clipping-induced impulse noise and nonlinear distortion within the QAM band, which was generated by the AM channels operating at a much higher RF level [1, 10]. Although the BER performance can be improved by operating the 256-QAM channels at higher QAM signal levels, the increased QAM channel loading may cause the degradation of AM channel quality. The difference in transmission performance between the 64-QAM and the 256-QAM channels in the presence of 79 AM channels can also be illustrated using the QAM modulation index. Figure 10.6 shows the BER of the 64/256-QAM channels with and without the in-line EDFAs as a function of the QAM modulation index per channel. The QAM signal level for the transmission link with the in-line EDFAs to obtain an equal BER is found to be as much as 2.5-dB higher than the signal level without the in-line EDFAs. This is expected, of course, considering the added ASE noise from the in-line EDFAs. However, no other QAM channel impairment due to the amplifiers is observed under

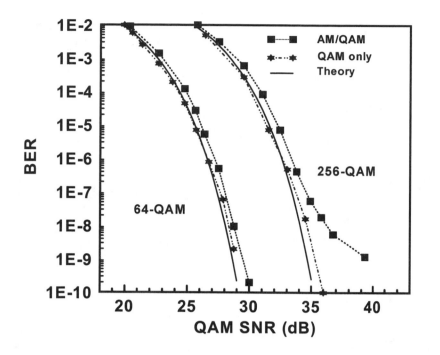

Figure 10.5 Measured uncoded BER of the 64-QAM and 256-QAM channels with and without the AM channels. The theoretical performance for only the 64-QAM and 256-QAM channels is also shown for comparison. After Ref. [8] (© 1996 IEEE).

the set operating condition. Figure 10.6 also indicates that at uncoded BER of 10^{-9}, the 256-QAM channels require almost three times the QAM modulation index of the 64-QAM channels operating at the same low BER. As the QAM modulation level increases from 64 to 256, the minimum distance between QAM clusters on the constellation map decreases, and thus the 256-QAM signals become more sensitive to intersymbol interference due to impulse noise and other impairments.

Do the transmitted 256-QAM channels affect the AM channels?

To answer this question, the AM CNR with the CSO and CTB distortions of the AM channels were measured as a function of the 256-QAM modulation index for the present system with four 256-QAM channels as shown in Figure 10.7. The AM modulation index remained fixed, while the QAM modulation index was increased up to 6% per channel. Notice that the AM CNR starts to deteriorate at a 256-QAM modulation index greater than about 2.5% per channel.

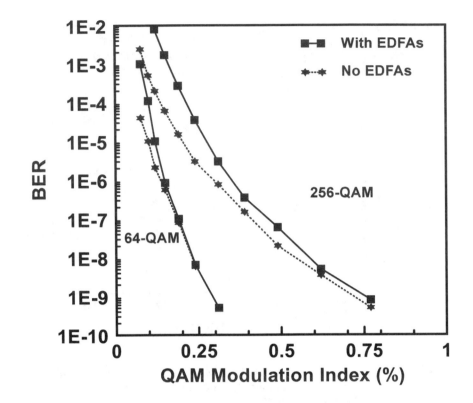

Figure 10.6 Measured uncoded BER of the transmitted 64-QAM and 256-QAM channels as a function of the QAM modulation index with and without the in-
line EDFAs. After Ref. [8] (© 1996 IEEE).

Figure 10.8 shows the calculated AM CNR (without in-line EDFAs) versus the 256-QAM
modulation index for a hybrid AM-QAM lightwave system with (A) 79 AM channels and four
256-QAM channels, (B) 79 AM channels and 33 QAM channels, and (C) 40 AM channels and
72 QAM channels. The model takes into account only the effect of the nonlinear clipping distortion at the laser transmitter [10]. As the QAM modulation index is increased, the four QAM
channels behave like additional AM channels, increasing both the observed CSO and CTB distortions. Thus, the observed AM CNR degradation (Figure 10.6) is mainly due to the increased
CSO and CTB distortions and not the additional clipping noise from the QAM channels.
However, in a fully loaded hybrid AM-QAM video transmission system [Figure 10.8, B], the
AM CNR degradation due to clipping distortion from both the AM and QAM channels appears
at about 3% 256-QAM modulation index per channel. The impact of the QAM channels on the

AM CNR is best demonstrated by case (C), in which the number of AM channels is reduced by one-half, while the number of QAM channels doubles. Notice the steeper AM CNR degradation at the QAM modulation index of 3% due to the increased nonlinear clipping distor tion. These results demonstrate that even in a fully loaded hybrid AM/256-QAM video lightwave transmission system operating under normal conditions, the effect of the QAM channels on the AM channels is essentially negligible. By using an FEC scheme, such as ITU-T J.83 Annex A, which provides a 4.8-dB coding gain at BER of 10^{-9}, the nonlinear distortion effect on the AM channels is reduced since the 256-QAM signal levels can be kept relatively low compared to the AM signals ($\approx$ −6-dBc).

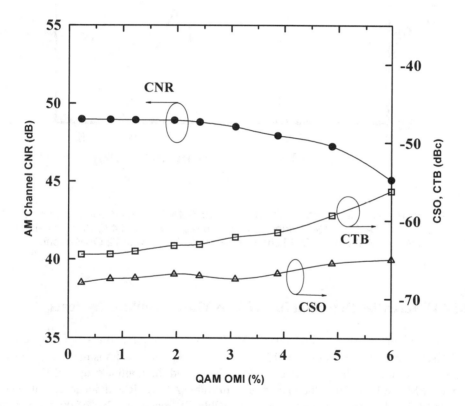

Figure 10.7 Measured AM CNR (solid circles) and CSO (triangles) and CTB (squares) distortions as a function of the 256-QAM modulation index. After Ref. [8] (© 1996 IEEE).

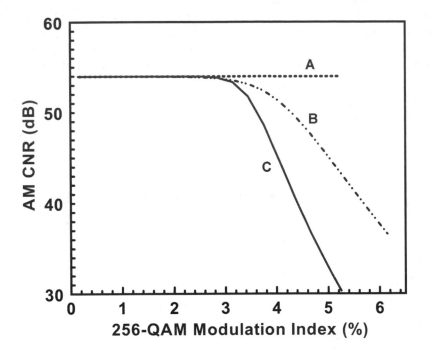

Figure 10.8 Calculated AM CNR versus the 256-QAM modulation index for a hybrid
AM/256-QAM video system with (A) 79 AM with 4 QAM channels, (B) 79
AM with 33 QAM channels, and (C) 40 AM with 72 QAM channels.

10.1.4 Differential Detection in AM-VSB Video Trunking Systems

As explained previously, multichannel AM-VSB video trunking links typically use EM-DFB laser transmitters operating at 1550-nm for a long-distance transmission ($\geq$ 100-km). However, the noise built up in the cascaded EDFAs and the nonlinear optical fiber effects, such as SPM and EPM, limit the maximum achievable AM CNR with an acceptable level of CSO and CTB distortions as measured at the cable TV receiver. One of the techniques to improve the AM CNR without degrading the CSO and CTB distortions is called the differential detection method [7, 11]. Figure 10.9 shows the differential detection scheme in a 50-km multichannel AM transmission system using a 1550-nm EM-DFB laser transmitter.

As we learned in Chapter 4, under the proper biasing condition, a dual-output LiNbO$_3$ MZ intensity modulator produces two signals with the same intensity, but 180° out of phase with each other. Using two optical receivers at the other end of the fiber, the RF outputs are combined in phase as shown in Figure 10.9. If the total fiber span of the two links is less than (c/n·Ω), where c/n is the speed of light in the fiber and Ω is the highest RF frequency, then the RF carriers are added coherently, increasing by 6-dB. However, the noise is added incoherently, increasing by 3-dB. Thus, the resultant CNR is increased by only 3-dB. The CSO and CTB distortions are canceled in the combined RF signal. Figure 10.10 shows the measured AM CNR versus the AM channel frequency after 100-km of standard SMF with two cascaded EDFAs using (a) a single receiver and (b) dual receivers. Below about 300 MHz, the AM CNR is improved by about 3-dB as expected. However, above 300-MHz, the AM CNR is improved by as much as 4.5-dB. This additional improvement of 1.5-dB in the AM CNR at high cable-TV frequencies is due to the cancellation in the phase-to-amplitude noise, which is correlated. According to Piehler et al., the differential detection scheme also reduced the measured CSO distortion, which was induced by dispersion and SPM, by as much as 20-dB compared with the same link using a single receiver [11]. Figure 10.10 (b) shows the measured AM CNR versus the channel frequency for 80 AM video channels, which were transmitted over 100-km of standard SMF at a low phase-modulation index ($\beta \leq 3$).

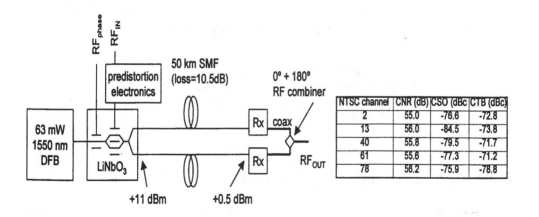

NTSC channel	CNR (dB)	CSO (dBc)	CTB (dBc)
2	55.0	-76.6	-72.8
13	56.0	-84.5	-73.8
40	55.8	-79.5	-71.7
61	55.6	-77.3	-71.2
78	56.2	-75.9	-78.8

Figure 10.9 Differential detection scheme in a 50-km multichannel AM transmission system using a 1550-nm EM-DFB laser transmitter. The measured CNR, CSO, and CTB distortions are shown in the table. After Ref. [7] (© 1996 Optical Society of America).

Thus, the advantage of the differential detection scheme, which is similar to the distortion cancellation by a push-pull amplifier, is to boost the AM CNR by 3-dB or more without degrading the CSO and CTB distortions at long-distance lightwave links. However, two optical receivers and one RF combiner must be used on two separate fibers.

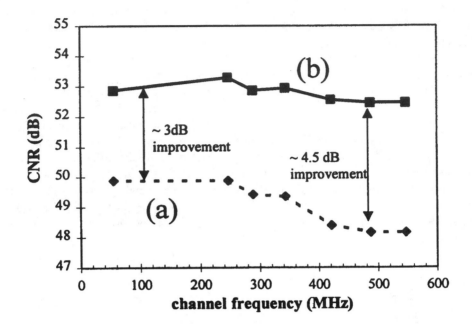

Figure 10.10 Measured AM CNR as a function of the AM channel frequency for 100-km optically amplified links with (a) single receiver, and (b) dual receivers with differential detection. After Ref. [7] (© 1996 Optical Society of America).

10.1.5 SPM and EPM Effects in AM Video Trunking Systems

As we learned in the previous chapter, CSO distortions induced by SPM and EPM effects degrade the performance of multichannel AM video links using 1550-nm EM-DFB laser transmitters. The impact of the SPM and EPM effects is particularly important when designing a long-distance (>100-km) AM video link with cascaded EDFAs. The CSO distortion can be still be approximated by Equation (9.67), where

$$\overline{z^2(L)} = (L - L_{eff})/\alpha \tag{10.1}$$

is replaced by

$$\overline{z^2(L)} + \left[1 + e^{-\alpha L/N}\right] \cdot \sum_{i=1}^{N-1} \overline{z^2}\left[(N-i)\cdot L/N\right]$$

where N is the number of EDFAs, and L/N is the fiber span between adjacent EDFAs. Figure 10.11 shows the measured, numerically calculated, and analytical calculation of the CSO distortion at cable TV channel 78 (worst case) as a function of the total fiber span, assuming AM video transmission link with equal span EDFAs. The fiber span between adjacent EDFAs was 60-km with launch optical power to the EDFA of 12-dBm. The optical

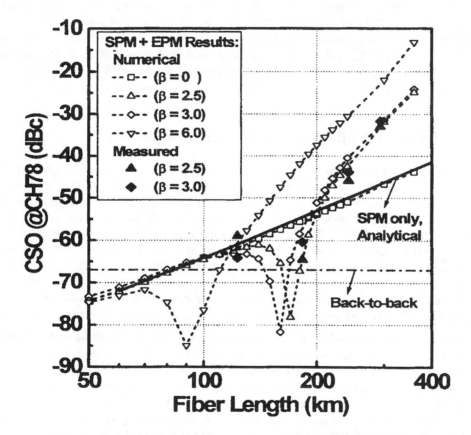

Figure 10.11 Measured, numerically calculated, and analytical CSO distortion at channel 78 as a function of the total fiber length in multichannel AM video trunking link, where the fiber span between adjacent EDFAs is 60-km, and with launched optical power at each EDFA of 12-dBm. After Ref. [12] (© 1999 IEEE).

modulation index per channel was 3% (78 AM channels) at operating wavelength of 1561.1-nm. A single-tone phase modulation at 1.9-GHz was used. The numerical calculations included both SPM- and EPM-induced CSO distortions and were based on a numerical solution to the envelope wave equations (Equations 9.62 and 9.63) using the split-step Fourier transform method. The accuracy of the numerical results was set to differ by less than 1-dB from the analytical results of the CSO distortion when the phase-modulation index was equal to 0 ($\beta = 0$).

Figure 10.11 indicates that when the phase-modulation index is relatively small ($\beta \leq 3$), and the total transmission distance is relatively short (e.g., < 120-km), the analytical results agree with the numerical and experimental results. At longer distances and/or larger phase-modulation index, the worst-case CSO distortion is determined by the combined effect of SPM and EPM. In other words, the predicted CSO distortion analysis based only on the SPM effect significantly underestimates the actual measured CSO distortion. Previous work on the SPM-induced CSO distortions was based on perturbation analysis of the envelope-wave equation and did not include the dependence of frequency, β, and L in the phase-modulation to intensity-modulation conversion, as well as carrier compression.

10.1.6 Polarization Effects in AM Video Trunking Systems

Large polarization-dependent CSO distortions can be generated in AM video trunking links with cascaded EDFAs, since each EDFA uses several optical isolators [13]. This is in spite of the fact that there may not be a measurable polarization-dependent distortion generated from a single EDFA. Figure 10.12 shows the measured CSO distortions versus the rotation angle of linearly polarized light input to cascaded EDFAs when (A) a direct-modulation scheme, and (B) an external-modulation scheme are adopted. The tested EDFAs use type B isolators for the input ports and type A isolators for the output ports. Notice that the CSO distortion degrades by more than 25-dB with 11-cascaded EDFAs compared with the no EDFA case under the direct-modulation scheme. In contrast, the CSO distortion varies by about 5-dB when the external-modulation scheme is used with a maximum extinction ratio.

However, it should be pointed out that the input polarization state to the $LiNbO_3$ MZ modulator under the minimum insertion condition is different from that under the maximum extinction condition. As a result, the CSO distortion shows a large degradation (> 15-dB) with cascaded EDFAs as a function of the rotation angle of the linearly polarized light under external modulation.

The CSO distortion in multichannel AM/QAM video transmission link with cascaded EDFAs fluctuates with time. Figure 10.13 shows the measured CSO distortion with and with 11-cascaded EDFAs during a 20-minute period. This is due to the fact that the polarization state of the propagating light fluctuates within each EDFA, resulting in a fluctuating CSO distortion.

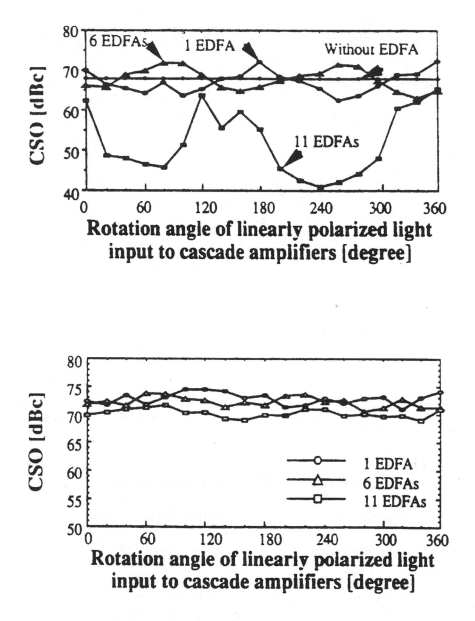

Figure 10.12 Measured CSO distortions versus the rotation angle of linearly polarized light input to cascaded type-γ EDFAs for (A) direct-modulation scheme (top), and (B) external-modulation scheme with maximum extinction ratio (bottom). After Ref. [13] (© 1994 IEEE).

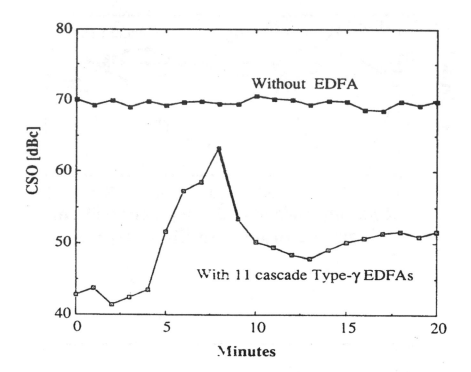

Figure 10.13 Measured CSO distortion level versus time with and without eleven cascaded EDFAs using direct modulation. After Ref. [13] (© 1994 IEEE).

10.1.7 Gain Tilt Distortion in AM Video Trunking Systems

A variation of the optical fiber amplifier gain with wavelength at a given optical input power level is called gain tilt. As we have learned in Chapter 6, EDFA can be used as a post amplifier in 1550-nm DM-DFB laser transmitter or as an in-line power amplifier to boost up the optical power. AM-VSB video lightwave trunking links using a 1310-nm DM-DFB laser transmitter with an in-line Praseodymium doped fiber amplifier (PDFA) are also included. In such laser transmitters, the wavelength-dependent gain of the EDFA interacts with the frequency chirping of the DM-DFB laser, primarily through the phase-to-intensity conversion, to enhance the CSO distortions [14–16]. Various authors have used both a simple and a rigorous analysis to explain the gain tilt distortion phenomenon. To derive an expression for the CSO distortion induced by gain tilt, let us follow the analysis of Clesca, et al. [14]. It is assumed that a single CW carrier modulates the DFB laser current, and the EDFA input optical power and wavelength are given by

$$P_{in}(t) = P_0[1 + m\sin(2\pi ft)] \qquad (10.3)$$

$$\lambda(t) = \lambda_0 - \Delta\lambda \cdot \sin(2\pi ft) \qquad (10.4)$$

where P_0 is the average input optical power, m is the optical modulation index, f is the modulation frequency, λ_0 is the average signal wavelength, and $\Delta\lambda$ is the wavelength variation due to modulation. Notice that the input optical power and wavelength are in phase and 180° out of phase with the current modulation. Since the frequency scale of the EDFA gain dynamic is much smaller than the modulation frequency f, the EDFA gain can be expanded in a Taylor series as follows:

$$G[P_{in}(t), \lambda(t)] \cong G(P_0, \lambda_0) + \left[\frac{dG(P_0, \lambda)}{d\lambda}\right]_{\lambda_0} \cdot [\lambda(t) - \lambda_0] \qquad (10.5)$$

where $dG/d\lambda$ at λ_0 is the EDFA gain tilt, namely, the wavelength-dependent gain as experienced by the modulated signal. Higher order terms have been neglected. The EDFA output power is simply the input optical power times the gain. Thus, using Equations (10.3), (10.4), and (10.5), the EDFA output power is given by

$$P_{out}(t) = P_{in}(t) \cdot G[P_{in}(t), \lambda(t)] \cong$$

$$G(P_0, \lambda_0) \cdot P_0 + G(P_0, \lambda_0) \cdot P_0 m\sin(2\pi ft) + \frac{1}{2}\left[\frac{dG(P_0, \lambda)}{d\lambda}\right]_{\lambda_0} \cdot \Delta\lambda m P_0 \cos(4\pi ft)$$
$$(10.6)$$

The ratio of the second-order harmonic distortion to the carrier (2HD/C) can easily be obtained from Equation (10.6). Recall from Chapter 2, to obtain the CSO distortion (relative to the carrier), one needs to multiply the 2HD/C by the CSO product count (N_{CSO}) and add 6-dB. Thus, the gain-tilt-induced CSO distortion expressed in dB is given by

$$CSO = 20 \cdot \log\left[\sqrt{N_{CSO}} \frac{\left[\dfrac{dG(P_0, \lambda)}{d\lambda}\right]_{\lambda_0} \cdot \Delta\lambda}{G(P_0, \lambda_0)}\right] \qquad (10.7)$$

The CSO distortions can be improved or degraded, depending on the phase of the EDFA gain tilt distortions relative to the distortions generated by the DFB laser transmitter. Clesca, et al., [14] have measured CSO distortion degradations at the EDFA output for a positive gain tilt, and CSO distortion improvement for a negative gain tilt. One way to avoid the CSO distortion induced by EDFA gain tilt is by using an externally modulated DFB laser transmitter (see Section 4.4.2).

10.2 The Problem with the Current HFC Networks

As we learned in Chapter 1, the broadband HFC networks with tree-and-branch architecture enabled robust transmission of the broadcast AM-VSB video channels through standard SMF to the multiple optical fiber nodes. In addition, cable TV operators were able to consolidate the deployment of headend equipment using fiber-optic rings to connect between the master headend and secondary headends or hubs in major metropolitan markets. Thus, cable TV operators were able to cut cost and further improve the quality and availability of the original broadcast services. The development of QAM modulators and low-cost QAM receivers with MPEG-2 digital video sources enabled the cable operators to offer tens of new digital video services along with the traditional AM-VSB channels. The rapid deployment of 750-MHz HFC networks also allowed the cable TV operators and some telecom services to provide competitive access to various businesses in major markets.

In the middle of the 1990s, the existing HFC network architecture started to evolve in a new direction. This evolution can be attributed to the following forces in the marketplace:

(A) An increased demand for high-speed data access in residential areas.
(B) The delivery of highly targeted interactive digital services.
(C) An increased competition from various telecommunication and DBS service operators.
(D) Improved cable TV network scalability with reduced life cycle cost.
(E) Improvements in fiber-optic technology and cable TV network management.

These requirements and market forces are demanding that the cable operators rethink their existing HFC network architecture. Fig. 10.12 shows, for example, the HFC network architecture used by the cable TV operators in a major metropolitan area. The signals from the master headend to the primary and secondary hubs are transmitted over SMF using 1550-nm externally modulated (EM) DFB laser transmitters. The composite signals are a mixture of traditional broadcast analog signals with MPEG compressed digital video. These hubs may house SONET equipment with the associated ADMs or some proprietary technology as well as modems, routers, and servers for high-speed data. The optical signals at 1550-nm are converted to RF signals and then back to optical signals for transmission to various optical-fiber nodes typically using 1310-nm DM-DFB laser transmitters.

The problem with this network architecture is that the secondary hubs had grown very rapidly to become very large facilities that are expensive to maintain since they house both passive and active systems. Another problem with this architecture is the upgrading cost of an existing one-way 450-MHz HFC network for premium video and data services. This is because it required not only additional installation of fiber-optic cables for two-way transmission, but also architectural changes to the HFC network with additional equipment installations at the primary or secondary headends [17].

Recently, DWDM technology has emerged as the most significant and important option for the cable TV operators to solve their current bandwidth bottleneck in their HFC network architecture. The next section discusses the DWDM network architecture for both downstream and upstream transmission.

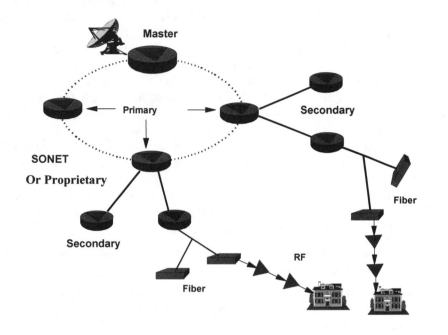

Figure 10.14 Conventional architecture of hybrid fiber/coax access network in a large metropolitan area.

10.3 Downstream DWDM Access Network Architecture

Simplified block diagrams of the downstream DWDM network architecture are shown in Figures 10.1 and 10.15 [18, 19]. The primary hubs are connected to the master headend, as shown in Figure 10.1, by self-healing primary fiber rings using SONET or some proprietary technology that can support bidirectional transmission. Up to 33 contiguous 64/256-QAM channels operating at RF center frequencies from 550-MHz to 750-MHz are loaded on each of the eight EM-DFB laser transmitters operating at International Telecommunication Union

(ITU) standard wavelengths in the primary hub. The ITU standard wavelength grid for fiber-optic telecommunication networks spans from 1520.25-nm to 1565.4961-nm with 100-GHz separation (i.e., about 0.8-nm) between adjacent wavelengths. Notice that each of the downstream DWDM laser-transmitters in Figure 10.15 may be loaded with different MPEG-2 video streams. The 1 x 8 multiplexed optical signal is transmitted to the primary or secondary hubs through, for example, 40-km of a conventional SMF. At the primary or secondary hubs, which are sometimes called optical transition nodes (OTN), the optical signal carrying the 64/256-QAM channels is amplified using an in-line EDFA with, for example, +17-dBm optical output power, and then optically 1 x 8 demultiplexed. The OTN, which is shown within the dotted line in Figure 10.15, consists of primarily passive optical components (see Section 3.2). Notice that only the EDFAs at the OTN require a temperature-controlled environment, which can be provided even at remote locations. The resultant optical signal can be multiplexed with typically 79 AM-VSB video channels, which are transmitted over a separate 40-km standard SMF using an EM-DFB laser transmitter with two different in-line EDFAs. Each primary hub can be connected to about five secondary hubs. From the secondary hubs, the optical signals can be transmitted to about 16 fiber nodes with each fiber node for every 1000–1200 homes passed. To further reduce cost, a 1 x 2 optical coupler can replace the 1 x 2 AM/QAM Mux at the OTN or secondary hub. The detected RF frequency spectrum at the output of the optical receiver is similar to the RF spectrum from a single-wavelength HFC network carrying both AM and QAM channels. There are three main advantages to this DWDM network architecture, as follows:

(A) Elimination of the need for an optical-to-RF-to-optical conversion at the primary or secondary hubs in order to deliver narrowcast video services.

(B) Increased network scalability and flexibility. Using, for example, dynamic wavelength allocation and/or add-drop wavelength multiplexing, new digital services can be added at one wavelength and/or remove at another wavelength in a specific serving area.

(C) Reduced operation, maintenance, and upgrade costs at the secondary hubs. This is possible since most of the downstream DWDM technology is optically passive, except for the EDFAs. Various cable TV operators anticipate about a 30% reduction in the cost of DWDM technology as well as 1550-nm laser transmitters in the next several years. In addition, only the EDFAs at the secondary hubs need a temperature-controlled environment.

Increased network flexibility enables cable TV operators to deliver the desired mix of narrowcast digital services and broadcast analog services to the targeted service area. For example, one of the main cable TV operators in the U.S. is planning to initially deploy one narrowcast laser transmitter per 10,000 homes passed, and incrementally add wavelength-specific laser transmitters as the subscribers' take rates increase [17]. At a take rate of 40% or more, it is expected that one dedicated narrowcast transmitter may be deployed for each 1000 homes passed in the network.

The disadvantages of this DWDM network architecture are as follows:

(A)   Increased initial deployment cost at the secondary hub since the subscribers' take rates for narrowcast services may be below the break-even point for cable TV operators.

(B)   Increased secondary-hub complexity as compared with the traditional HFC network architecture design.

As the DWDM network architecture becomes more mature with higher subscriber take rates for the premium digital services, the number of AM-VSB channels is expected to shrink to 30 channels or less, while the number of 64/256-QAM channels is expected to grow to more than 70 channels. In this case, a 1 x 4 AM/QAM mux can replace the 1 x 2 AM/QAM mux (Figure 10.15) to enable a wider mix of interactive digital services that are carried by different wavelength-specific WDM laser transmitters.

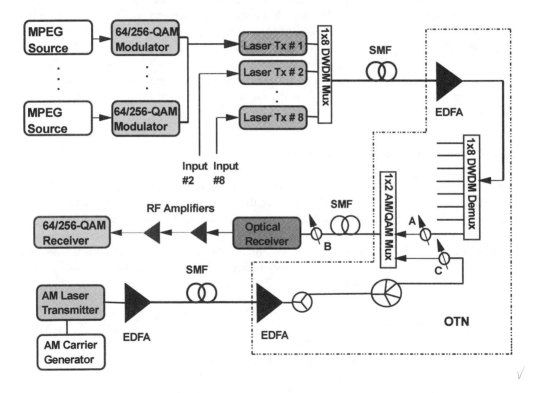

Figure 10.15 Basic downstream DWDM network architecture. The OTN is shown within the dotted line. After Ref. [19] (© 1999 IEEE).

10.4 Upstream DWDM Access Network Architecture

The return-path network architecture is also modified by the DWDM technology. Figure 10.16 shows a simplified block diagram of the upstream DWDM network architecture. It is assumed that each subscriber is connected to the closest fiber node through a coaxial cable plant. At the fiber node, a low-cost uncooled DFB laser transmitter operating at either a 1310-nm or a 1550-nm wavelength band will be used to transmit the high-speed data in the 5–42-MHz band to the secondary hub. These low-cost upstream laser transmitters typically have an optical link budget of 5-dB at 1550-nm or 7-dB at 1310-nm. The DWDM laser transmitters, which are located at the secondary or primary hubs, are typically operating in 1550-nm wavelength band with an optical link budget of 8-dB. Using a 1 x 4 DWDM multiplexer, the optical signal is transmitted, for example, through 40-km of a standard SMF to the primary hub. At the primary hub, the optical signal is amplified, for example, with a 13-dBm in-line EDFA, and optically demultiplexed to four different optical receivers as shown in Figure 10.16. Each cable subscriber can send high-speed data to the headend using a time-division-multiple access (TDMA) scheme according to the DOCSIS protocol standard [20]. Thus, the DWDM upstream network architecture with the TDMA scheme can provides at least four-fold traffic capacity over a standard HFC network architecture with TDMA. Two recently developed and field-deployed return-path technologies, namely, the frequency-stacking scheme (FSS) and the digitized return-path transport are reviewed in the following sections.

10.4.1 Frequency-Stacking Scheme

The demand for more upstream bandwidth from an increasingly larger number of cable subscribers generates additional pressure on the cable operators to provide larger return-path capacity and/or larger allocated bandwidth per subscriber. It turns out that the return-path network traffic capacity can be significantly expanded using the FSS in combination with TDM [18,19, 21]. Figure 10.16 shows an optimized upstream DWDM network architecture with FSS at both the secondary hub and fiber nodes. The FSS system typically consists of four main components as follows:

- Return-path upconverter, which is sometimes called frequency stacker, is a multistage phase-locked upconverter that translates the return-path frequency bands (5–42-MHz) to higher RF frequency bands. For example, Figure 10.17 shows the composite RF spectrum of four upconverted return-path frequency bands (5–42-MHz) using, for example, a Motorola FSS system. A reference pilot tone is generated at a frequency of 370-MHz in the FSS in order to synthesize the four-band stack. The pilot tone is transmitted along with the upconverted signal.

- High-performance return-path laser transmitter, typically an isolated DFB laser with TEC, with optical output power of 5-dBm or more. It should be pointed out that the laser RIN is required to be sufficiently low for robust transmission.

- Block converter receiver, which is an optical receiver with wide dynamic range for power control of both the optical and the RF signals. The transmitted pilot tone is utilized in the block down-converter unit to synchronize the down-conversion process, thus removing any frequency offset errors.

- Block downconverter, which is sometimes called frequency destacker, is controlled by the locally generated pilot tone and is responsible for desynthesizing the composite RF spectrum into four separate return-path frequency bands. The composite RF signal is used to drive each of the DWDM upstream laser transmitters at the secondary hub.

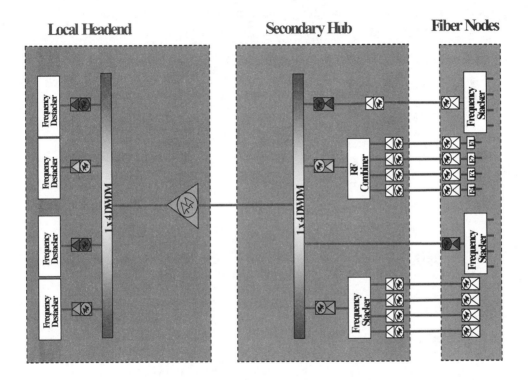

Figure 10.16 DWDM upstream network architecture with frequency stacking at both the fiber nodes and at the secondary hub. After Ref. [18].

The composite output RF signal from each optical receiver is transmitted to the block down-converter unit, which extracts the four separate 5–42-MHz upstream bands. Then, each of the high-speed data bands can be routed and transmitted to return-path burst receivers, and/

or other network control and management equipment. It should be pointed out that the FSS is best utilized at the optical fiber nodes instead of the secondary headends in the upstream DWDM network architecture.

Let us look more closely at the combined FSS/DWDM expansion process. A single shared traditional return-path 37-MHz segment provides 74-kHz bandwidth per home, assuming a 500-home-passed node, for example. The implementation of frequency stacking (4 band) increases the shared segment to 148-MHz, thereby increasing the return-path bandwidth per home to 296-kHz. The implementation of the combined scheme of frequency stacking and DWDM increases the return-path bandwidth segment to 32 times the return-path bandwidth segment, or 1184-MHz, thereby increasing the return bandwidth per home to 2.368-MHz. To meet subscribers' demands for more bandwidth beyond the 32 times bandwidth expansion factor, the cable operator can further reduce the number of homes passed for each optical fiber node.

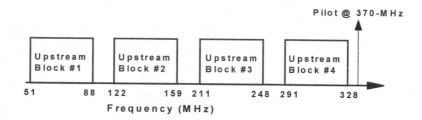

Figure 10.17 The composite RF spectrum of four upconverted upstream frequency blocks (5–42-MHz). After Ref. [19] (© 1999 IEEE).

10.4.2 Digital Return-Path Transport

There are two important factors that make the replacement of analog return-path transport with a digital technology desirable. These factors are a simplification of the reverse-path alignment, afforded by digital components, and a lowering of the return-path system cost [18]. These factors have been enabled by the continuous improvement in the digital component performance while their cost is declining, driven by high-volume applications. Figure 10.18 shows the basic element of a digital return-path in a DWDM network. The composite analog return-path waveform is converted to a sequence of digital words by an analog-to- digital converter (ADC) operating at a typical sampling rate of 100-MHz with 8–

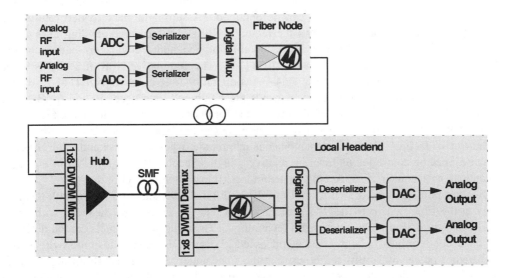

Fig. 10.18 Block diagram of digitized return path at the fiber node in DWDM network architecture. After Ref. [22] (© 2001 Optical Society of America).

12 bits of resolution. The parallel digital words are converted to a serial bit stream with the appropriate synchronization at the fiber node in order to recover the transmitted serial bit stream at the optical receiver output in the local headend. Using, for example, two 12-bit ADC (as shown in Fig. 10.18), which are currently being deployed by cable TV operators, the laser transmitter is modulated by an approximately 2.5-Gb/s TDM data stream (OC-48). The TDM data stream is optically multiplexed at the hub, and then demultiplexed at the local headend, where it is digitally demultiplexed and deserialized before it is presented to the digital-to-analog converter (DAC). The performance of the ADC is typically limited by its quantization noise, clipping noise, and distortions. Given the number of required resolution bits and the sampling frequency, the ADC noise density per unit bandwidth can be easily calculated from its SNR for a full-scale sine wave input. In analog fiber-optic links, the corresponding noise source is the laser RIN. The performance of various components, particularly laser transmitters, in return-path cable TV networks is typically characterized in terms of the noise-power ratio (NPR) curve. The NPR is simply the signal-to-noise plus the inband clipping-and-distortion ratio. The NPR measurement loads the tested device with a Gaussian noise containing a notch at a test frequency. The depth of the notch is measured with respect to the noise power as is varied over the operating range of the tested device [22, 23]. The NPR curve is bound by Gaussian noise at low input signal levels and by clipping noise and distortions at high input signal levels relative to the optimum input level for a full-scale sine wave. The dynamic range (DR) of the component under test captures the range of

input levels between the Gaussian noise limit and the clipping-noise limit for which greater or equal to a specific NPR value can be obtained. Assuming a fully loaded return-path DWDM network, Table 10.1 shows the required NPRs and the corresponding DRs at the optical receiver output in the local headend using the FSS and digital return-path technologies with 10-bit and 12-bit ADCs and DACs [24]. The cumulative NPR of 37-dB is based on the following factors. Assuming a 64-QAM operation, one needs C/(N + I) of 27-dB for robust transmission. If four upstream bands are combined, one needs to add 6-dB to the cumulative NPR. An additional 4-dB are needed for operational margin to guarantee the 27-dB at the subscriber's home. The required large DR at the local headend is primarily attributed to two factors. First, the instantaneous magnitude of ingress noise and other interference cannot be controlled and can be relatively high. Second, STBs and CMs in the cable plant can acts as interferers, forcing upstream laser transmitters into clipping.

Several points can be drawn from Table 10.1. First, 12-bit digitized return-path technology at the fiber node provides the necessary NPR and DR at the local headend with about 2-dB margin over a 10-bit return-path technology. The return-path signals can be digitized at the secondary hubs (12-bits), and an uncooled 1310-nm DFB laser transmitter with a typical output power of 3-dBm can be used at the fiber node. Second, the 10-bit digitized return-path technology does not meet the necessary NPR and DR at the local headend. It should be pointed out that the required NPR and DR at the fiber node must be achieved over a wide operating-temperature range (−20°C to +65°C).

The digitization of the return-path with TDM in a DWDM network has a significant advantage over analog fiber links. When the return-path analog signals are digitized, the DWDM network becomes transparent to them. In other words, the digital return-path signal can be efficiently transmitted throughout the network without a noticeable degradation.

Table 10.1 Performance requirements for different return-path technologies.
After Ref. [24].

Technology	NPR/DR (Node to SH)	NPR/DR (SH to PH)	Total NPR/DR (including RF)
DWDM (from fiber node)	40-dB/11-dB	42-dB/11-dB	37-dB/11-dB
DWDM+FSS (from fiber node)	38-dB/11-dB		37-dB/11-dB
DWDM+FSS (from SH)	42-dB/11-dB	42-dB/11-dB	37-dB/11-dB
Digitized at Node (12-bits)	38-dB/11-dB		37-dB/11-dB
Digitized at SH (12-bits)	40-dB/11-dB	42-dB/11-dB	37-dB/11-dB
Digitized at Node (10-bits)	36-dB/11-dB		Insufficient

Furthermore, digital signal-processing techniques can be used to minimize signal degradation before the ADC. Thus, the use of digital return-path technology in DWDM network architecture is very promising since it allows cable TV operators to reduce their network upgrade costs while maintaining network transparency and flexibility. As the prices of the digital components continue to decline, driven down by high volumes, while their performance improves, the deployment of digital return-path technology becomes more and more attractive.

References

1. S. Ovadia and C. Lin, "Performance Characteristics and Applications of Hybrid Multichannel AM-VSB/M-QAM Video Lightwave Transmission Systems," *IEEE Journal of Lightwave Technology* **16**, 1171–1186 (1998).

2. C. Lin, K. P. Ho, H. Dai, S. K. Liaw, and H. Gysel, "High-Capacity Digital and Analog Video Trunking and Distribution Based on Hybrid WDM Systems," *OFC '97 Technical Digest*, paper ThS3.

3. W. I. Way, M. M. Choy, A. Yi-Yan, M. Andrejco, M. Saifi, and C. Lin, "Multichannel AM-VSB Television Signal Transmission Using an Erbium-Doped Optical Fiber Power Amplifier," *IEEE Photonics Technology Letters* **1**, 343–345 (1989).

4. H. Bulow, R. Fritschi, R. Heidemann, B. Junninger, H. G. Krimmel, and J. Otterbach, "Analog Video Distribution System with Three Cascaded 980 nm Single-Pumped EDFA's and 73 dB Power Budget," *IEEE Photonics Technology Letters* **4**, 1287–1289 (1992).

5. B. Clesca, P. Bousselet, J. Aug'e, J. P. Blondel, and H. F'evrier, "1480-nm Pumped Erbium-Doped Fiber Amplifiers with Optimized Noise Performance for AM-VSB Video Distribution Systems," *IEEE Photonics Technology Letters* **6**, 1318–1320 (1994).

6. W. Muys, J. C. van der Plaats, F. W. Willems, H. J. van Dijk, J. S. Leong, and A. M. J. Koonen, "A 50-Channel Externally Modulated AM-VSB Video Distribution System with Three Cascaded EDFA's Providing 50-dB Power Budget Over 30 km of Standard Single-Mode Fiber," *IEEE Photonics Technology Letters* **7**, 691–693 (1995).

7. C. Y. Kuo, D. Piehler, C. Gall, J. Kleefeld, A. Nilsson, and L. Middleton, "High-Performance Optically-Amplified 1550 nm Lightwave AM-VSB CATV Transport System," *OFC '96 Technical Digest* **2**, paper WN2.

8. H. Dai, S. Ovadia, and C. Lin, "Hybrid AM-VSB/M-QAM Multichannel Video Transmission Over 120-km of Standard Single-Mode Fiber with Er-Doped Optical Fiber Amplifiers," *IEEE Photonics Technology Letters* **8**, 1713–1715 (1996).

9. S. Ovadia, "CNR Limitations of Er-Doped Optical Fiber Amplifiers in AM-VSB Video Lightwave Trunking Systems," *IEEE Photonics Technology Letters* **9**, 1152–1154 (1997).

10. Q. Shi and S. Ovadia, "Effects of Clipping-Induced Impulse Noise in Externally Modulated Multichannel AM/QAM Video Transmission Systems," *IEEE Transactions on Communications* **46**, 1448–1450 (1998).

11. D. Piehler, J. Kleefeld, C. Gall, C. Y. Kou, A. Nilsson, and X. Zou, Technical Digest, Optical Amplifiers and their Applications, Monterey, paper SaB2 (1996).

12. M. C. Wu, C. H. Wang, and W. I. Way, "CSO Distortions Due to the Combined Effects of Self- and External-Phase Modulations in Long Distance 1550-nm AM-CATV Systems," *IEEE Photonics Technology Letters* **11**, 718–720 (1999).

13. K. Kikushima, Ko-ichi Suto, H. Yoshinaga, and E. Yoneda, "Polarization Dependent Distortion in AM-SCM Video Transmission Systems," *IEEE Journal of Lightwave Technology* **12**, 650–657 (1994).

14. B. Clesca, P. Bousselet, and L. Hamon, "Second-Order Distortion Improvements or Degradations Brought by Erbium-Doped Fiber Amplifiers in Analog Links Using Directly Modulated Lasers," *IEEE Photonics Technology Letters* **5**, 1029–1031 (1994).

15. C. Y. Kuo and E. E. Bergmann, "Second-Order Distortion and Electronic Compensation in Analog Links Containing Fiber Amplifiers," *IEEE Journal of Lightwave Technology* **10**, 1751–1759 (1992).

16. J. Ohya, H. Sato, M. Mitsuda, T. Uno, and T. Fujita, "Second-Order Distortion of Amplified Intensity-Modulated Signals with Chirping in Erbium-Doped Fiber," *IEEE Journal of Lightwave Technology* **13**, 2129–2135 (1995).

17. O. Sniezko, "Video and Data Transmission in Evolving HFC Networks," OFC '98 Technical Digest, paper WE4 (1998).

18. O. Sniezko, "Reverse Path for Advanced Services – Architecture and Technology," *NCTA Technical Papers*, 11–30 (1999).

19. S. Ovadia, "Advanced Upstream CATV Access Technologies and Standards," *Digest of the IEEE/LEOS RF Photonics for CATV and HFC Systems*, 35–36 (1999).

20. CableLabs, *Data-Over-Cable Service Interface Specifications (DOCSIS): Radio Frequency Interface Specification*, SP-RFIv1.1-I03-991105 (1999).

21. T. Brophy, R. Howald, and C. Smith, "Bringing Together Headend Consolidation and High-Speed Data Traffic in HFC Architecture Design," *NCTA Technical Papers*, 189–204 (1999).

22. O. J. Sniezko and T. Werner, "Return-Path Active Components Test Methods and performance Requirements," Conference on Emerging Technologies Proceedings, 263–294 (1997).

23. L. West, "The Effect of Preamplifier Compression on Measured Noise Power Ratios," *IEEE Photonics Technology Letters* **12**, 924–926 (2000).

24. S. Ovadia, "Advanced Return-Path Cable TV Access Technologies," *OFC '01 Technical Digest* (2001).

CHAPTER 11

DATA-OVER-CABLE INTERFACE
SPECIFCATIONS (DOCSIS) PROTOCOL

The data-over-cable interface specifications (DOCSIS) 1.1 standard is the primary protocol in the US for high-speed data transmission using CMs operating over HFC network. The purpose of this chapter is to explain the operation principles of the downstream and upstream physical (PHY) layer (Sections 11.2 and 11.3) as well as the *medium access control* (MAC) layer of the DOCSIS protocol. Sections 11.5 to 11.8 discuss the MAC layer format, features, and operation. In particular, quality of service and an upstream packet fragmentation are discussed in Section 11.8, and the interaction between the CM and the CM termination system (CMTS) is discussed in Section 11.9. In addition, the different random access and contention resolution methods are reviewed in Section 11.6.

11.1 DOCSIS Communication Protocol

The primary function of the CM system is to transparently transmit IP packets between the cable TV headend and the subscriber. The CM and CMTS operate as IP and logical-link-control (LLC) hosts as defined in the IEEE 802 standard [2] for communication over cable TV network. Figure 11.1 shows the defined DOCSIS protocol stack on the RF interface. The physical (PHY) layer consists of the *physical media dependent* (PMD) sublayer and the downstream *transmission convergence* (TC) sublayer. The data-link-layer (DLL) above the PHY layer consists of three sub-layers as follows: (A) medium-access-control (MAC) layer, (B) link security sublayer, and (C) logical link control (LLC) sublayer. The MAC defines the procedure and parameters in which the CM can interact with the CMTS in order to support various services on both downstream and upstream channels. The Link Security sublayer provides the necessary security for the transmitted IP packets [3]. The network layer protocol, which operates above the LCC sublayer, is IP version 4 and migrating to IP version 6 [4]. In the network layer, both the CM and CMTS must support *address resolution protocol* (ARP) over DIX link-layer framing [5,6]. Above the network layer, there are various network management and operation protocols including:

- Simple network management protocol (SNMP) for network management

- Trivial file transfer protocol (TFTP) is a file transfer protocol, which is used for downloading software and configuration information to a remote location [7]

- Dynamic host configuration protocol (DHCP), which is used to assign IP addresses and passing configuration information to hosts on a TCP/IP network [8]

- Time of Day Protocol, which is used to obtain the time of day.

Figure 11.2 shows the data forwarding through a CM and a CMTS. The data forwarding through the CMTS may be transparent bridging, or may employ network-layer forwarding (routing and IP switching). The data forwarding through the CM is link-layer transparent bridging. It should be pointed out that the data forwarding between the upstream and downstream channels within the MAC layer is different from a conventional local-area-network (LAN) forwarding because the upstream channels are point-to-point transmission on a shared media, while downstream channels are point-to-multipoint transmission on a shared media. In addition, the RF characteristics of upstream channels are significantly different from the downstream channels.

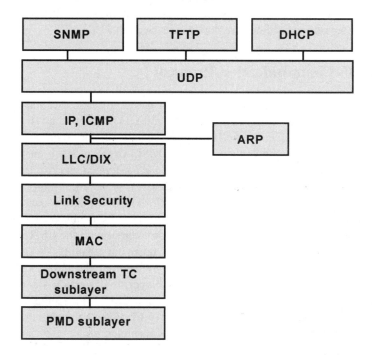

Figure 11.1 DOCSIS communication protocol stack on the RF interface. After Ref. [1].

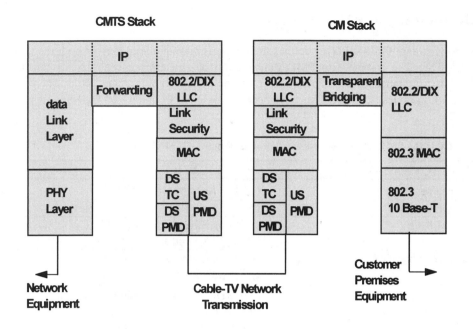

Figure 11.2 Data forwarding through a CM and a CMTS. After Ref. [1].

There are various forwarding rules for the CM and the CMTS, which are similar to the IEEE 802.1D standard, but with some modifications as described in DOCSIS 1.1 protocol standard [1]. The forwarding rules are beyond the discussion here, and further reading is encouraged [9].

11.2 Downstream PHY Layer

The DOCSIS protocol standard defines the procedure and requirements for high-speed data transmission between the subscriber's CM and the cable-modem-termination-system (CMTS) over HFC cable TV network. The PHY layer of DOCSIS protocol stack for downstream data transmission is divided into two sublayers as follows:

- PMD sublayer (both downstream and upstream)
- Downstream TC sublayer, which is sandwiched between the PMD sublayer the MAC layer, exists only for downstream data transmission

The next two subsections discuss the technical specifications of each of these sublayers.

11.2.1 Downstream PMD Sublayer

The downstream PMD sublayer defines the RF transmission parameters of both the CM and the CMTS. The downstream PMD transmission parameters are specified to conform to ITU-T Recommendation J.83 Annex B for low-delay video applications [10]. These downstream RF parameters for the CMs have been discussed in Chapter 7 on QAM modems, and

Table 11.1 CMTS RF output parameters according to DOCSIS 1.1 standard.

CHARACTERISTICS	SPECIFICATION
Center Frequency Range (f_c)	91 MHz to 857 MHz $\pm$ 30 kHz[1]
RF Output Level	50-dBmV to 61 dBmV (adjustable)
Symbol Rate	5.056941 Mbaud for 64-QAM
	5.360537 Mbaud for 256-QAM
Spectrum Shaping Filter	SRRC with $\alpha \cong 18\%$ (64-QAM)
	SRRC with $\alpha \cong 12\%$ (256-QAM)
Phase Noise	1–10 kHz: −33 dBc DS noise power
	10–50 kHz: −51 dBc DS noise power
	50 kHz to 3 MHz: −51 dBc DS noise power
Output Return Loss	> 14 dB up to 750 MHz
	> 13 dB above 750 MHz
Total Discrete Spurious In-Band	< −57 dBc for $f_c \pm 3$ MHz
In-Band Spurious and Noise	< −48 dBc
Adjacent Channel Spurious and Noise	< −58 dBc in 750 kHz from $f_c \pm 3$ MHz to $f_c \pm 3.75$ MHz
	< −62 dBc in 5.25-MHz from $f_c \pm 3.75$ MHz to $f_c \pm 9.0$ MHz[2]
	< maximum of −65-dBc or −12-dBmV in 6 MHz from $f_c \pm 9.0$ MHz to $f_c \pm 15$ MHz
Spurious and Noise in Other Channel	< −12 dBmV in each 6-MHz channel
Output Impedance	75-ohms
Connector	F connector per IPS-SP-406
CMTS Master Clock Frequency	10.24 MHz $\pm$ 5-ppm over 0°C to 40°C temperature range and up to 10 years
CMTS Timestamp Jitter	< 500 ns

---

[1] $\pm$ 30-kHz was specified to allow up to 25-kHz frequency offset due to the QAM RF upconverters at the headend.
[2] Excluding up to 3 spurs, where each spur must be < −60-dBc when measured in 10-kHz band.

thus will not be repeated here. However, let us briefly review the CMTS RF modulated output requirements. The CMTS must transmit to the CM according to the standard, IRC, or HRC frequency plans, but it is not required to transmit 64/256-QAM channels below 91-MHz. The rationale was to set aside a frequency gap (42 to 91-MHz) to minimize potential performance degradation due to nonlinear distortions and spurious noise leaking from upstream to the downstream frequency band.

Table 11.1 summarizes the CMTS RF output parameters according to DOCSIS 1.1 standard. The total in-band spurious and noise specification in Table 11.1 includes all the discrete spurious noise, interference peaks, AWGN, and undesired downstream QAM transmitter products. In order to reduce the effect of adjacent channel interference (ACI) on the transmitted QAM channel, the relative magnitude of the spurious noise outside the desired channel is reduced from less than −58-dBc to less than −65-dBc, depending on the frequency location of the spurious noise. For discussion of the upstream spurious output from a CM, see Section 11.3.4.

11.3 Upstream PHY Layer

The upstream PMD sublayer specifies the use of frequency-division-multiple access (FDMA) and time-division-multiple access (TDMA) methods in a burst mode with two modulation formats, QPSK and 16-QAM, operating at different symbol rate and channel bandwidths. Before discussing the different burst attributes and parameters, let us review the building blocks of the upstream CM transmitter and their requirements.

The various signal-processing building blocks of the upstream QAM burst transmitter are shown in Figure 8.4. The first building block separates the input packets into codewords, where each packet is equal to the number of information bytes in a codeword. This step is called blocking the data. The next building block is the FEC encoder. Note that the upstream PHY layer FEC encoder for a CM is different than the one used in a digital STB (see Section 8.4.1). It consists of R-S coding, bit-to-symbol ordering, and randomizer. The R-S codes over GF(256) can be selected from T = 1 to T = 10 with the following R-S generating polynomial:

$$g(X) = (X + \alpha^0)(X + \alpha^1)...(X + \alpha^{2T-1}) \tag{11.1}$$

where the primitive α is 0x02. The primitive polynomial for the R-S code over GF(256) is given by:

$$p(X) = X^8 + X^4 + X^3 + X^2 + 1 \tag{11.2}$$

The codeword sizes that the upstream CM transmitter can provide range from 18 bytes (16 information bytes plus two parity bytes for T = 1) up to a maximum of 255 bytes. Note that the uncoded word size can be as small as one byte.

The input serial bit stream to the R-S encoder, which was received from the MAC layer in the CM, is mapped into R-S symbols at the encoder. The first bit of the stream is mapped into the MSB of the first R-S symbol. The MSB of the first R-S symbol out of the encoder is mapped into the first bit of the serial stream before it is fed to the randomizer.

After the bit-to-symbol mapping, the symbols are randomized or scrambled to ensure equal probabilities to all the QPSK and 16-QAM constellation points. Figure 11.1 shows the randomizer structure using linear shift registers. The randomizer generating polynomial is given by:

$$g(X) = X^{15} + X^{14} + 1 \qquad\qquad (11.3)$$

The randomizer works as follows. First, the randomizer's registers are cleared and loaded with the seed value at the beginning of each burst. Then, the first output bit from the randomizer is calculated by XOR, the seed value with the first bit of data of each burst, which is the MSB of the first data symbol. The seed value of the randomizer, which can take an arbitrary value, is calculated in response to the *upstream channel descriptor* (UCD) message from the CMTS, which will be discussed later. The next building block, after the FEC block, prepend a variable-length preamble to the coded data symbols.   The preamble length

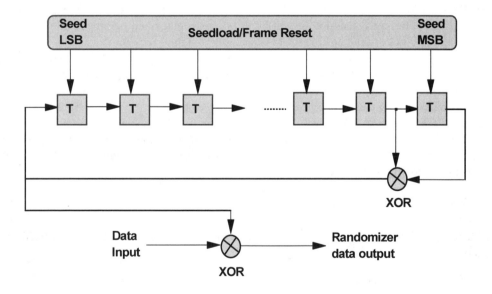

Figure 11.3 Randomizer structure in upstream CM transmitter. After Ref. [1].

is up to 1024 bits for both QPSK and 16-QAM modulation formats, and its value is programmable and must be an integer number of modulation symbols. Thus, the maximum preamble length is 512 QPSK symbols and 256 16-QAM symbols. The first bit of the preamble pattern, which is the also the first bit into the symbol mapper, is designated by the preamble value offset (PVO) to identify the bits to be used for the preamble value. The maximum length of the PVO is 1022 bits. The preamble length and values are calculated based on the transmitted UCD message from the CMTS.

The next building blocks in the upstream CM transmitter, namely, the symbol mapper, SRRC filters, and modulator, are similar to the one already described in Chapter 8 and will not be repeated here. In general, the upstream transmission specifications are divided into three main categories as follows: (A) channel parameters, (B) burst profile attributes, and (C) unique user parameters. Let us review the upstream channel parameters for CM as specified by DOCSIS 1.0 standard [1].

11.3.1 Upstream Channel Parameters and Requirements

The upstream channel parameters include the allowed symbol rates, channel center frequency, and the 1024-bit preamble. The Table 11.2 shows the five different symbol rates with the corresponding upstream channel bandwidths. Notice that the upstream channel bandwidth is the −30-dB bandwidth and not the −3-dB bandwidth as typically used in communication systems. The upstream spectrum is shaped by the SRRC filter with a roll-off factor of 25%. Thus, according to Table 11.2, the maximum transmission rate is 10.24-Mb/s using a 16-QAM modulation format. The required burst profile attributes as well as timing from an upstream CM transmitter are discussed in the next section.

Table 11.2 Upstream symbol rates and channel bandwidths.

Upstream Symbol Rate	Upstream Channel Bandwidth
160 kbaud	200 kHz
320 kbaud	400 kHz
640 kbaud	800 kHz
1.28 Mbaud	1.6 MHz
2.56 Mbaud	3.2 MHz

11.3.2 Burst Profiles

To successfully transmit upstream packets, the CM must generate each burst at the appropriately allocated time. All the burst profile attributes are specified via the burst descriptors in the UCD, and are assigned to the CM for upstream transmission in the upstream *bandwidth allocation map* (MAP) by the CMTS. A burst length of 0 mini-slots in the upstream channel profile means that the burst length is variable on that channel for the particular burst type. Tables 11.3 and 11.4 summarize the burst profile attributes as well as the unique burst parameters for each subscriber.

Table 11.3 Burst profile attributes according to DOCSIS 1.1 standard.

Burst Profile Attributes	Specification
Upstream Modulation Format	QPSK or 16-QAM
Differential Encoding	On or Off
Preamble Length	Up to 1024 bits
Preamble Offset Value	0 to 1022
FEC	T = 0 to 10 (0 implies no FEC)
FEC Codeword Information Bytes (k)	Fixed or Shortened, 16 to 253 (FEC on)
Randomizer Seed	15 bits
Randomizer Setting	On or Off
Maximum Burst Length	0 to 255 (mini-slots)
Guard Time	5 to 255 symbols
Last Codeword Length	Fixed or Shortened

The CM ranging process provides the correct timing offset such that the transmitted packets are aligned to the proper mini-slot boundary. Since the timing delays between the CM and the CMTS are essentially constant, any additional time delay must be accounted by the CMTS in the guard time of the transmitted packets to synchronize the upstream TDMA transmission. The ranging offset parameter in Table 11.4 is the time-delay correction applied by the CM to the CMTS, and it is essentially equal to the round-trip delay of the CM from the CMTS. In return, the CMTS provides feedback timing offset correction to the CM,

Table 11.4 Upstream unique burst parameters for each user according to DOCSIS 1.1

User Unique Parameters	Specification
Upstream Power Level Range	+8 dBmV to +55 dBmV (16-QAM) +8 dBmV to 58 dBmV (QPSK)
Power Level Increment	1 dB
Channel Frequency Offset	± 32 kHz in 1-Hz increments, but with ± 10-Hz implementation
Ranging Offset	0 to $(2^{16}-1)$ in increments of 6.25-μs/64
Burst Length Variations	1 to 255 mini-slots
Transmit Equalizer Coefficients	Up to 64 coefficients
Equalizer Coefficient Length	4 bytes (2 real and 2 imaginary)

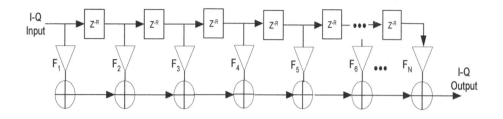

Figure 11.4 Pre-equalizer structure in upstream CM transmitter. After Ref. [1].

based on successful reception of one or more bursts with an accuracy of 1/2 symbol and timing resolution of 1/64 of the frame tick increment ($6.25\text{-}\mu s/64 = 97.65625\text{-ns} = 1/10.24\text{-MHz}$). In other words, the timing resolution is equal to the inverse of the CMTS clock frequency. The CM responds to the ranging offset correction with a resolution of 1 burst symbol and with accuracy of $\pm 0.25\text{-}\mu s \pm 1/2$ symbol, relative to the mini-slot boundaries.

Adaptive pre-equalization, which is also known as precoding, in the upstream CM transmitter is a well-known method to combat the effect of linear distortions in the transmission channel [11, 12]. DOCSIS 1.1 specifies the use of up to 64 fractional-spaced complex taps with 4 bytes for each complex tap coefficient. The total number of taps per symbol varies from 1 to 4. The upstream channel characteristics are learned by the burst receiver in the CMTS, which sends the updated pre-equalizer coefficients for the CM transmitter using the ranging response message. The ranging response message is transmitted by the CMTS in response to the received ranging request (RNG-REQ) from the CM. The CM transmits initially and periodically the RNG-REQ as requested by the CMTS to determine network time-delay to adjust its power level. The ranging and synchronization process will be further discussed in this chapter.

Prior to making an initial ranging request and whenever the upstream channel frequency or symbol rate changes, the CM must initialize the coefficients of the pre-equalizer to a default setting in which all coefficients are zero except the real coefficient of the first tap (i.e., F1). During initial ranging, the CM must compensate for the delay (ranging offset) due to a shift from the first tap to a new main tap location of the equalizer coefficients sent by the CMTS. The pre-equalizer coefficients are then updated though the subsequent ranging process (periodic station maintenance). The main tap location cannot change during periodic station maintenance by the CMTS. Equalizer coefficients may be included in every RNG-RSP message, but typically they occur only when the CMTS determines that the channel response has significantly changed. The frequency of equalizer coefficient updates in the RNG-RSP message is determined by the CMTS. The CM must normalize the pre-equalizer coefficients in order to guarantee proper operation (such as not to overflow or clip). The CM must also compensate for the gain (or loss) due to any coefficient scaling performed locally.

11.3.3 Burst Timing

The DOCSIS standard requires a CM to switch between different burst profiles without a reconfiguration time between consecutive bursts. However, for the following burst profile parameters: (A) output power, (B) modulation format, (C) symbol rate, (D) offset frequency, (E) channel frequency, and (F) ranging offset, the CMTS has to allocate at least 96 symbols reconfiguration time to the CM.   Figure 11.5 shows the burst timing convention according to DOCSIS 1.1 protocol standard. The ramp-up and ramp-down times represent the time spreading of the transmitted symbols beyond their nominal symbol duration Ts (= 1/Rs) due to the spectral shaping filter at both the upstream CM transmitter and CMTS receiver. In this figure, ten-symbol ramp-up time with eight-symbol ramp down time are shown with no guard band and assuming no timing errors. Notice that the timing is referenced to the symbol center of the first symbol of each burst. Thus, the 96-symbols reconfiguration time is measured from the last symbol center of one burst to the first symbol center of the following burst. If the output power of the CM is changed by more than 1-dB or less than 1-dB, the reconfiguration time is increased to 96 symbols plus 10-μs and 96 symbols plus 5-μs, respectively. It should be pointed out that the output power of the upstream CM transmitter is

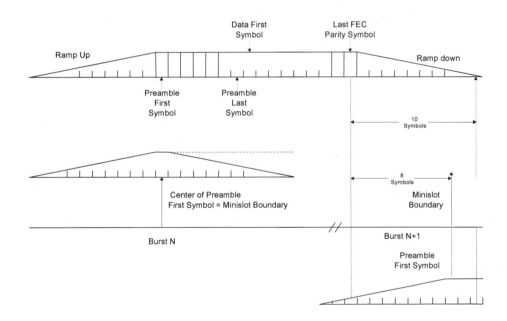

Figure 11.5 Upstream burst timing convention. After Ref. [1].

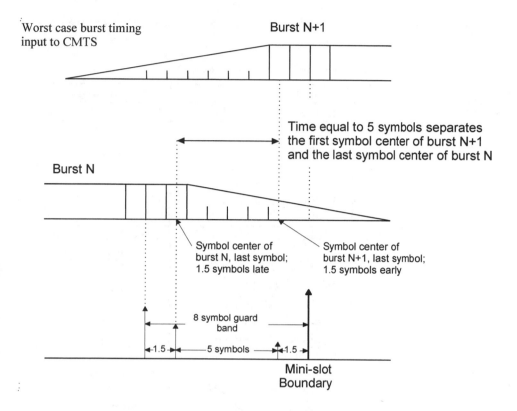

Worst case burst timing
input to CMTS

Burst N+1

Time equal to 5 symbols separates
the first symbol center of burst N+1
and the last symbol center of burst N

Burst N

Symbol center of
burst N, last symbol;
1.5 symbols late

Symbol center of
burst N+1, last symbol;
1.5 symbols early

8 symbol guard
band

1.5   5 symbols   1.5

Mini-slot
Boundary

Figure 11.6 Worst case upstream burst timing. After Ref. [1].

required to be essentially constant (less than ± 0.1-dB) within any given TDMA burst. If the
channel frequency is changed, the reconfiguration time is increased to 96 symbols
plus 100-ms.

Figure 11.6 shows the worst-case burst timing transmission for a CM according to DOC-
SIS 1.1 standard [1]. Burst N arrives 1.5 symbols late, while the following burst (N+1)
arrives 1.5 symbols early, reducing the symbol guard band to 5 symbols, which is the minimum time separation between consecutive bursts required by DOCSIS 1.1 protocol.

11.3.4 Upstream Spurious Power Outputs

Due to the harsh upstream channel characteristics, there is a need to improve the CM trans-

Table 11.5 Upstream spurious emission specifications for a CM.

Parameter	Transmitting Burst	Between Bursts
Inband Spurious Power	−40 dBc	Max {−72 dBc, −59 dBmV}
Adjacent Band Spurious Power	−45 dBc @ 160-kBaud[1] −42 dBc @ 160-kBaud[2] −39 dBc @ 160-kBaud[3] −45 dBc @ other symbol rates[4] −44 dBc @ 1280-kBaud[5] −41 dBc @ 2560-kBaud[6]	Max {−72 dBc, −59 dBmV}
≤ 3 Carrier-Related Frequency Bands	−47 dBc	Max{−72dBc, −59 dBmV}
Spurious Power within 5–42-MHz	−53 dBc @ 160-kBaud −50 dBc @ 320-kBaud −47 dBc @ 640-kBaud −44 dBc @ 1280-kBaud −41 dBc @ 2560-kBaud	Max{−72 dBc, −59 dBmV}
CM Integrated Spurious Power Limits (in 4-MHz)	Max {-40 dBc, -26 dBmV} in 42–54-MHz −35 dBmV in 54–60-MHz −40 dBmV in 60–88-MHz −45 dBmV in 88–880-MHz	 −26 dBmV −40 dBmV −40 dBmV Max {−45 dBmV, −40 dBc}
CM Discrete Spurious Power Limits	Max {−50 dBc, 36 dBmV} in 42–54-MHz −50 dBmV in 54-880-MHz	 −36 dBmV −50 dBmV

[1] Assuming adjacent channel symbol rates of 160-kBaud, 320-kBaud, and 640-kBaud, with measurement intervals from 20-kHz to 180-kHz, 40–360-kHz, and 80–720-kHz relative to channel edge.
[2] Assuming adjacent channel symbol rate of 1280-kBaud, with a measurement interval from 160-kHz to 1440-kHz relative to channel edge.
[3] Assuming adjacent channel symbol rate of 2560-kBaud, with a measurement interval from 320-kHz to 2880-kHz relative to channel edge.
[4] Assuming adjacent channel symbol rates of 160-kBaud, 320-kBaud, and 640-kBaud, with the same measurement intervals as in footnote 1.
[5] Assuming adjacent channel symbol rate of 1280-kBaud, with measurement interval from 160-kHz to 1440-kHz relative to channel edge.
[6] Assuming adjacent channel symbol rate of 2560-kBaud, with a measurement interval from 320-kHz to 2880-kHz relative to channel edge.

mission fidelity. Consequently, this rationale led the DOCSIS 1.1 standard to impose streak limits on the upstream and downstream spurious power levels at different frequency bands relative to the CM transmitted burst power level. The transmission performance of a CM operating at a given symbol rate may be degraded due to adjacent channel spurious emissions from other CMs operating at different symbol rates. Table 11.5 specifies the required in-band, discrete, and integrated spurious power levels from a CM during burst transmission (active state) and between bursts (idle state). The spurious power measurement is done in the adjacent channel interval, which corresponds to the adjacent channel symbol rate. However, the measurement interval does not start at the channel edge, but at a frequency spacing of 12.5% of the adjacent symbol rate relative to the channel band edge. This is because the channel bandwidth is the bandwidth measured at −30-dB, while the symbol rate bandwidth is the bandwidth measured at −3-dB. For example, the 200-kHz channel bandwidth has only a 160-kHz noise bandwidth or 20-kHz excess bandwidth on each side of the channel band. Thus, the spurious measurement interval is specified to start from 20-kHz away from the channel edge up to 180-kHz. The measurement of the upstream spurious emissions in the 5–42-MHz band starts at initial distance of 220-kHz from the channel band edge for a symbol rate of 160-kBaud, and it is repeated until at increasing frequency spacing from the channel until the upstream band edge is reached. For these specifications, the spurious measurement interval starts beyond the adjacent channel. Consequently, 200-kHz (channel bandwidth) is added to 20-kHz, which is initial starting frequency from the channel edge. In addition, the CM must control its spurious output levels immediately before and after the TDMA burst; namely, it's burst on and off transients. This on/off spurious output levels must be less than 100-mV, and occur not faster than 2-μs of constant slewing when the CM is operating at +55-dBmV or more. For each 6-dB drop in output power level from 55-dBmV, the transient voltage must be reduced by a factor of 2, down to 7-mV at +31-dBmV or below.

11.3.5 Burst Frame Structure

The CM can operate in either a fixed-length codeword mode or in a shortened last codeword mode. A codeword consists of FEC parity bytes and payload data bytes, where the minimum codeword length (number of information bytes) is 16 bytes. These two modes are illustrated in Figure 11.7. In the fixed-length codewords, the preamble starts at the mini-slot boundary, and after all the data are encoded, zero-fill occurs in the last codeword to reach the assigned k-bytes per codeword. The zero-fill continues until no additional fixed-length codewords can be inserted before the end of the last allocated mini-slot in the grant. In the shortened last codeword mode, let k' (k' < k) be the number of remaining information bytes in the last codeword. We assumed that the full-length burst was partitioned into k-bytes length codewords. Also, let k" (k' < k") be the number of burst bytes plus zero-fill bytes in the shortened last codeword. In this mode, the CM zero fill (k"−k') bytes until the end of the mini-slot allocation, which in most cases are next to the mini-slot boundary. In example 2 (Figure 11.7), the packet length satisfies the following relation: k + k' ≤ k + k" ≤ 2k. It should be pointed out that if after the zero-fill of additional codewords, each with k-bytes, there are less than 16 bytes remaining in the allocated grant of the mini-slot, then the CM must not create a shortened codeword.

Example 1. Packet length = number of information bytes in codeword = k

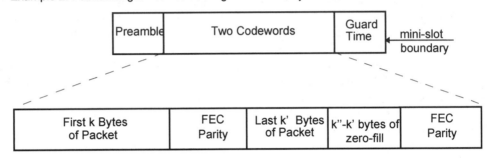

Example 2. Packet length = k + remaining information bytes in 2nd codeword = k + k'

Figure 11.7 Burst frame format examples with fixed and flexible burst length packets. After Ref. [1].

11.4 Downstream Transmission Convergence Sublayer

The purpose of the downstream TC sublayer is to provide a common interface between the PMD sublayer and the MAC layer for future multiplexing/de-multiplexing of MPEG-2 video and data over the PMD sublayer. The downstream bit-stream is defined as continuous stream of 188-byte long MPEG packets [13].   The format of MPEG packet carrying DOC-SIS data is shown in Figure 11.8. The 4-byte MPEG header identifies the content of the MPEG payload as either carrying DOCSIS MAC payload or digital video payload. The MPEG header format conforms to the MPEG standard, but its use is restricted to prevent the inclusion of an adaptation_field in the MPEG packets.

MPEG TC Header (4-bytes)	pointer_field (1-byte)	DOCSIS Payload (183/184-bytes)

Figure 11.8 Format of MPEG packet according to the DOCSIS protocol.

The MPEG payload portion of the MPEG packet carries the DOCSIS MAC frames. The first byte following the MPEG header, which is called the pointer_field, is present if the payload_unit_start_indicator (PUSI) bit of the MPEG header is set. The pointer_field indicates the number of bytes in the transmitted packet that the CM decoder must skip immediately after the pointer_field for the beginning of a DOCSIS MAC frame. The pointer_field is also present to indicate any stuff_byte(s) following the MAC frame. Stuff_bytes, which have the value of 0xFF, are used within the DOCSIS payload to fill possible gaps between contiguous DOCSIS MAC frames. This value (0xFF) was chosen as an unused value for the first byte of the DOCSIS MAC frame. Note that MAC frame may begin anywhere within the MPEG packet, and several MAC frames may exist within one MPEG packet. The MAC frames may be concatenated one after the other or may be separated by optional stuffing bytes. If a long MAC frame is used such that it spans multiple MPEG packets, the pointer_field of the succeeding frame points to the byte following the last byte of the tail of the first frame.

Another important point is related to the multiplexing of MPEG packets carrying DOCSIS data with digital video data. The CMTS is required not to transmit MPEG packets with the DOCSIS PID consisting of only stuffing bytes. Instead, null MPEG packets should be transmitted. The reason for this requirement is to preserve the timing information that exists in the DOCSIS MAC during the multiplexing operation.

11.5 Media Access Control (MAC) Layer

The MAC layer in the DOCSIS 1.1 protocol standard over HFC networks defines the procedure and parameters in which the CM can interact with the CMTS in order to support various services on both downstream and upstream channels. The primary features of the MAC layer protocol are as follows:

- Upstream bandwidth allocation controlled by the CMTS
- Upstream allocation of mini-slots to the CM by the CMTS
- Dynamic mix of contention and reservation-based upstream transmit opportunities
- Bandwidth efficiency through support of variable length packets
- Support of quality of service (QoS) including:
  1. Support of bandwidth and latency guarantees

2. Packet classification
3. Dynamic service establishment
- Extensions provided for security at the data link layer
- Support of a wide range of data rates

Before we discuss the MAC layer frame formats, let us introduce the concept of service flow, which is an essential part of the DOCSIS MAC protocol. A service flow is a MAC-layer transport service, which provides unidirectional transport of packets between the CM and the CMTS. The service flow also shapes, polices, and prioritizes traffic according to QoS traffic parameters, which are defined for the flow. A service flow ID (SFID) is a 32-bit identifier, which is assigned to a service flow by the CMTS. The CMTS can assign one or more SFIDs to each CM, depending on the requested service flows from the particular CM. This unique mapping can be negotiated between the CMTS and the CM during CM registration or via dynamic service establishment. With an active or admitted upstream service flow, the CMTS assigns the so-called service ID (SID) in addition to the SFID. The SID provides the necessary means for the CMTS to implement the requested upstream QoS. Multiple SIDs allow advance CM to support multiple service classes and to be interoperable with basic CMs, which support only a single SID. The length of the SID parameter is 14 bits.

The second parameter is called mini-slot, which is a unit of granularity for upstream transmission opportunities. A mini-slot is $2^K 6.25$-μs ($K = 1,\ldots,7$). Notice that one mini-slot corresponds to the inverse of the smallest symbol rate, namely, 160-kHz. The minimum mini-slot is 12.5-μs, and has two time ticks.

11.5.1 MAC Frame Format

Since upstream request packets and data packets are transmitted in a contention manner, the MAC frame, which is a variable-length packet, is the basic unit of data exchanged at the data link layer (DLL) between the CM and the CMTS over the cable TV network. Figure 11.9 shows a generic MAC frame format. Preceding the MAC frame is either a PMD sublayer overhead (for upstream transmission) or MPEG TC sublayer header (for downstream transmission). The PMD sublayer overhead indicates to the MAC layer the start of a burst at the mini-slot boundary, where the MAC frame is required to start. Thus, the MAC layer can account for the PMD overhead when upstream bandwidth allocation is done (see Section 11.4.2). It should be pointed out that the FEC overhead is spread throughout the MAC frame and is assumed to be transparent to the data stream. The MAC frame consists of a MAC header and an optional variable-length data protocol-data-unit (PDU).[7] The MAC header uniquely identifies the contents of the MAC frame, including the presence of data PDU in the MAC frame. Four different types of data PDUs are supported: (A) variable-

---

[7] PDU is a unit of data specified in a layer protocol and consisting of protocol control information and layer user data.

length packets (18 to 1518-bytes), (B) ATM packets (n x 53-bytes), (C) MAC specific header (no PDU), and (D) reserved PDU for future use.

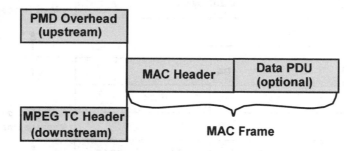

Figure 11.9 Generic MAC frame format according to the DOCSIS protocol.

Many MAC messages incorporate the use of type-length-value (TLV) fields. The TLV represents encoding of three fields, in which the first field indicates the type of element (1-byte), the second field indicates the length of the element (1-byte), and the third field indicates the value. This encoding allows new parameters to be added and correctly interpreted by the CM and CMTS. If the CM or CMTS do not recognize the new parameters, they must skip them and not handle them as errors.

11.5.2 MAC Management Messages

Figure 11.10 shows the MAC header and the MAC management message header formats, which are common across all the MAC management messages. The first byte of the MAC header is the frame control (FC) field, which defines the type and format of data protocol data unit (PDU). Four different types of data PDUs that are supported are (A) variable-length packets (18 to 1518-bytes), (B) ATM packets (n x 53-bytes), (C) MAC specific header (no PDU), and (D) reserved PDU for future use. The MAC specific header can be further divided into five categories as follows: (I) timing header, (II) MAC management header, (III) request frame, (IV) fragmentation header, and (V) concatenation header. The MAC specific headers are used for specific functions including downstream timing and upstream ranging or power adjustments, bandwidth requests, and fragmentation and concatenation of multiple MAC frames.

Timing Header:
This specific MAC header, which is followed by a packet data PDU, is used to transport the *Global Timing Reference* in the downstream channel for synchronization all the CMs. CM MAC synchronization is achieved if the CM has received two SYNC messages within the maximum SYNC interval (= 200-ms). Each SYNC message contains the CMTS timestamp,

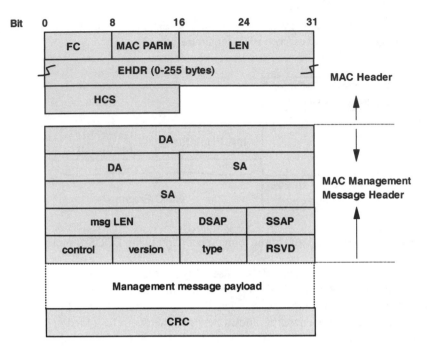

Figure 11.10 MAC header and MAC management header fields.

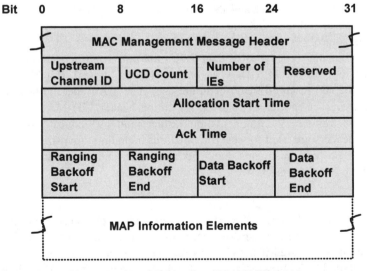

Figure 11.11 Upstream bandwidth allocation MAP format.

which tells the CM when exactly the CMTS has transmitted the message. The CM then compares the actual time the message was received with the CMTS timestamp, and adjusts its local clock accordingly. In this way, the CM can time its transmission to accurately arrive at the CMTS in the cable headend (see Section 11.6). For upstream transmission, the timing header is used as part of the ranging request (RNG-REQ) message, which is needed to adjust the time and power of the CM.

The FC field is followed by 3 bytes of MAC control. The MAC_PARM field of the MAC header is used for different purposes, depending on the FC Field. If the EHDR_ON = 1, then the MAC_PARM field is used as the extended-header-length (ELEN), which can vary up to 240-bytes. For concatenated MAC header, the MAC_PARM field represents the number of concatenated MAC frames. For request MAC header (REQ), the MAC_PARM field represents the amount of requested bandwidth. The 2-bytes LEN field is used in most cases to indicate the length of the MAC frame. However, for the REQ case, this field is used to indicate the CM SID.

The optional extended header (EHDR) field, which provides extension to the MAC frame format, is used to add support for additional functions such as data link security and frame fragmentation. The header check sequence (HCS) field is a 16-bit CRC that is used to detect errors in the entire MAC header, particularly in the presence of collisions. The HCS is calculated using the following CRC-CCITT generating polynomial [14]:

$$g(X) = X^{16} + X^{12} + X^5 + 1 \tag{11.4}$$

MAC Management Header:
This is a specific MAC header that is used to support MAC management messages. The Destination Address (DA) can be a specific CM unicast or multicast address, and the Source Address (SA) is the address of the source CM or CMTS. The "msg LEN" field is the total length of the MAC message from DA to the end of the MAC management message payload. For the other fields in the MAC management header, the reader is encouraged to see the DOCSIS 1.1 protocol specifications.

MAC Management Payload:
There are many different types of MAC management messages with variable lengths, which are briefly outlined as follows:

- CMTS Timestamp: 32-bit counter, which is locked to the CMTS 10.24-MHz master clock.
- Upstream channel descriptor (UCD): The UCD, which is transmitted periodically by the CMTS, defines the upstream channel characteristics. The UCD parameters include the upstream channel ID, the downstream channel ID on which this message was transmitted, the number of mini-slots in the upstream channel, counter for any changes in the CM configuration, and channel type-length-value (TLV) parameters.

- Upstream bandwidth allocation MAP (MAP): The MAP format, which is generated by the CMTS, is shown in Figure 11.11. The upstream bandwidth allocation MAP has a fixed length header and various housekeeping parameters such as the upstream channel ID, the number of information elements (IEs) in the MAP, UCD counter, and a number of important time references. The allocation start time (Alloc Start Time) parameter shows the effective start time from the CMTS initialization (counted in mini-slot) for the first entry in this MAP. The parameter Ack Time in the MAP indicates the latest time from the CMTS initialization (in mini-slots) processed in the upstream that generated a Grant, Grant Pending, or Data Ack messages. Ranging Backoff Start and Ranging Backoff End indicate the initial and final backoff windows for initial ranging contention (see Section 11.5). Data Backoff Start and Data Backoff End parameters indicate the initial and final backoff windows for contention data and requests.

As the name indicates, IEs define individual grants or deferred grants in the transmitted MAP from each CM. Each IE is a 32-bit quantity, where the most significant 14-bits represent the SID (for unicast, multicast, or broadcast services), the middle 4-bits represents interval usage code (IUC), and the last 14-bits represent the mini-slot offset. Since all the CMs must scan all their IEs, it is important that the IEs be short. A null IE is used to terminate the list. The DOCSIS 1.1 protocol standard defines the following IEs:

- Request IE: It provides an upstream interval for all CMs to request upstream bandwidth for data transmission.
- Request/Data IE: It provides upstream interval for either bandwidth or short data packets. For example, under "light" network loading, the allocation algorithms in the CM may use this IE to provide for "quick" data contention.
- Initial Maintenance: It provides an upstream interval for new CMs to join the cable TV network (ranging opportunity). This interval is equal to the round-trip propagation delay between the CM and CMTS plus the transmission time of the ranging request message.
- Station Maintenance IE: It provides an interval in which some stations are expected to perform some aspects of routine network maintenance, such as ranging or power adjustments.
- Short and Long Data Grant IEs: A data grant IE provides an opportunity for a CM to transmit upstream one or more data PDUs.
- Data Acknowledge IE: It provides an acknowledgement that a data PDU was received. The CM has to request this acknowledgement within the data PDU.
- Null IE: It terminates all actual allocations in the IE list. All Data Acknowledge and Data Grant Pending IEs must follow the Null IE.

There are also various request and response messages that are used with the MAC management message payload, which will be briefly introduced as follows:

1. Ranging Request (RNG-REQ):

A CM at initialization must transmit a ranging request and periodically as requested by the CMTS to determine cable TV time delays and requests for output power level adjustments.

2. Ranging Response (RNG-RSP):

A ranging response must be transmitted by the CMTS in response to RNG-REQ. This response message must include TLV encoded parameters such as SID, upstream channel ID, ranging status, timing/power/frequency adjust information. In addition, if the CM transmitter employs a pre-equalizer, then the adjustments of the equalization coefficients are transmitted in the downstream channel.

3. Registration Request (REG-REQ):

A CM at initialization must transmit a registration request after receipt of a CM parameter file. This request includes initialization SID as well as the CM configuration file settings such as downstream frequency, upstream channel ID, network access control object, downstream packet classification, upstream service flow, class of service, baseline privacy configuration, and others.

4. Registration Response (REG-RSP):

A registration response must be transmitted by the CMTS in response to receive REG-REQ. The response includes the SID from the corresponding REG-REQ as well as TLV encoded parameters such as service flow, payload header suppression, classifier and classifier error set, service flow set, payload header suppression error set, service not available, modem capabilities, and vendor-specific data.

5. Registration Acknowledge (REG-ACK):

The CM in response to REG-RSP to the CMTS must transmit a registration acknowledgment. This message confirms the acceptance of the quality-of-service (QoS) parameters of the flow by the CM as reported by the CMTS.

6. Upstream Channel Change Request (UCC-REQ):

An upstream channel change request must be transmitted by the CMTS to cause a CM to change the upstream channel on which it is transmitting.

7. Upstream Channel Change Response (UCC-RSP):

An upstream channel change response must be transmitted by a CM in response to a receive UCC-REQ to indicate to the CMTS that it has received the request and it is complying with the UCC-REQ.

11.6 Random Access and Contention Resolution Methods

Before we can discuss the primary contention resolution methods for upstream transmission by a CM, let us review the conventional random access protocols that were used in early cable TV data networks. Some of these random access methods are still used by many cable TV operators to access their digital STBs.

11.6.1 Random Access Methods

The earliest random access method, which was developed at the University of Hawaii to interconnect several of its computers with a communication satellite, is the Aloha protocol [15]. In the Aloha system, a station can transmit at any time, encoding its transmission with error detection code. Then, the station listens for time duration equal to maximum roundtrip propagation time on the network for an acknowledgment from the receiver. In case of collisions, the station receives a negative acknowledgment, and it re-transmits the message after a random time delay. After repeated negative acknowledgments, the station stops. It can easily be shown that the Aloha protocol is quite inefficient, achieving a maximum traffic throughput of 18% [16]. To improve the maximum throughput of the Aloha system, the slotted Aloha (S-Aloha) protocol was developed. In the S-Aloha system, a central clock is used to synchronize all the stations by broadcasting a sequence of synchronization pulses. The transmission time is divided into uniform slots, and each station can begin to transmit only at the slot boundary. This simple modification reduces the rate of collisions by half, since collisions only occur when two or more stations attempt to transmit within the same time slot. Thus, the maximum traffic throughput for S-Aloha is increased to $1/e$ or 36.8%. DAVIC 1.2 standard uses the S-Aloha protocol to manage the upstream transmission over HFC networks [17]. The DAVIC upstream 64-byte slot consists of a 4-byte unique word (UW), 53-bytes of data payload, 6-bytes of R-S parity, and 1-byte of guard time. Thus, there are 500 and 3000 slots per second for upstream data rates of 256-kb/s and 1.544-Mb/s, respectively, in the DAVIC 1.2 standard. Appendix A summarizes the upstream and downstream PHY layer parameters between DAVIC 1.2 and DOCSIS 1.1 specifications.

To improve over the Aloha and S-Aloha access schemes, let us review the access methods that were developed in the 1980s to interconnect computers over a local area network [18, 19]. Let us assume that a station began to transmit multiple packets over a local area network. However, it may take some time before other stations know about the first transmitting station. Thus, one or more stations may also try to transmit multiple packets, resulting in a collision with the previously transmitted packet, and neither of the packets is getting through. If the propagation time to the other stations is much smaller than the packet transmission time, and all the stations are listening before talk (LBT) when a given station begins to transmit, then collisions can be avoided since these stations know it almost immediately. Thus, collisions can occur only when a station has begun transmitting within the propagation time delay interval. This approach is sometimes called carrier-sensing multiple access (CSMA) or LBT method. Using the CSMA method raises two important questions. First, what should a station do when the network is found to be busy? There are two common algorithms to address this problem as follows:

- Non-persistent: If the network is busy, the station has to wait a random amount of time and try to transmit again.
- p-Persistent: If the network is busy, delay by a fixed time slot and try to transmit again with a probability of p ($0 \leq p \leq 1$). A special case of this method is 1-persistent, where a station continues to listen when the network is busy, and transmit as soon as the network becomes available. If collision occurs due to lack of acknowledgment, the station waits a random amount of time and tries to transmit again ($p = 1$).

There are drawbacks to each method. In the non-persistent method, the network capacity is not efficiently utilized since the network remains idle following a station transmission. In the 1-persistent method, a collision is guaranteed when two or more stations are waiting to transmit at the same time. A compromise is achieved in the p-persistent method by setting p low such that $n \cdot p < 1$, where n is the number of stations waiting to transmit, resulting in a collision and unnecessary delays.

The CSMA method can be improved by adding two more transmission rules as follows:

- When a collision is detected, the transmitting station aborts the transmission and sends a short jamming signal to inform all he other stations that a collision has occurred.
- After the jamming signal, the station tries to re-transmit after a random time-delay using CSMA. The time-delay before the nth attempt is a uniformly distributed random number from 0 to 2n -1, where the re-transmission time delay unit is 512 bits.

This is essentially the CSMA with collision-detection (CSMA/CD) algorithm, which is the most commonly used access method in today's Ethernet local-area networks. The maximum Ethernet packet size is 1536 bytes, which consists of 8-byte preamble, 14-byte header, 4-byte parity, and 1500-byte payload. The baseband physical layer of the Ethernet using CSMA/CD was issued as the IEEE802.3 standard [20], while the broadband physical layer for a two-way cable TV system as provided as supplements to IEEE 802.3 standard [21]. In the early 1980s, various two-way cable TV systems were implementing the CSMA/CD method. In the CSMA/CD method, the amount of unused bandwidth is reduced to the time it takes to detect a collision. However, broadband Ethernet has a severe drawback over the HFC network. For example, due to time-delay constraint of 10-Mb/s Ethernet, the maximum distance of a station to the headend is required to be less than 2-km.

To overcome this limitation, two MAC protocols have been proposed and standardized for contention resolution of data over cable. The p-persistent with binary exponential backoff algorithm [22] has been selected for the DOCSIS 1.0/1.1 standard, and the ternary-tree algorithm for IEEE 802.14 [23]. In our discussion here, we are primarily focusing on the p-persistent binary exponential backoff algorithm since the DOCSIS 1.1 is the only widely used standard for CMs in North America.

11.6.2 p-Persistence with Binary Exponential Backoff Algorithm

The contention resolution method for upstream transmission, which is used by DOCSIS 1.1 standard, is based on a truncated binary exponential backoff algorithm, where the initial backoff window (W_i) and the maximum backoff window (W_m) are controlled by the CMTS at the cable headend [22]. In order for a station to enter the contention resolution process, it must set its internal backoff window equal to W_i. Then, the station must select random number X within its backoff window, which is between 0 and R, where R is given by:

$$R = 2^{\max[W_i + K - 1, W_m]} - 1 \qquad\qquad (11.5)$$

where K is the number of collisions. For example, for $W_i = W_m = 4$, the backoff window is between 0 and 15. If the station selects X = 7, then it has a total of seven contention opportunities.

After the contention transmission, the station must wait for feedback information from the CMTS, which could be a data grant or data acknowledge in the upstream allocation MAP. The contention resolution process is completed once the station receives the feedback information. However, if there is no feedback information or if the station determines that the contention transmission was lost due to collision, the station must increase its backoff window by a factor of 2, as long as it is less than maximum backoff window (W_m). The station must select another random number within its new backoff window and repeat the process as described above. The maximum number of contention resolution re-tries is 16, and it is independent on the initial and maximum backoff windows as defined by the CMTS.

It should be pointed out that the p-persistent contention resolution algorithm is a free-access algorithm, since it allows new stations to content for mini-slots at any time. This type of algorithm is sometimes called a non-blocking algorithm. In contrast, the IEEE802.14 contention resolution protocol based on the ternary tree algorithm is essentially a blocking algorithm. It precludes new stations contending for mini-slots from entering into an ongoing contention resolution process. Each protocol has its own advantages and disadvantages. The p-persistent algorithm cannot guarantee the contention resolution duration, since new stations can always join the ongoing resolution process. In other words, there is no upper bound on the longest resolution time for successful transmission.

The contention resolution in the MAC protocol of the IEEE 802.14 standard uses the ternary-tree algorithm with variable entry persistence. Performance comparison results between the ternary-tree and the p-persistent contention resolution algorithms in terms of the mean access delays and throughput indicate that the ternary-tree-blocking algorithm outperforms both blocking and non-blocking p-persistence algorithms [23].

11.7 MAC Layer Protocol Operation

As we have learned in Section 11.4, a stream of mini-slots defines the upstream channel in the DOCSIS protocol, which is referenced to the CMTS clock. The upstream bandwidth allocation MAP, which is a MAC message transmitted by the CMTS on a downstream channel, specifies the granted mini-slots for particular stations to transmit, the available mini-slots for contention transmission, and other mini-slots as an opportunity for new stations to join in. An upstream transmit opportunity is defined as any mini-slot in which the CM may be allowed to start a transmission. Each mini-slot is number relative to a master reference clock maintained by the CMTS, which distributes the clocking information to the CMs using the synchronization (SYNC) packets. The allocation MAP, which is a variable-length MAC management message, defines the upstream transmit opportunities for a given CM on a given channel. The allocation MAP consists of a fixed-length header and a variable number of information elements (IEs), which defines the allowed usage for the granted mini-slots. Each IE consists of a 14-bit SID, a 4-bit type code, and a 14-bit starting offset parameter. There are four different types of SID as follows: (A) Broadcast SIDs intended for all the CMs, (B) Unicast SID intended for a particular CM or a particular service within that CM, (C) Multicast SID intended for a group of CMs, and (D) Null SID addressed to no CM. The maximum size of an allocation MAP is limited to 240 IEs, and it is bounded to no more than 4096 mini-slots.

Before we explain the MAC message exchange mechanism between the CM and the CMTS for upstream transmission, we will assume that initial CM ranging between the CM and the CMTS has been established. The CM initialization is discussed in the next section. Figure 11.12 illustrates the MAC message exchange example. The CMTS transmits a MAP in a downstream channel at time t_1 with effective starting time of t_3. This time difference ($t_3 - t_1$) is needed to compensate for downstream propagation delay to allow the CM to receive the MAP message, processing time at the CM, and upstream propagation delay. At t_2, the CM receives the transmitted MAP and scans it for request opportunities. The CM has to request bandwidth from the CMTS for upstream data transmission using the Request IE in the MAP message. Thus, the CM sends the Request IE to start at t_5, and calculates a new time t_6 as a random offset based on the Data Backoff Start value (e.g., the initial backoff window starting time) in the most recently transmitted MAP. At t_4, the CM transmits a request to the CMTS for a particular number of mini-slots needed to accommodate the required data packet data units (PDUs). Notice that t_4 is selected such that the transmitted request will arrive at the CMTS at t_6. If, for example, due to collision with requests from other CMs, the CMTS does not receive the transmitted MAP, the CM must perform the binary exponential backoff algorithm and re-try to transmit. At t_6, the CMTS received the transmitted request from the CM and schedules it for upstream data PDU transmission in the next MAP. At this point the scheduling algorithm of the CMTS determines which request to grant, depending on various factors such as class of service requested and any competing requests from other CMs. It should be pointed out that if the CMTS scheduler cannot grant the CM request in the next MAP message, it must either reply with a zero-length grant in that MAP or discard the request by not giving a grant. In addition, the CMTS must indicate to the CM that the request is still pending by continuing to reply with a zero-length grant in the MAP message until the request can be granted. At t_7, the CMTS transmits an allocation MAP to the CM with an effective starting time t_9, and with a data grant starting time t_{11}. When the CM receives the transmitted MAP at time t_8, it scans the MAP for its data grant and then transmits its data PDU at t_{10} to arrive at the CMTS at t_{11} as shown in Figure 11.12.

Multiple upstream and downstream channels can be transmitted to and from a single CM. If multiple upstream channels are associated with a single downstream channel, then the CMTS must transmit one allocation MAP per upstream channel.

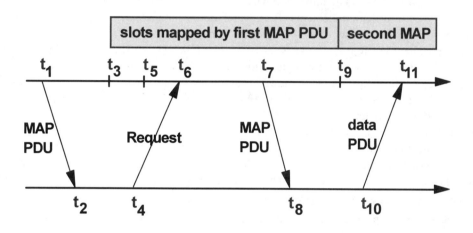

Figure 11.12 Example of MAC protocol between CM and CMTS. After Ref. [1].

11.8 Quality of Service (QoS) and Fragmentation

Several new concepts related to QoS have been introduced in the DOCSIS 1.1 protocol standard as compared with the DOCSIS 1.0 standard [24]. Classifying the transmitted packets over the DOCSIS MAC interface into service flows provides enhanced QoS. As we have learned earlier, a service flow defines a unique unidirectional mapping of packets between the CM and the CMTS to provide a particular QoS. A service flow is characterized by a set of QoS parameters such as latency or end-to-end delay, timing jitter, and throughput assurances. Both the CM and CMTS provide QoS by shaping, policing, and prioritizing the traffic according to the QoS parameter set defined for the service flow. The SFID exists for all service flows, while SID exists only for admitted or active upstream service flows.

The following subsection discusses several basic concepts related to QoS, including packet classification and types of service flows.

11.8.1 Basic Concepts and Operation

The first important concept is the classification of QoS parameter set or service flows. The QoS parameter set for a given service flow can be divided into three different categories as follows:

Provisioned QoS Parameter Set:
This set of QoS parameters appears in the configuration file of the CM and is present during its registration. The provisioned QoS parameter set is defined when the service flow is created in the CM registration process. Notice that if the service flow is created dynamically, the provisioned QoS parameter set is null.

Admitted QoS Parameter Set:
The CMTS and possibly the CM are reserving resources for this set of QoS parameters. The primary reserved resource is upstream bandwidth, but may also include the required memory or a time-based resource to activate the service flow. The admitted QoS parameter set can be thought as a subset of the provisioned QoS parameter set.

Active QoS Parameter Set:
This QoS parameter set defines the actual service that is being provided to the service flow. Only an active service flow may forward packets. The active QoS parameter set can be thought of as a subset of the admitted QoS parameter set. It should be pointed out that SID exists only for admitted or active upstream service flows.

One can also think in terms of three different types of service flows. The provisioned service flow is provisioned through the configuration file in the CM, while the admitted service flow has reserved resources by the CMTS that are not active. Only for the active service flow, the CMTS has committed resources by its QoS parameter set such as allocation upstream bandwidth.

Another important concept is the concept of Classifiers and packet classification within the MAC layer. A Classifier is a set of matching criteria applied to each transmitted packet over the cable TV network. The packet matching criteria can be, for example, destination IP address, a reference to a service flow, or scheduling priority. If the packet matches the specified packet matching criteria, then it is delivered on the referenced service flow. Downstream Classifiers are applied by the CMTS to the packets it is transmitting, while upstream classifiers are applied at the CM and may be applied by the CMTS in order to police the classification of upstream packets. Packet classification by the CMTS or CM is done in a table format, and consists of multiple Classifiers. Each Classifier has a priority field, which determines the search order for the Classifier. The highest priority Classifier must be applied first. The Classifier priority is used for ordering the application of Classifiers to packets. Classifiers can be added to the packet classification table using either management operations such as configuration file and registration, or via dynamic operation such as dynamic signaling.

The CMTS approves or denies every change to the QoS parameter set and Classifiers associated with a given service flow using a logical function called *authorization module*. The authorization module defines the range and acceptable values of the admitted and active QoS parameter sets that can be used. Such changes include requesting an admission control decision to set, for example, the admitted QoS parameter set, and requesting activation of a service flow. There are two types of authorization models in which the authorization module functions. In the static authorization model, the authorization module receives all registration messages and stores the provisioned status of all the service flows. Admission and activation requests for these provisioned service flows are permitted, as long as the active QoS parameter set is a subset of the admitted QoS parameter set and the admitted QoS parameter set is a subset of the provisioned QoS parameter set. The static authorization model can be thought of as a nested set of envelopes, with the smallest envelope defining the active QoS parameter set. Thus, in the static model, all the possible services are defined in the initial configuration of the CM. In the dynamic authorization model, the authorization module not only receives all registration messages, but also communicates them through a separate interface to an independent policy server. The policy server provides advance notice of upcoming admission and activation requests, and specifies the correct authorization action to the authorization module. The authorization module checks all the admission and activation requests from the CM. Only those admission and activation requests from the CM that are signaled in advance by the policy server are permitted. All other requests are either refused by the authorization module or sent for real-time query to the policy server. In the dynamic authorization model, it is possible to change the parameters of an existing service flow using the so-called *dynamic service change request* (DSC-REQ).

For any given service flow, its QoS parameter set may be explicitly specified, including all the traffic parameters, or indirectly by specifying a Service Class Name, which is referred to a set of traffic parameters.

11.8.2 Upstream Service Flow Scheduling

In order for the DOCSIS protocol to support constant bit rate (CBR) services such as voice over IP (VoIP) packets, the CMTS needs to schedule a steady stream of grants. These grants are referred to as unsolicited grants because there is no need for the CM to request them. The CMTS may reserve specific upstream opportunities for real-time traffic flows by providing data grants at periodic interval to the service flow. The term *unsolicited grant service* (UGS) is a type of upstream flow scheduling service that is used to map CBR traffic onto a service flow. Upstream flow scheduling services are associated with service flows through the use of SID. Each service flow may have multiple Classifiers, where each Classifier may be associated with a unique CBR media stream. When a service flow is provisioned for UGS, the Nominal Grant Interval is selected to be equal to the packet interval in the CBR application. In other words, the size of the grant is chosen to satisfy the bandwidth requirements for the CBR application.

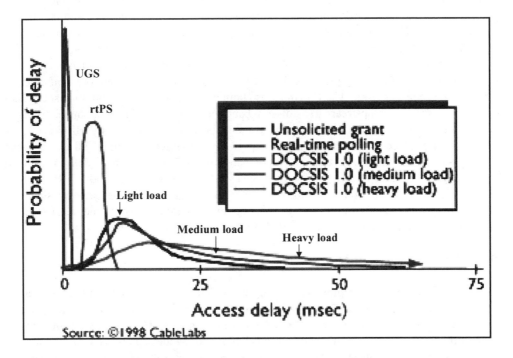

Figure 11.13 Probability of time delay versus the access delay (ms) for (A) unsolicited grant service, (B) real-time polling service, and (C) best effort service capabilities using DOCSIS 1.0 with light, medium, and heavy traffic load. After Ref. [26].

Besides UGS, there are other types of upstream service flow scheduling services, including real time polling service (rtPS), UGS with activity detection (UGS-AD), non-real time polling service (nrtPS), best effort (BE) service, and committed information rate (CIR) service. Let us briefly describe each of these scheduling services. The rtPS flows periodically receive transmission opportunities regardless of network congestion, and they release their transmission reservations to other service flows when inactive [25]. The CMTS polls rtPS SIDs on a periodic basis for upstream transmission traffic usage by using unicast request opportunities. The CM must use both unicast request opportunities as well as unsolicited data grants for upstream transmission.

For active service flows, the CMTS provides periodic unicast grants, but when the service becomes inactive, the CMTS must provide periodic unicast request opportunities. The problem is that the algorithm for detecting when a service flow is changing from an active to inactive state is dependent on the CMTS implementation. In the UGS-AD, one is trying to conserve system resources by taking advantage of the reduced latency of UGS for active service flows, and increase the efficiency of rtPS for inactive service flows. The UGS-AD is particularly tailored for VoIP applications, where meeting packet latency is essential. In this application, the CM must not send packets during the inactive state. This requirement implies that the CMTS has to support silent detection and the CM silent suppression.

The goal of nrtPS was to reserve upstream transmission opportunities for non-real-time traffic flows. The CMTS typically polls nrtPS SIDs on a periodic or random time interval (about one second or less). The CM must use both contention requests and unicast requests opportunities to obtain upstream transmission bandwidth. In the BE service, one is trying to provide efficient service to the upstream traffic on a best-effort basis. In the BE service, the CM must use all contention and unicast request opportunities as well as all unicast data transmission opportunities. Figure 11.13 compares the probability of time delay versus the access delay (in msec) for the following services: (A) UGS, (B) rtPS, and (C) BE service using DOCSIS 1.0 with light, medium, and heavy traffic load [26]. Other services such as a CIR service can be configured in many different ways, including a combination of some of the previously discussed types of scheduling services.

11.8.3 Upstream Fragmentation

The goal of the upstream fragmentation concept, which was introduced to the DOCSIS 1.1 protocol standard by both the Broadcom and Nortel corporations, is to improve the upstream traffic latency, particularly to support CBR services such as VoIP. The CMTS initiates an upstream packet fragmentation when it grants bandwidth to the requesting CM with a grant size smaller than the requested bandwidth. This is known as a Partial Grant. Fragmentation is encapsulation of only a portion of a MAC frame within a fixed-size fragmentation header and a fragment CRC. A single PDU or concatenated PDUs are encapsulated in the same way. At the CMTS, the fragment packets are processed similarly to non-fragmented packets except when baseline privacy is used; it starts just after the fragmentation header. The CMTS supports two different modes for packet fragmentation. In the first mode, which is called *multiple grant* mode, the CMTS allows multiple outstanding partial grants for any given SID. The CMTS can break a request up into two or more grants in a single MAP or over successive MAPs, and calculates the required additional overhead in the remaining partial grants to satisfy the request. If the CMTS cannot grant the remainder in the current MAP, it must send a grant pending (i.e., zero-length grant) in the current MAP and in all subsequent MAPs to the CM until it can grant additional upstream bandwidth. If there is no grant or grant pending in the subsequent MAPs, the CM must re-request for the packet payload remainder. To prevent the CM from being out of synchronization, the CMTS has to make sure that the packet payload remainder is at least greater than a mini-slot.

The second mode of the CMTS to perform fragmentation is called the *piggyback* mode, where it is assumed that the CMTS does not retain any fragmentation state. Thus, there is only one outstanding partial grant, and the CM must insert the payload remainder into the piggyback field of the fragment header. The CM calculates how much of a frame can be transmitted in the granted bandwidth, and forms the packet fragment to send it. Using the piggyback field in the fragment extended header, the CM requests the necessary bandwidth for the payload remainder to be sent. In this mode, the CM must track the physical-layer and fragmentation overhead of the payload remainder to be sent. At the CMTS, once the fragment HCS is correct, the payload fragment is ready for re-assembly with the original MAC frame, and then forward to the appropriate destination. From a practical perspective, the piggyback mode is easier to implement than the multiple-grants mode since the CMTS does not have to keep track of all the outstanding multiple partial grants for any given SID.

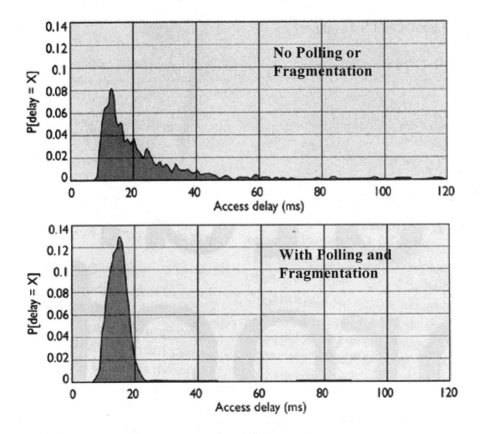

Figure 11.14 Time delay performance of CBR sources with and without polling and packet fragmentation. After Ref. [27].

Each CMTS vendor, based on an implementation specific algorithm, can select one or a combination of these operating modes, depending on the specific services the CMTS is supporting.

The advantage of using both polling and packet fragmentation when CBR and BE traffic are mixed is illustrated in Figure 11.14 [27]. Ten cable modems, each with a 64-Kb/s constant bit rate (CBR) source, modeled low latency voice traffic. Additional 44 CMs with packet sizes distributed over a range from 64 bytes to 1,500 bytes modeled BE traffic. The average bit rate for the best effort sources was 16-Kb/s. Notice that without polling and fragmentation, voice packets can occur time delays in excess of 100-ms. In contrast, the voice packets access time delays are upper bounded around 25-ms. However, there are still some delay variations, resulting from the frame-to-frame shifts in where a packet will lie within a particular frame.

11.9 CM and CMTS Interaction

The interaction between the CM and the CMTS can be divided into five basic categories as follows: (A) initialization, (B) authentication, (C) configuration, (D) authorization, and (E) signaling. The CM initialization procedure can be divided into the following steps:

1. Scan for a downstream channel and establish synchronization with the CMTS.
2. Obtain transmit parameters from the UCD message.
3. Perform ranging.
4. Establish IP connectivity.
5. Establish time of day.
6. Transfer operational parameters.
7. Perform registration.
8. Establish security association if provisioned to run Baseline Privacy.

When a CM is shipped from the manufacturer, the following information is sent:

- A unique IEEE 802 48-bit MAC address that can be used to identify the CM to the CMTS or to other provisioning servers during initialization
- Security information that can be used to authenticate the CM to the security server and authenticate the responses from the security and provisioning servers

Let us explain in more details each of the CM initialization steps.

1.   Scanning and Synchronization:
The first step of the CM initialization is to scan and acquire the downstream channel. A downstream signal is considered valid when the CM has achieved the following steps: (A) synchronization of the QAM symbol timing, (B) synchronization of the FEC framing, (C)

synchronization of the MPEG packetization, and (D) recognition of timing synchronization (SYNC) MAC message. As we have noted earlier, the SYNC messages must be periodically transmitted by the CMTS to establish MAC layer timing.

2.   Obtain Upstream Parameters:

After synchronization has been established, the CM must wait for the UCD message from the CMTS in order to retrieve the transmission parameters for a possible upstream channel. It should be pointed out that the CMTS periodically broadcasts on the cable TV network all available upstream channels using UCD messages. The CM must determine whether it can use a given upstream channel from the UCD parameters. If yes, the CM extracts the upstream channel parameters, and then it must wait for the next SYNC message and extract the upstream mini-slot timestamp from this message. The CM must also wait for the upstream bandwidth allocation MAP for the selected channel, and then may begin to transmit.

3.   Ranging and Automatic Adjustments:

After receiving the upstream bandwidth allocation MAP, the CM transmits a ranging request (RNG-REQ) in a contention mode to the CMTS using a temporary SID = 0. The CMTS receives recognizable ranging packets, allocates temporary SID, sends ranging response, and adds the temporary SID to its poll list. The CM is using a temporary SID before it is registered with the CMTS and is changing the upstream channel. When the CM receives the ranging response, it stores the temporary SID and adjusts its local parameters such as output power level, timing information, and upstream channel ID.

4.   Establish IP Connectivity:

After adjusting its local parameters, the CM must transmit *dynamic host configuration protocol* (DHCP) request to the server at the cable TV headend to obtain an IP address and any other parameters, which are needed to establish IP connectivity. The DHCP protocol is typically used to assign network-layer IP addresses. The server DHCP response must contain all the necessary configuration parameters for the CM.

5.   Establish Time of Day:

Both the CM and the CMTS need to have the proper time of day and date for time-stamping of various events, which can be retrieved by the management system. The CM creates its current local time by retrieving the time from the server, which is adjusted by the timing offset from the DHCP response.

6.   Transfer of Operational Parameters:

After successful DHCP operation, the CM must download its configuration file using trivial file transfer protocol (TFTP). TFTP is an Internet protocol for transferring files without requiring user names or passwords, and it is typically used for automatic downloads of data and software. If the downloaded configuration file has different upstream and/or downstream channels from what the CM is currently using, then the CM must not send a registra-

tion request message to the CMTS but must redo its initial ranging using the configured upstream and/or downstream channels.

7.   Registration:
To register with the CMTS, the CM must send a registration request with its assigned SID
and all the downloaded configuration file settings as follows:

- Downstream frequency configuration setting
- Upstream channel ID configuration setting
- Network access control object
- Downstream packet classification configuration setting
- Class of service configuration setting
- Upstream service flow configuration setting
- Downstream service flow configuration setting
- Baseline Privacy configuration setting
- Maximum number of Classifiers
- Privacy enable configuration setting
- Payload header suppression
- TFTP server timestamp
- TFTP server provisioned modem address
- Vendor specific information configuration setting
- CM MIC configuration setting
- CMTS MIC configuration setting

In addition, the CM must send the CM ID assigned by the vendor and the CM IP address as
well as the encoding capabilities of the CM. The CM must wait for a registration response
from the CMTS, authorizing it to transmit data to the network.

8.   CM Security Initialization:
After the registration process is completed, the CM can initialize cable TV network security
operation if it is provisioned to run Baseline Privacy [3]. The CM is provisioned to run
Baseline Privacy if its configuration file includes a Baseline Privacy configuration setting.

References

1.   CableLabs, *Data-Over-Cable Service Interface Specifications (DOCSIS): Radio Fre-
     quency Interface Specification*, SP-RFIv1.1-I02-990731 (1999).

2. IEEE Standards Department, IEEE Project 802.14/a Draft 3 Revision 3, *Cable TV Access Method and Physical Layer Specification*, October (1998).

3. CableLabs, *Data-Over-Cable Service Interface Specifications, Baseline Privacy Plus Interface Specification*, SP-BPI-W02-981228 (1998).

4. C. Huitema, *Ipv6: The New Internet Protocol*, Prentice Hall PTR (1996).

5. Ethernet Protocol Version 2.0, Digital, Intel, Xerox (1982).

6. D. Plummer, Ethernet Address Resolution Protocol: Or Converting Network Protocol Addresses to 48-bit Ethernet Address for Transmission on Ethernet Hardware, November (1982).

7. K. Sollings, *The Trivial File-Transfer Protocol*, Revision 2, Internet Engineering Task Force RFC-1350, July (1992).

8. R. Droms, *Dynamic Host Configuration Protocol*, Internet Engineering Task Force RFC-2131, March (1997).

9. ISO/IEC 10038, ANSI/IEEE Standard 802.1D: Information technology - Telecommunication and Information Exchange between Systems; Local Area Networks Media Access Control Bridges (1993).

10. ITU-T Recommendation J.83 Annex B, *Digital Multi-program Systems for Television Sound and Data Services for Cable Distribution*, April (1997).

11. H. Harashima and H. Miyakawa, "Matched-Transmission Technique for Channels with Intersymbol Interference," *IEEE Transaction on Communications* **COM-20**, 774–780 (1972).

12. M. Tomlinson, "New Automatic Equalizer Employing Modulo Arithmetic, " *IEE Electronics Letters* **7**, 138–139 (1971).

13. *MPEG Video Compression Standard*, Edited by J. L. Mitchell, W. B. Pennebaker, C. E. Fogg, and D. J. LeGall, Chapman and Hill, New York (1997).

14. ITU-T Recommendation X.25, *Interface between Data Terminal Equipment and Data Circuit-Terminating Equipment for Terminals Operating in the Packet Mode and Connected to Public Data Networks by Dedicated Circuit*, March (1993).

15. N. Abramson, "The ALOHA System, Another Alternative for Computer Communications," Proceedings, Fall Joint Computer Conference AFIPS vol.37, 281–285 (1970).

16. B. Sklar, *Digital Communications Fundamentals and Applications*, PTR Prentice Hall, Englewood Cliffs (1988).

17. Digital Audio-Visual Council (DAVIC), *Lower Layer Protocols and Physical Interfaces*, DAVIC 1.2 Specification Part 8 (1997).

18. W. Bux, "Local-Area Subnetworks: A Performance Comparison," *IEEE Transactions on Communications* **COM-29**, 1465–1473 (1981).

19. W. Stallings, "Local Network Performance," *IEEE Communications Magazine* **22**, 27–36 (1984).

20. IEEE, *Carrier Sense Multiple Access with Collision Detection (CSMA/CD) Access Method and Physical Layer Specifications*, American National Standard ANSI/IEEE Standard 802.3 (1985).

21. IEEE, *Broadband Medium Attachment Unit and Broadband Medium Specifications, Type 10 Broad36 (Section 11)*, Supplements to ANSI/IEEE Standard 802.3 (1988).

22. R. Citta and J. Xia, "Adaptive P-persistive Algorithm with Soft Blocking for Contention Resolution," IEEE 802.14/97-037, March (1997).

23. N. Golmie, D. H. Su, G. Pieris, and S. Masson, "Performance Evaluation of Contention Resolution Algorithms: Ternary Tree versus p-Persistence," IEEE 802.14/96–241, October (1996).

24. CableLabs, *Data-Over-Cable Service Interface Specifications (DOCSIS): Radio Frequency Interface Specification*, SP-RFI-I04-980724 (1998).

25. R. Rabbat and K.-Y. Siu, "QoS Support for Integrated Services over CATV," *IEEE Communications Magazine* **1**, 64–68 (1999).

26. F. Dawson, "Is IP Telephony in Your Future?" *Communications Engineering & Design Magazine* **11** (1998).

27. T. Quigley and D. Hartman, "Future Proofing MCNS Data-Over-Cable Protocol," *Communications Engineering & Design Magazine* **3** (1998).

CHAPTER 12

DIGITAL SET-TOP BOX SOFTWARE
ARCHITECTURE AND APPLICATIONS

So far, we have learned about the key system technologies that enable the current and emerging two-way cable TV networks. In the final chapter, we review first the STB software architecture, including the real-time operating systems; the middleware layer; and applications. The different types of STB applications are divided into native applications and TV-based and Internet-based applications and are discussed in more detail. The native applications, which include electronic program guides (EPGs) and parental control, are discussed in Section 12.2. The TV-based applications include pay-per-view (PPV), enhanced broadcast, interactive and targeted advertisements, and video-on-demand (VOD), which are discussed in Section 12.3. The Internet-based applications, which include webcasting, e-mail, e-commerce, home banking, education, gaming, and others, are discussed in Section 12.4. Section 12.5 discusses the ongoing customized integration of TV-based and Internet-based applications. It should be pointed out that since there are many different middleware and applications vendors, which are operating on different platforms, the review here of the software architecture and applications is intended not to be exhaustive but to provide the reader with different examples.

12.1 Digital Set-Top Box Software Architecture

In Chapter 8, our discussion focused primarily on the hardware architecture of the digital STB. The software architecture of the STB is a critical part of the STB operation, enabling cable operators to offer new services to their subscribers. Let us review the basic building blocks of the STB software architecture. A generic block diagram of the software architecture of a digital STB is shown in Figure 12.1 [1,2]. It consists of the following layers: (A) hardware abstraction layer, (B) native middleware layer, (C) real-time operating system (RTOS) layer, (D) standard application program interface (API) layer, and (E) native and high-level application layers.

The hardware abstraction layer consists of software drivers to control the various underlying hardware devices in the STB such as cable modem, cable TV tuner, QAM transceiver, and others. Any design upgrade of an existing hardware device or the addition of new hardware components may require the STB manufacturer to modify the device driver. In the

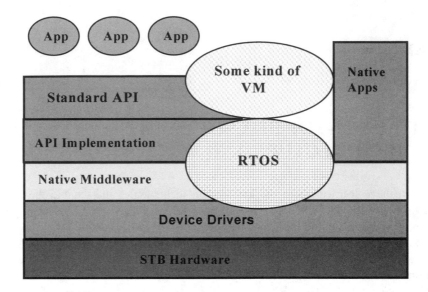

Figure 12.1 Generic software architecture block diagram of digital STB. After Ref. [1].

RTOS layer, the kernel, which is typically stored in flash memory and can be as large as several megabytes, is first loaded when the STB is powered-up. The RTOS kernel is typically responsible for managing the STB memory resources, real-time applications, and high-speed data transmission. Another important feature of the RTOS kernel is its support of multithreading and multitasking, namely, different portions of a program or different programs can be simultaneously executed [2]. In addition, the RTOS performs other important functions such as managing the network communication stacks, graphics, audio and video subsystems, and conditional access system, as well as service acquisition download, system information management, and return-path transmission.

The native middleware layer isolates the native applications and the various APIs from the underlying hardware devices, allowing these applications to operate transparently across the network. In other words, the software vendor may benefit from having the same middleware environment on different STB models, allowing these applications to be easily ported. This can significantly reduce the development complexity of new applications, cutting their development cycle. The standard APIs, such as the ATSC DTV application software environment (DASE), form the necessary interface between a specific RTOS and high-level applications and provide the environment to write applications that are independent of a specific STB hardware or RTOS vendor. The API implementation, on the other hand, is most likely RTOS specific (see Section 12.1.1).

The flexibility of the software architecture enables the cable operators to add new services. The different STB applications can be categorized in terms of orthogonal groups of applications. Common groups of applications are, for example, native versus non-native or STB-independent applications, broadcast versus resident applications, and passive versus interactive applications [1,3,4]. Resident applications are programs, which typically reside in the STB flash memory, and are downloaded from the headend by the cable operators to the subscriber STB. The RTOS must have a loader to enable the cable operator to upgrade resident applications, download new applications, and/or download patches to the existing RTOS. When multiple applications are running simultaneously, it is essential to prevent possible conflicts as well as to provide a consistent look and feel among these applications. This goal is a function of the middleware implementation and will be discussed in Section 12.1.2. Broadcast applications are typically programs that are synchronous and related to the video content with which they are broadcast. Such broadcast applications, for example, include stock tickers, sports and weather information, and enhanced advertisement. Interactive applications can have local interactivity where the same content is sent to all the STBs and the user makes a choice locally, or transactional applications where the user actually interacts with the headend server (i.e., VOD, web browsing, and e-commerce). These types of applications are discussed in more detail in Section 12.2.

12.1.1 Real-Time Operating System (RTOS)

The operating system (OS) is one of the key components of the digital STB software that enables it to function. Since there are many different types of STB, the OS should be as independent as possible from the hardware devices, shielding the various applications running on top of it. Furthermore, unlike the PC, a digital STB can have multiple operating systems to enable the widest range of applications over the cable TV network. The OS of the STB must be fast, functioning in a real-time environment, and handling sophisticated tasks in a limited memory size and in a smaller hardware system than a desktop PC. The primary RTOS responsibilities include:

- Controlling the use of the CPU and memory management
- Controlling video, graphics, and data outputs to TV, hard drives, and other devices
- Controlling smooth operation of different applications
- Controlling communications to other devices such as cable modems or PCs
- Managing STB hardware functions and data storage, and scheduling real-time tasks

Currently, there is no standard RTOS for digital STB, and some of the most common RTOSs are Microsoft Windows CE [5], PowerTV [6], Java OS [7], VxWorks [8], and Microware David OS-9 [9]. Let us very briefly discuss the features and functionality of two popular OSs, namely, Windows CE and PowerTV.

Microsoft Windows CE is a modular RTOS designed to run on variety of TV-centric devices such as digital STBs and hand-held devices such as a Pocket PC. In order to lure cable operators to use its software, Microsoft has developed the so-called TV platform adaptation kit (TVPAK), which consists of Windows CE and a software headend server called Microsoft TV server [2]. The modularity of Windows CE means that software developers can customize the RTOS for a specific hardware platform using Windows CE platform builder of development tools. In order to connect the various Windows CE modules, Microsoft uses dynamic link libraries (DLLs), where a DLL is a library of executable functions or data that can be used by Windows CE applications.

To accelerate their TVPAK strategy, Microsoft has added various drivers to its Windows CE (version 2.12) to support the underlying STB hardware components. The STB hardware components include CMs, smart cards for conditional access and e-commerce, IP telephony module, IEEE 1394 bus, hard disk drives, and cable TV tuner. Another useful feature of Windows CE is its registry, which is a small-footprint database used by the operating system to store the various STB configurations. From a cable operator perspective, the digital STB must be able to communicate securely and efficiently over the cable TV network. Consequently, Microsoft has provided support for a variety of communications and security protocols. Windows CE is one of the possible OSs that can run on Motorola/GI DCT5000+ advanced digital STBs.

The PowerTV OS, which has been developed by PowerTV since 1994, is specifically designed for the digital STB environment [6]. This means that it exists in a low-cost ROM or flash memory that can easily be upgraded and customized by the cable operators through network downloads to deliver integrated and secure enhanced TV services. The PowerTV RTOS consists of the following layers:

- Hardware-independent porting layer, which contains all the necessary device drivers for the STB
- PowerCore layer that provides the base functionality required by the applications, including a multitasking PowerKernel designed to operate in a small memory footprint
- High-level processing functions built on top of the PowerCore are encapsulated as software PowerModules, providing efficient and hardware-independent application services. The set of PowerModules enables cable operators to offer customized network-delivered applications and content.

In addition, the PowerTV platform provides cable operators with the necessary STB middleware to enable Internet-based applications. PowerTV RTOS has been running on Scientific Atlanta Explorer 2000/3000 digital STBs.

12.1.2 Set-Top Box Middleware

As mentioned earlier, the STB middleware is an important link between the native applications and APIs and the underlined hardware devices. Currently deployed digital STBs support a wide range of enhanced interactive applications, including web browsing and video and audio streaming. In order to create applications that are not limited to proprietary hardware, developers have started to write middleware interpreters that are capable of decoding many different types of applications, independently from the hardware. The most popular web content formats are hypertext markup language (HTML), JavaScript language, and Java bytecodes. The Java language is a general-purpose object-oriented programming language with similar syntax to C/C++, but omits many of the features that make C and C++ complex and confusing. It was designed to support multiple host architectures and to allow secure delivery of software components. The compiled Java code had to survive transport across networks, operate on any client, and assure the client that it was safe to run. The interesting capabilities of Sun's Hot Java browser to embed Java programs, which are known as applets, inside HTML pages were demonstrated before most other web browsers added support for the Java JM and applets. The applets are transparently downloaded into the HotJava browser along with the HTML pages in which they appear. Before being accepted by the browser, applets are carefully checked to make sure they are safe. Like HTML pages, compiled Java programs are network- and platform-independent, and behave the same way regardless of where they come from, or what kind of machine they are being loaded into and run on. Consequently, web browsing was no longer limited to a fixed set of capabilities.

The essential component of the Java technology, which is responsible for Java portability, is called the *Java virtual machine* (JVM) [10]. The Java source code (*.java files) are compiled into bytecodes (*.class files), which are interpreted by the JVM. A java class is a collection of data and related methods that operate on the data. The JVM does not know the Java programming language except for a particular file format called the class file format. It

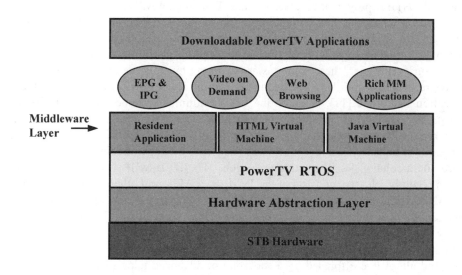

Figure 12.2 Architecture of PowerTV STB software showing the middleware layer with the various applications. After Ref. [2].

contains JVM instructions (or bytecodes) and a symbol table, as well as other ancillary information. For secure delivery, the JVM imposes strong format and structural constraints on the code in the class file. However, any language with functionality that can be expressed in terms of a valid class file can be hosted by the JVM. Personal Java (P-Java), which is a Java API and application development environment, consists of a modified JVM, a set of core libraries, and optional libraries that can be used as needed.

Another related popular virtual machine is called JavaScript virtual machine. JavaScript is a scripting language rather than a programming language, which uses less complex rules, and it is easier to learn and use than Java. Since JavaScript code is typically embedded within an HTML document, it can perform more complex tasks than HTML alone. New extensions to JavaScript are being developed by middleware vendors to provide an enhanced TV viewing experience. It should be pointed out that JavaScript is the only interpreted language in which the source code files are executed directly in the virtual machine and stored in the STB.

The emerging API standardization efforts want to eliminate porting of applications to all the different RTOSs in a given STB. The middleware platform must have a common set of APIs to make downloadable content and applications interoperable among different STBs. In other words, the ultimate goal of the software vendor is to develop applications once using standard APIs, which are STB hardware and RTOS independent. The STB is only required to support a chosen VM such as JVM and a set of APIs such as Java APIs. This increases the STB deployment cost because the JVM and the API implementation layer must be included on top of the native middleware in the STB software architecture as shown in Figure 12.1.

Currently, various companies are offering middleware platforms that are mostly proprietary solutions. However, these solutions are evolving towards the emerging open standard middleware, which is split between two main efforts. One effort, which is led by the DVB organization in Europe [11], is called the DVB Multimedia Home Platform (DVB-MHP) standard. The DVB-MHP standard, which intends to enable a wide variety of home terminals such as a PC, an integrated TV, and a digital STB, is responsible to define user requirements for enhanced cable TV services (i.e., dual broadcast audio/video and Internet), and provide a solution for interactive television. The DVB-MHP 1.0 specification, which has been recently released, consists of the JVM, Java APIs, and other APIs such as Java TV APIs. MediaHighway middleware, which has been developed by Canal+ Technologies and internationally indorsed by many countries, provides a powerful and flexible solution in this category [12]. The latest version of this software is called MediaHighway+, which is integrating the open technical specifications of DVB-MHP middleware standard.

A second alternative effort led by a subgroup of advanced television systems committee (ATSC) in the US, which is known as T3/S17, is defining the software architecture standard for digital STB called DTV DASE [13]. DASE has two environments as follows: (1) a procedural environment (called Application Execution Engine), which is generally written in Java, and (2) a declarative environment (called Presentation Engine), which is for declarative application content (HTML) [14]. The T3/S17 group has selected Sun's Java TV APIs

[15], P-Java APIs [16], Java Media Framework APIs [17], and Home Audio/Video Interoperability (HAVi) User Interface APIs [18] as the standard development environment for digital interactive services. The Java TV APIs, which have also been adopted by other standards organizations such as DVB, provide a high-level, protocol-independent, common set of APIs for applications to be used on different hardware platforms. The complete set of DASE APIs is expected to be implemented on top of P-Java platform, and it is also expected that TV applications written in a Java programming language will be able to run on digital STBs that comply with the DASE effort.

In addition to the standard efforts, there is also a consortium effort, consisting of US-based companies, including Microsoft, Intel, Disney, DirecTV, and others, and it is called advanced television enhancement forum (ATVEF) [19]. The ATVEF group is proposing to use HTML virtual machine and JavaScript virtual machine for the middleware standard. ATVEF version 1.1 specification, which has been released last year, defines the fundamentals for creating HTML-enhanced TV content for reliable delivery to many different access terminals such as digital STBs across a wide variety of networks, including terrestrial (i.e., over the air), cable, satellite, and the Internet.

To illustrate the middleware platform concepts that have been discussed so far, let us look at the following example. Figure 12.2 shows a block diagram of the PowerTV software architecture, which has been implemented on Explorer 2000/3000 digital STB by Scientific Atlanta [6]. Notice that the middleware layer contains both JVM and HTML virtual machine APIs. The use of HTML virtual machine allows the PowerTV enabled STB to process web content which has been broadcasted over the cable TV network. The downloadable PowerTV applications are applications that can be used, for example, to control other devices on the home network. It is expected that both Scientific Atlanta and Motorola will support ATVEF content delivery on their digital STBs.

Finally, let us mention the proprietary middleware efforts such as OpenTV and Liberate TV Navigator for DTV by Liberate Technologies [20]. OpenTV software environment for digital STBs consists of a middleware platform, a set of authoring tools, a library of APIs, and utilities, which can fit into a small-footprint configuration (350–560 kB flash memory), for cost-effective integration into interactive TV applications. The addition of O-code VM, which translates the OpenTV scripting language into native instructions, shields the OpenTV applications from the hardware layer. Recently, OpenTV has begun to implement jointly with Panasonic a fully compliant DVB-MHP extension to their STB software.

Liberate Technologies, which was originally founded by Oracle and Netscape and was known as Network Computers, Inc., has developed a platform solution, which integrates Internet-based applications with TV-based content to digital STB subscribers. Liberate TV Navigator platform integrates HTML extensions for TV with JavaScript 1.2 content. The Liberate middleware platform, which is competing with Microsoft's WebTV solution, has been installed on various STBs, including Motorola's DVi5000+ to allow European subscribers access to a wide variety of applications.

As mentioned earlier, the middleware implementation determines the feel and look of the content display for a specific digital STB. This can be achieved by using (1) native master application to manage all the concurrent applications when they are launched, and assign priority for sharing hardware resources, (2) a set of implementation APIs for all the downloadable and broadcast applications, and (3) non-native resident application to control the TV display.

   There are many different ways to classify the various STB applications. For reasons that will become clearer later on, let us classify the various STB applications into three main groups as follows:

(A) Native applications, which are specific to a given STB, such as Electronic Program
     Guides (EPGs)
(B) TV-based interactive applications, which include local and server interactivity, such as
     pay-per-view (PPV) and video-on-demand (VOD)
(C) Internet-based applications, which include a wide range of applications from web surf-
     ing to e-commerce

From here to the end of this chapter, these applications are reviewed.

12.2 Native Applications

Native applications, which typically reside above the STB native middleware layer as shown in Figure 12.1, are STB specific applications that do not require the use of the RTOS or APIs. EPGs and parental control are examples of native applications, and are discussed in the next subsections.

12.2.1 Electronic Program Guides (EPGs)

As the possibility of 500 or more TV channels with potentially hundreds or even thousands of additional web sites becomes closer to a reality, the necessity of software tools to help navigate through large number of options becomes clear. EPGs became available in the ear ly 1970s and provided consumers with onscreen guides of the channel line-ups. The EPG development in the 1980s and early 1990s had a limited functionality with mostly text-based onscreen display due to hardware limitations of the analog STBs. With the development of interactive digital STBs, a new type of advanced multimedia EPGs, which are sometimes called interactive program guide (IPG), were developed that changed the consumer TV experience. Some of the most popular EPGs are TV Guide [21] and StarSight [22], running on the most widely deployed digital STBs such as Motorola's DCT2000/5000+ and Scientific Atlanta's Explorer 2000/3000. The basic EPG provides onscreen display of program lineup and episode information for every channel for the next seven days. The viewer typically navigates around the program grid menu using the arrow keys on his remote control, and can

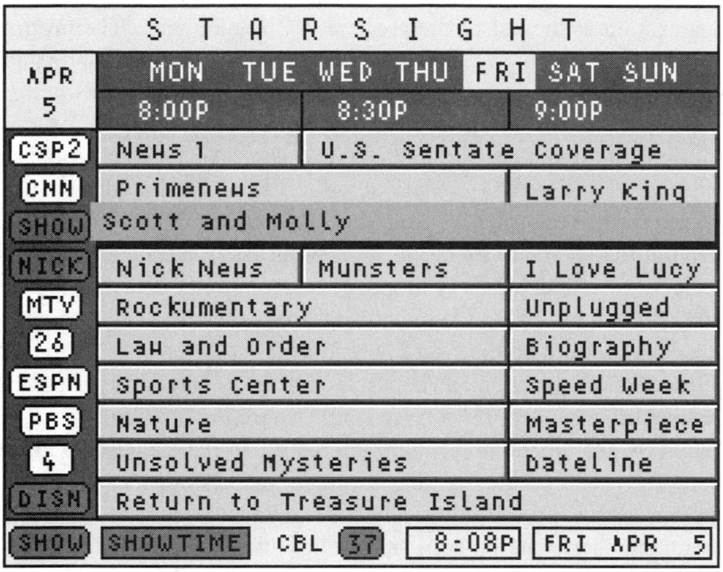

```
          S   T   A   R   S   I   G   H   T
  APR     MON   TUE  WED  THU  FRI  SAT  SUN
   5      8:00P          8:30P          9:00P
 CSP2   News 1        U.S. Sentate  Coverage
 CNN    Primenews                   Larry King
 SHOW   Scott and Molly
 NICK     Nick News    Munsters      I Love Lucy
 MTV    Rockumentary                Unplugged
 26     Law and Order               Biography
 ESPN   Sports Center               Speed Week
 PBS    Nature                      Masterpiece
  4     Unsolved Mysteries          Dateline
 DISH   Return to Treasure Island
 SHOW   SHOWTIME   CBL  37    8:08P   FRI APR  5
```

Figure 12.3 StarSight electronic program guide features where the selected program is highlighted in the program menu. Courtesy of Gemstar TV Guide International.

receive immediate information for any available program, including show description, main actors, and rating. Figure 12.3 illustrates, for example, StarSight EPG with the available program menu within the selected time slot, and with the desired program highlighted [22].

Advanced EPGs offer new features in addition to the basic EPG features, and can be configured for personalized TV viewing as well as provide detailed program information in multiple languages. For example, StarSight EPG let the viewer arrange the TV channels in the program grid menu (available channels versus time) such that the favorite channels are placed first, and delete those channels that the viewer never wants to watch. Furthermore, StarSight lets the viewer turn on his or her VCR, record the selected programs, and turn it off. Advanced EPGs may also include additional features such as alerting the viewer before his selected program is shown, control of the disk storage device within the digital STB, and searching capabilities for a particular episode on a specific time and date.

The EPG information has exploited the digital characteristics of MPEG-2 video streams. Special control data structures, which help to identify the transmitted programs within a multiplexed stream or among different multiplexed streams, have been defined by ITU/ISO MPEG-2 system specifications (ISO-13818 section 1) [23]. A packet identifier (PID) field in the header identifies each video or audio stream. The PID streams, which form virtual channels, are multiplexed and transmitted, for example, within a 64-QAM channel. A special reserved PID of 0 is used only for the periodic transmissions of the so-called program assignment table (PAT). The PAT contains a list of the available program numbers for the transmitted multiplex as well as the list of the associated PID numbers. The program map table (PMT), which is transmitted on an assigned PID in the PAT, lists descriptors about the program, stream type, and PIDs for each stream that is associated with that program as well as descriptors for each stream. Since some of the transmitted programs may be encrypted, PID 1 is reserved for the conditional access table (CAT). The CAT lists the associated entitlement management messages (EMMs) or the decryption keys for the addressable digital STBs that are allowed to receive the specific programs that were paid for by the subscriber.

It should be pointed out that there are two EPG extensions to this MPEG-2 system specification, which have been specified by the European Telecommunications Standardization Institute (ETSI)/DVB [24] and by ATSC [25]. Unfortunately, these extensions are not compatible with each other and will not be discussed here [26].

The ongoing development of advanced EPGs is still in its early stages. The integration of web-based applications with enhanced TV services will require the next-generation EPGs to become more personalized, intelligent, and highly interactive to reduce the increasing complexity and offer more TV friendly viewing. For example, the EPG can group TV channels based on a common content such as sporting events or children programs. Other features may include a personalized agent who is familiar with your viewing habits and interests can scan the available TV and web programs when asked. By interacting with the intelligent agent, the viewer can reduce the list of choices according to his preferences. The personalized agent can also alert you to future programs based on your viewing schedule and your interaction with him. Such advanced interactive EPGs will surely transform the TV viewing experience from a passive one to an interactive and exciting one.

The discussion on EPGs is not complete without mentioning intellectual property rights. Gemstar TV Guide International, in the US, holds various key patents for the functionality, look, and feel of the basic and advanced EPGs. For example, Gemstar has the patent rights for the typical TV program grid design (see Figure 12.3). Since this grid look is familiar with many consumers, various software vendors and STB manufacturers have licensed the patent rights rather than engage in an expensive legal battle with Gemstar International.

12.2.2 Parental Control

Another application related to the EPG application is called parental control. The parental control service allows viewers to control access to certain programs or channels that are viewed at the subscriber's location. The parental control service, which can be launched on the TV screen using a lock or a secret PIN number, allows the subscribers to specify the category of programs to be restricted, such as adult movies, as well as how much Can be spent when ordering a non-restricted PPV event. The parental control of certain TV programs is typically accomplished by highlighting the selected program rating from the on-screen menu and pressing the lock button. Figure 12.4 shows, for example, a parental control screen with no lock on the selected channel. To keep the lock secret, which is only known to several family members, the cable operators typically advise the subscriber to change the lock periodically, say every three months.

Figure 12.4 A parental control screen showing the selected title and the lock status.

12.3 TV-Based Interactive Applications

The TV-based applications can be divided into two main groups. The first group of TV-based interactive applications, which enables the user's local interactivity, includes applications such as PPV, impulse PPV (IPPV), and parental control, which are discussed in Section 12.3.1. In addition, there are non-native applications such as enhanced broadcast and interactive and targeted advertisements, which may require an increasing level of local user interactivity and are discussed in Section 12.3.2. The second group of TV-based applications, which require server interactivity, includes applications such as VOD and are discussed in Section 12.3.3. Although Near VOD (NVOD) application requires only weak server interactivity, it is also discussed here due to its close relationship with VOD.

12.3.1 Pay-Per-View (PPV) and Impulse PPV (IPPV)

PPV and IPPV enable consumers to watch a broadcast program such as a movie or a sporting event by purchasing it within a designated time window. The cable operators typically allow the viewer to purchase the desired event from several weeks before the starting time of the event up to 15 minutes after the event starts. Different PPV menus have a different look and feel in terms of the screen layout and appearance. The PPV menu may be organized, for example, according to movie titles, titles by time, or according to the type of programs such as children, adult, and sport. Each cable operator can insert his logo on the PPV screen and further customize the PPV menu. The concept of PPV application was implemented as call-ahead PPV (CAPPV) service, where the subscriber had to call ahead using a toll-free number before receiving authorization for the requested event. The IPPV application allowed the subscriber to watch the requested programs and then pay for them up to a maximum pre-determined credit. The calling was done automatically using a low-speed (9600-kb/s) telephone modem built into the STB. Figure 12.5 shows an example of an IPPV menu selected according to the desired time slot, where the viewer is reviewing his order for the movie "Batman Forever" rated PG-13. After the viewer has purchased the desired movie, the billing information such as the time, date, title, and charge for the movie is transmitted (not real-time and low-speed) to the billing system server located at the cable headend through the return OOB channel. The subscriber billing information is typically stored in a

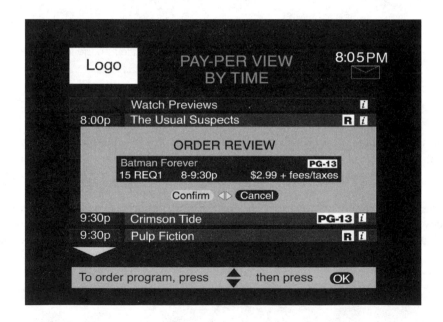

Figure 12.5 Pay-per-view menu selected according to desired time slot.

non-volatile memory such as flash memory in the STB. Some international cable operators allow subscribers to use the so-called prepaid smart cards to purchase the requested PPV events. It should be pointed out that some cable operators might have cumbersome and/or noisy return-path network architecture, which require an expensive upgrade. Consequently, IPPV applications cannot be implemented, and the cable operators have to resort to using low-speed telephone modems or call-ahead options.

12.3.2 Enhanced Broadcast, Interactive, and Targeted Advertisements

Enhanced broadcast (EB) and interactive and targeted advertisements are non-native STB applications that are associated with the audio/video stream and are enabled via the subscriber's local interactivity. EB, which is targeted by DVB-MHP and ATSC-DASE, is represented by additional information (i.e., text, graphics, and some level of interactivity) sent to the subscriber along the audio/video stream to enhance the broadcast video program. For example, such information may be sports scores of an ice hockey team playing, which can be displayed on the screen when the subscriber pushes a specific button on his remote control. Other examples may include a customized stock ticker during a CNN business news program or an additional text/press release during an evening news program. Some of these features are also being targeted by ATVEF [19].

Interactive and targeted advertisement applications, which offer potentially lucrative opportunities to many businesses, require higher user's interactivity than EB, depending on the particular implementation of these applications. For example, during a commercial break, the user may click on a specific button on his remote control and either request further information on the displayed product or service, or even purchase it. The application may allow the subscriber to print a coupon of the advertised product or service in the convenience of his home, or connect him to the web for e-commerce transactions (see Section 12.4.3). In addition, different subscribers may see different commercials during the same program break, depending on their location and/or personal preferences. For example, some people who live in Miami may prefer to see commercials in Spanish rather than English, and some people who live in Phoenix may not want to see winter tire ads during the winter months.

12.3.3 Video-on-Demand (VOD) and Near VOD (NVOD)

One of the primary applications of digital TV is VOD and Near VOD (NVOD) [2]. In the VOD system, the consumer can purchase a program on demand using the STB remote control and view it when he wants to. In other words, VOD system offers the viewer the convenience of renting a movie from the cable operators without having to leave his home. Figure 12.6 shows, for example, the interactive VOD grid menu by Diva, where the viewer can search, purchase, and view the desired movie from the available movie menu [27]. VOD

Figure 12.6 Interactive video-on-demand grid menu, showing the available movie
categories to be selected. Courtesy of DIVA.

is sometimes called "pull" technology because the media server "pulls-in" the subscriber's
requests over the cable TV network. In the NVOD system, a limited number of programs
start on different digital channels with a fixed time interval (i.e., every 30 minutes) such that
the consumer can easily choose the desired time and date to view his selected program. Currently, there are various vendors who offer commercially scalable VOD system solution,
including Diva [27], SeaChange International [28], and Vivid Technologies. Generally
speaking, the VOD system architecture consists of high-performance video servers, which
have a large library of movies, connected to a network management system and a billing
system at the cable TV headend. For example, the DIVA DVS5000 video server consists of
multiple processor modules that stream movies from disk drive arrays over VOD networks.
Each processor module uses a twelve-disk redundant array of independent (or inexpensive)
disks (RAID) storage. The DVS 5000 includes a 34-Gb/s non-blocking MPEG switch that
can be configured up to 16 ports, and can interconnect server modules to form video server
clusters. A DVS 5000 cluster can deliver anywhere from 500 to over 20,000 3.4-Mb/s
streams. The different MPEG-2 video streams are then encrypted, multiplexed, and coded
for 64-QAM or 256-QAM transmission over the cable TV network to the subscriber's digital
STB. In practice, it is cheaper and simpler for the cable operators to offer NVOD service
rather than VOD service, since the NVOD system can be implemented using low-cost PC-

based digital media servers and there is no need for high-throughput real-time MPEG-2 switching and decoding. Furthermore, some NVOD vendors offer a scalable NVOD solution to VOD, depending on the cable operator penetration rates and number of subscribers.

12.4 Internet-Based Applications

Internet-based applications are perhaps the most important emerging application class for the digital STB. In fact, web-based applications are converging with broadcast and interactive TV-based applications [29]. Before discussing the different web-based applications, let us discuss how web-based content is implemented to enhance broadcast TV content. For the broadcast analog channel, the web content is carried in the VBI according to North American basic Teletext specification (NABTS) [30] and Internet standard IP over VBI [31]. For the broadcast digital channel, there are several options. An extension to the ATSC *program and system information protocol* (PSIP) supports data broadcast as defined by ATSC T3/S13, specifying how IP data is carried in the MPEG-2 TS [32].

In the next section, we provide a brief overview, which is not intended to be exhaustive, of the various web-based applications for a digital STB, including webcasting, e-mail, e-commerce, home banking, education, and gaming.

12.4.1 Set-Top Box Web Browsers

Internet-based applications are the most important emerging application class for a digital STB. Similar to PC-based web browsers, STB-based browsers allow subscribers to receive multimedia web content displayed on their TV. However, the web browsers in the STB environment are operating with some constraints such as limited memory size and CPU processing power. As a result, it may not be possible to support all the different forms of web content such as 3D graphics or a virtual reality content. The Internet access of the STB is accomplished by connecting to a file server at the headend through the cable TV network. The file server listens and responds to requests from the subscriber STB browser using *hypertext transfer protocol* (HTTP). In addition, the web server is responsible for other important tasks such as protecting the web content from unauthorized users, and sending requests to other web servers on the Internet. To help browse the web, the digital STB often incorporates Proxy server software. The Proxy server acts as a gateway with firewall-class security between the cable TV network and the public Internet, while providing easy-to-use management features. The Proxy server offers distributed (hierarchical and array-based) web caching, scalability, fault tolerance, resulting in improved network response time and efficiency by as much as 50%. Some Proxy servers provide network filtering, allowing cable operators and subscribers to restrict access to certain web site containing inflammatory or inappropriate content.

The security between the digital STB and the Proxy server is typically established using *secure sockets layer* (SSL) protocol, which is supported by most web browsers using 40-bit or 128-bit keys for data encryption [34]. The SSL protocol runs above the TCP/IP layer and below the high-level application protocols such as HTTP and IMAP. The transmitted data is encrypted using a public key, but can only be decrypted using a private key. The SSL server authentication allows a user to confirm the identity of a server. This feature may be critical when the STB subscriber, for example, is sending his credit card information over the cable TV network and wants to check the receiving server identity. Furthermore, the SSL authentication running on the user's STB allows a server to confirm the users' identity. This confirmation may be important, for example, to a bank sending financial information to its client. The SSL protocol uses both public-key and symmetric-key encryption. First, the server authenticates itself to the STB, or the STB can authenticate itself to the server using public-key encryption. Then, symmetric keys are created for rapid encryption, decryption, and tamper detection during the session between the server and the STB.

Figure 12.7 Web browsing through a standard TV using a digital set-top box. Courtesy of Interactive Channel.

An interesting and important feature of the web browser, in contrast to the VOD application discussed in Section 12.2.3, is called *webcasting* ("push" technology). Set-top webcasting is the process of distributing multimedia content by the web server to the STB subscriber over the cable TV network. The reason it is called push technology is because the server "pushes" the multimedia content to the user rather than waiting for his request. In practice, the push of the multimedia content is triggered by either the user or the cable operator, and arrives only as a result of the user request.

In webcasting, the STB user subscribes to his favorite "channels" that are frequently accessed. A channel consists of the main web page and any related web pages. The set-top browser periodically monitors the selected channels and notifies the user when a new content is available. As the updates arrive, they are either stored for you to view the first time you click on a "channel" or, if a channel is already active, the information is presented to you immediately. Microsoft has developed a channel specification called *channel definition format* (CDF) that permits the subscribed STB user to automatically receive new multimedia content posted on the selected channel. The CDF file, which is typically written in extensible markup language (XML), contains a list of all the necessary uniform resource locators (URLs) to point to the selected multimedia content on the Internet [35]. Using some type of a scheduling tool, the cable operators can decide how often the multimedia content needs to be updated on the Internet [2].

12.4.2 E-mail

The use of electronic mail or e-mail has become very popular with the recent explosive use of the Internet. E-mail is a non-real-time electronic communication between computers that are connected, for example, via a LAN or WAN. It is a fast and economical way for anyone with a PC and a modem to exchange information with friends and colleagues around the globe. Thus, the e-mail information, which may include multimedia-rich content, promotes the virtual office concept, where people can communicate with each other wherever they are and whenever they want. Since e-mail is cheaper and faster than making phone calls or sending faxes, the use of e-mail is expected to grow dramatically to more than a billion global users within three years.

E-mail application can be viewed as part of the STB client/server family of applications. In the client/server computing model, the client STB interacts with the cable operator e-mail servers equipped with a firewall at the cable headend. Every subscriber receives a unique e-mail address and mailbox, which is stored and managed for billing purposes at the e-mail server. The e-mail server acts as a gateway to the Internet using common protocol standards such as *simple message transfer protocol* (SMTP) and *secure multipurpose Internet mail extensions* (S/MIME). As part of the TCP/IP protocol stack, the SMTP is a reliable and efficient mail transfer protocol, which specifies how the messages should be transmitted from the sender to the recipient 's mailbox, for only plaintext files. Then, the STB user can retrieve the messages using either a *post-office protocol* (POP) or an *Internet message access protocol* (IMAP). The newest version of POP is POP3, which is used by most e-mail applications and can be used with or without SMTP [36]. The latest version of IMAP is IMAP4, which is similar to POP3, but supports additional features such as selecting which messages on the e-mail server to download to your STB [37].

The S/MIME protocol, which is a new version of the MIME protocol for secure transmission, allows subscribers to send non-ASCII messages such as graphics, audio, and video files over the public Internet network [38]. The S/MIME supports RSA public-key encryption (see Section 8.6.1). The advantage of the S/MIME protocol is that it will allow people to send secure e-mail messages to one another using different STB e-mail clients. A STB e-mail client is an application that runs on STB that enables one to send, receive, and organize his e-mail messages.

12.4.3 E-Commerce

Electronic commerce or e-commerce is one of the most significant emerging Internet applications. E-commerce is typically defined as buying and selling goods on the Internet. According to Forrester Research, the on-line Business-to-Consumer (B2C) e-commerce is predicted to reach $200 billion in consumer purchases by 2004. This kind of explosive growth is not only fueled by ease of use and broad range of available products, but also because it drives prices down, and there are no taxes for goods and services purchased on-line. Consequently, cable operators are faced with a compelling e-business opportunity to deploy new e-commerce applications. In this e-business model, cable operators are forming profit-sharing partnerships with various existing on-line merchants. Examples of popular e-commerce services over cable TV networks are [39]:

- Home catalog shopping
- Travel services (airline tickets, hotels, car rental)
- E-auction (i.e., art, antiques, jewelry)
- E-gambling
- On-line ticket sales (i.e., concerts, shows, and sports)
- Local fulfillment (flowers, stamps, food)

The key component of this e-business model is payment for the purchased on-line goods and services. The consumer can pay for his on-line transactions in two ways, namely, electronic cash (e-cash) or credit cards. E-cash is a new form of payment that allows subscribers to make electronic transactions over the cable TV networks. Currently, there are three main categories for e-cash as follows: anonymous cash, micro-payments, and smart cards. The anonymous cash method is just like real cash where the subscriber can spend cash anywhere in the world while being anonymous. An example in this category is eCash, which uses digital signatures based on public-key cryptography, provides authentication, non-repudiation, data integrity, and confidentiality [40]. In the eCash system, only the bank signs the payment data, and the same bank verifies its signature and the integrity of the payment data when a payment recipient submits an eCash payment for validation. The eCash software eliminates the complex key management issues associated with issuing or verifying digital signatures from millions of consumers or merchants.

The micro-payments method allows subscribers to make small payments just as one would use coins in the real world. Since the micro-payments method handles transaction typically up to $10, it is generally not economical for the merchant to use a credit card. For example, the Pay2See system uses a browser-like payment system, which handles specially encrypted web pages and lets the subscriber see them only if he pays.

Smart cards are electronic cards that can store a limited amount of cash in them. Since most digital STB have a built-in smart card reader, cable operators can download a fixed cash amount to the smart card for e-commerce transactions. Examples of such e-cash systems include BarclayCoin, NetFare, Mondex, and Smartcard Axis.

The use of credit cards for electronic transactions over the cable TV network is similar to in-person transactions. There are two main categories for credit card payments. The first category simply requires the subscriber to enter his credit card at the STB e-commerce application when he decides to make a purchase. The credit card information can be sent encrypted or unencrypted (not recommended) across the cable TV network to the merchant's web server. The second category of credit card payment involves the use of the so-called electronic or digital wallets. Electronic wallet, which is securely stored in the STB, is a software program that has the subscriber's credit card and billing information. This information is sent encrypted to the desired merchant once a secure connection has been established between the merchant and the subscriber. The advantage of electronic wallet is that the subscriber does not have to enter his credit card information every time he wants to purchase.

12.4.4 Home Banking, Education, and Gaming

Home banking, education, and gaming can also be viewed as part of the client/server family of applications. These applications, like e-commerce application, require the establishment of a secure means to transmit the sensitive subscriber information over the cable TV network. Home banking service is an important emerging new application typically via the cable TV network. This service offers both convenience and time saving opportunity to the STB subscribers. For example, it enables subscribers to obtain up-to-date information on their checking and saving accounts, and allows them to transfer funds between accounts and pay bills 24 hours a day, 7 days a week. The home banking service is also advantageous to the financial institution as well as the cable operator. From the financial institution's point of view, it lowers its operating costs, and allows it to offer new financial services to its customers. From the cable TV operator's point of view, it increases subscriber satisfaction and could generate additional revenues through adding additional subscribers and through targeted advertisement campaigns.

STB education is also a promising new application that is expected to have an explosive growth. Generally speaking, STB education is the delivery of STB educational materials to subscribers in the convenience of their homes. This application essentially creates a virtual classroom, where students at different locations receive specially broadcast audio and video content from a central location and can interact with each other and the teacher as if they were all in same physical classroom. The educational material may include, for example, university curriculums, training courses, or even lectures and speeches given by subject matter experts. To enable the STB education service, the cable operator has to provide the subscriber with a downloadable resident application to the STB, and must have some kind of database server that stores and manages the educational material. To reach the STB user, the educational content can be multiplexed with other MPEG-2 video stream to form a 64/256-QAM channel that is transmitted over the cable TV network.

The popularity of interactive video games has been increasing in recent years. However, most of these games are limited to a single user. With the evolution of cable TV networks toward DWDM high-capacity digital networks, multi-user interactive video games are expected to become very hot applications for many people. From the cable TV operator's perspective, it may be easier to implement than home banking or e-commerce applications. The operator can simply download one or more resident applications for specific games to the STB user over a dedicated gaming channel when the subscriber tunes in. The gaming application can be stored in the STB if it has enough flash or RAM memory. Then, the list of authorized games will be displayed on the subscriber's TV monitor, and he can select the desired games using his remote control. The selected game request is then forwarded to the game application server at the cable headend, which authorizes the request, and starts the game session. The game server also manages the game interaction among multiple users.

12.5 Integrated Set-Top Box Applications

In the previous sections so far, we have reviewed separately the different TV-based and Internet-based applications. These applications are sometimes called third-party applications since software vendors independently from the RTOS or STB vendors have developed them. In order to deliver enhanced TV and web content, several software vendors such as Microsoft, Liberate Technologies, and WorldGate Communications are developing a customized interactive menu that integrates the various STB applications. This important point is best illustrated by the following example. Figure 12.8 the shows the WorldGate home page menu, showing the available services, including web browsing, e-mail, and TV programming using a digital STB [41]. Using a wireless keyboard or a remote control, the subscriber logs in to the WorldGate home page menu that can be customized to his city or community. In addition to web surfing, the WorldGate menu supports the so-called Channel HyperLinking, which enables the viewer to go directly to the corresponding web site related to the viewed TV channel or advertisement by pressing a special key on the keyboard. This feature offers a smooth transition between TV and web-watched programming. Using the WorldGate menu, the viewer can activate parental control filters such as Surf Watch to assist parents who wish to block certain types of web sites such as those dedicated to gambling or guns. The WorldGate menu also offers a home banking capability, which allows subscribers up-to-date account information as well as safe direct bill payment. New emerging applications such as VOD and distance learning can also be integrated into the current interactive menu. Liberate TV Navigator is also offering similar features by integrating web browsing, e-mail, and an interactive program guide for TV programming for enhanced TV viewing.

Figure 12.8 WorldGate home page menu with accesses to standard TV programs, e-mail, as well as customized web browsing features. Courtesy of WorldGate Communications, Inc.

Emerging applications for digital STB should be able to support these technologies:

- Animated and virtual reality graphics
- High-definition TV (HDTV) display
- Video and audio streaming
- 3-D video display

As subscribers become more comfortable with the enhanced TV viewing, they will demand more advanced interactivity to TV programs using, for example, Channel HyperLinking. Advanced TV interactive applications (not limited to specific web content) will open the door to a whole new TV experience where many people will be able to interact with each other based on the viewed TV or webcasting programs.

References

1. P. Peterka, "Interactive TV Applications: Standard APIs for Digital TV Receivers," *NCTA Technical Papers*, 134–143 (1999).
2. G. OÆdriscoll, *The Essential Guide to Digital Set-Top Boxes and Interactive TV*, Prentice Hall PTR (2000).
3. R. F. Annibaldi, J. Buehl, and S. Johnson, "System Implementation of a Resident Application and User Interface for Digital Set-Tops," *NCTA Technical Papers* (1998).
4. R. W. Brown, "Pegagus Set-Top Terminal," *NCTA Technical Papers*, 24–31 (1997).
5. Product information on Windows CE can be found at http://www.microsoft.com/tv/
6. Product information on PowerTV RTOS can be found, for example, at http://www.powertv.com/product/product.html
7. Product information on Java OS and APIs can be found at http://www.sun.com/javaos.
8. Product information on VxWorks OS can be found at http://www.vxworks.com/
9. Product information on Microware DAVID OS-9 can be found at http://www.microware.com/ProductsServices/Technologies/david.html.
10. Java Virtual machine (JVM): http://java.sun.com/docs/books/vmspec/index.html.
11. See for example http://www.dvb.org/.
12. For more information on MediaHighway middleware platform see for example http://www.canalplus-technologies.com/media/mediahighway.htm.
13. ATSC T3/S17 DTV Application Software Environment (DASE): http://toocan.philabs.research.philips.com/misc/atsc/dase/.
14. For example, look at www.w3c.org.
15. Suns' Java TV: http://java.sun/com/products/javatv.
16. Personal Java: http://java.sun.com/products/personaljava/spec-1-1/pJavaSpec.html.
17. Java Media Framework (JMF): http://java.sun.com/products/javamedia/jmf/forDevelopers/playerguide/index.html.
18. Home Audio/Video Interoperability (HAVi): http://www.havi.org.
19. For information on ATVEF specifications, see for example http://www.atvef.com/index.html.
20. Liberate product information can be found at http://www.liberate.com.
21. See, for example http://www.tvguide.com/.
22. See, for example http://www.starsight.com/.
23. International Telecommunication Union (ITU) H.222/ISO 13818 Section 1, Generic Coding of Moving Pictures and Associated Audio Information: Systems (1997).
24. European Telecommunications Standardization Institute (ETSI) ETS 300468, Digital Broadcasting Systems for Television Sound and Data Services; Specification for Service Information (SI) in Digital Video Broadcast (DVB) Systems (1996).

25. A/65 Program and System Information Protocol (PSIP) for Terrestrial Broadcast and Cable, Advanced Television Standardization Committee, *Digital Television Standard* (1997).

26. D. Ruiu, "Electronic Program Guide Standards", in *Testing Digital Video*, Hewlett Packard (1997).

27. For information on Diva VOD system, see http://www.divatv.com/diva_home.html.

28. For SeaChange International VOD product, see http://www.schange.com/products/resvod.html.

29. R. W. Brown, "Where Web and TV Collide," *NCTA Technical Papers*, 95–102 (1998).

30. Joint EIA/CVCC Recommended Practice for Teletext: North American Basic Teletext Specification, ANSI/EIA-516-88 (1988).

31. The Transmission of IP Over The Vertical Blanking Interval Of A Television Signal, Internet draft of IPVBI working group (1998). http://www.ietf.org/internet-drafts/draft-ietf-ipvbi-tv-signal-00.txt.

32. A/90 Data Broadcast Standard, Advanced Television Standardization Committee (August, 2000).

33. Information Technology – Generic Coding of Moving Pictures and Associated Audio: Digital Storage Media and Control – ISO/IEC 13818-6 International Standard, ISO/IEC JTC1/SC29/WG11 MPEG96/N1300p1 (1996).

34. More information on the SSL protocol can be found at http://home.netscape.com/eng/ssl3/draft302.txt.

35. Tutorial on Microsoft's CDF standard can be found at http://msdn.microsoft.com/workshop/delivery/channel/cdf1/cdf1.asp?RLD=363.

36. For more information on POP version3 protocol standard, see for example http://www.cis.ohio-state.edu/htbin/rfc/rfc1939.html.

37. For more information on IMAP version4 protocol standard, see for example http://www.cis.ohio-state.edu/htbin/rfc/rfc2060.html.

38. For more information on S/MIME protocol standard, see for example, http://www.ietf.org/ids.by.wg/smime.html.

39. T. Wasilewski, "E-Commerce over Cable: Providing Security for Interactive Applications," *NCTA Technical Papers*, 122–133 (1999).

40. For more information on eCash, see for example http://www.ecash.net/.

41. WorldGate user guide, WorldGate communications, Inc. (1999).

APPENDIX A

COMPARISON OF DAVIC AND DOCSIS SPECIFICATIONS

In Chapter 11, we reviewed CableLabs DOCSIS protocol, which is the primary standard in the United States for high-speed data transmission. The Digital Audio Video Council (DAVIC) protocol standard is a competing standard, which is used mainly in Europe. In an attempt to bridge this gap, CableLabs has added Appendix N (EuroDOCSIS) to the DOCSIS 1.1 standard, which provides PHY layer specifications for CMs operating in European cable plants. The intent of this appendix is simply to highlight the primary PHY layer differences between the DAVIC and DOCSIS protocol specifications; accordingly, what is presented should not be viewed as a complete comparison. Table A.1 summarizes the primary PHY layer differences between DAVIC 1.2 and DOCSIS 1.1 specifications.

Table A.1 Comparison of DAVIC 1.2 and DOCSIS 1.1 Specifications.

Parameter	DAVIC 1.2	DOCSIS 1.1
OOB Modulation Format	Differentially encoded QPSK	N/A
OOB Transmission Rate	1.544 Mb/s	N/A
OOB Error Correction	R-S T = 3 (59,53)	N/A
OOB Link Layer(s)	ATM-like/ extended superframe	N/A
Inband Modulation Format	64/256-QAM	64/256-QAM
Transmitted Spectrum 64-QAM 256-QAM	Square-root raised cosine shaping $\alpha = 0.13$ (6-MHz) $\alpha = 0.15$ (8-MHz)	Square-root raised cosine shaping $\alpha = 0.18$ (6-MHz) for 64-QAM $\alpha = 0.12$ (6-MHz) for 256-QAM
Inband Information Rate	29.172/38.896 Mb/s (6-MHz) 38.236/50.981 Mb/s (8-MHz)	26.97035 / 38.8107 Mb/s (6-MHz)
Inband Error Correction 64-QAM 256-QAM	R-S (204,188) T = 8 R-S (204,188) T = 8	R-S T= 3 (128,122) +14/15 rate TCM R-S T = 3 (128,122) +19/20 rate TCM

Parameter	DAVIC 1.2	DOCSIS 1.1
Convolutional Interleaver	I = 12, J = 17 for 64-QAM I = 204, J = 1 for 256-QAM	Variable I = 8, 16, 32, 64, 128 J = 16, 8, 4, 2, 1 for both 64/256-QAM
CNR @ Receiver Input (only AWGN)	$\geq$30 dB @ BER = $1\cdot10^{-12}$ (64-QAM)[1] $\geq$36 dB @ BER = $1\cdot10^{-12}$ (256-QAM)[2]	$\geq$23.5 dB @ BER $\leq1\cdot10^{-8}$ (64-QAM) $\geq$ 30 dB @ BER$\leq1\cdot10^{-8}$ (256-QAM) at RF input -6 to $+15$ dBmV $\geq$ 33 dB @ BER$\leq1\cdot10^{-8}$ (256-QAM) @ RF input -15 to < -6 dBmV
Adjacent QAM Channel Level	$\leq$ 3 dB (same QAM channels) $\leq$ 6 dB (diff. QAM channels)	0 dB for both 64/256QAM
Adjacent Analog Channel Level	0 to 10 dBc for 64-QAM 0 to 6 dBc for 256-QAM	10 dBc for both 64/256-QAM
RF Input Level Range (Downstream)	-8 to 20 dBmV (64-QAM) $+2$ to 20 dBmV (256-QAM)	-15 to $+15$ dBmV (64/256-QAM)
Inband Link Layers	ATM	DIX or 802.3 over DOCSIS MAC
Assumed Set-top Addressing Architecture	48 bits	48 bits
Upstream Modulation Format	Differentially encoded QPSK	QPSK and 16-QAM
Upstream Frequency Band	8–26.5 MHz	5–42 MHz
Upstream Transmitted Spectrum	α = 0.30 for QPSK Square-root raised cosine shape	α = 0.25 for QPSK or 16-QAM Square-root raised cosine shape
Upstream Transmission Rate(s)	256 kb/s for Grade A 1.544 Mb/s for Grade B 3.088 Mb/s for Grade C	160, 320, 640, 1280, and 2560 kSym/s
Upstream Channel Spacing	200 kHz for Grade A 1 MHz for Grade B 2 MHz for Grade C	200, 400, 800, 1600, and 3200 kHz

[1] EuroDOCSIS specifies CNR of 25.5 dB or greater (post FEC) at BER=10^{-8} for 64-QAM.

[2] EuroDOCSIS specifies CNR of 31.5 dB (post FEC) at BER=10^{-8} at RF input from -15 dBmV to > -6 dBmV, and 34.5 dB or greater (post FEC) at BER=10^{-8} at RF input level from -6 dBmV to $+15$ dBmV for 256-QAM.

Parameter	DAVIC 1.2	DOCSIS 1.1
Upstream Transmitter Output Power	+25 to +53 dBmV (RMS) QPSK	+8 to 58 dBmV (QPSK) +8 to 55 dBmV (16-QAM)
Upstream Demodulator Input Power (in each carrier)	N/A	−16 to +14dBmV @ 160 kSym/s −13 to +17 dBmV @ 320 kSym/s −10 to +20 dBmV @ 640 kSym/s −7 to +23 dBmV @ 1280 kSym/s −4 to +26 dBmV @ 2560 kSym/s
Upstream CNR (Only AWGN)	≥ 20 dB @ BER $= 1 \cdot 10^{-6}$ (post FEC packet loss)	@ SER$=1 \cdot 10^{-6}$ (uncoded) ≤ 14.29 dB for QPSK ≤ 21.42 dB for 16-QAM
Upstream Error Correction	R-S T = 3 (59,53)	RS over GF(256) with T = 1–10
Upstream Data Link Packet Structure	UW = 4 bytes, Pay Load = 53 bytes, RS parity = 6 bytes, Guard band = 1 byte	FC = 1-byte, MAC_PARM = 1 byte, LEN=2 bytes, HCS = 2 bytes, Packet PDU = 18–1518 bytes
Upstream Link Layer(s)	ATM cell	Variable-length Ethernet/802.3
TDMA Synchronization Approach	OOB Extended Superframe synchronization bits	Time synchronization MAC management messages

N/A	Not Applicable
R-S	Reed-Solomon Code
FC	Frame Control: Identifies type of MAC header
MAC_PARM	Parameter field whose use is dependent on FC
LEN	The length of the MAC frame
HCS	MAC header check sequence
PDU	Protocol Data Unit
SER	Symbol Error Rate

APPENDIX B

INTERNATIONAL CABLE TV FREQUENCY PLANS

B.1 CCIR System B/G Frequency Plan

The North America cable TV frequency plans, namely, the standard, HRC, and IRC, were discussed in Chapter 2. In this appendix, we provide selected cable TV frequency plans commonly used in various parts of the globe. In particular, the CCIR system B/G cable TV plan, which is used in Western Europe, the CCIR system I, which is used in Great Britain, and the CCIR system D, which is used in China, are listed in the tables that follow. Notice that E-2 to E-12 channels (Table B.1) are 7-MHz wide, while all the channels from E-21 and above are 8-MHz wide. However, the sound carrier is fixed at 5.5-MHz above the picture carrier frequency for either the 7-MHz or 8-MHz wide channels.

Table B.1 CCIR System B/G Cable TV Frequency Plan Used in Western Europe.

Channel	Picture Carrier Frequency (MHz)	Sound Carrier Frequency (MHz)
E-2	48.25	53.75
E-3	55.25	60.75
E-4	62.25	67.75
E-5	175.25	180.75
E-6	182.25	187.75
E-7	189.25	194.75
E-8	196.25	201.75
E-9	203.25	208.75
E-10	210.25	215.75
E-11	217.25	222.75
E-12	224.25	229.75
E-21	471.25	476.75
E-22	479.25	484.75
E-23	487.25	492.75

Channel	Picture Carrier	Sound Carrier
E-24	495.25	500.75
E-25	503.25	508.75
E-26	511.25	516.75
E-27	519.25	524.75
E-28	527.25	532.75
E-29	535.25	540.75
E-30	543.25	548.75
E-31	551.25	556.75
E-32	559.25	564.75
E-33	567.25	572.75
E-34	575.25	580.75
E-35	583.25	588.75
E-36	591.25	596.75
E-37	599.25	604.75
E-38	607.25	612.75
E-39	615.25	620.75
E-40	623.25	628.75
E-41	631.25	636.75
E-42	639.25	644.75
E-43	647.25	652.75
E-44	655.25	660.75
E-45	663.25	668.75
E-46	671.25	676.75
E-47	679.25	684.75
E-48	687.25	692.75
E-49	695.25	700.75
E-50	703.25	708.75
E-51	711.25	716.75
E-52	719.25	724.75
E-53	727.25	732.75
E-54	735.25	740.75
E-55	743.25	748.75
E-56	751.25	756.75
E-57	759.25	764.75
E-58	767.25	772.75
E-59	775.25	780.75
E-60	783.25	788.75
E-61	791.25	796.75
E-62	799.25	804.75

Channel	Picture Carrier	Sound Carrier
E-63	807.25	812.75
E-64	815.25	820.75
E-65	823.25	828.75
E-66	831.25	836.75
E-67	839.25	844.75
E-68	847.25	852.75
E-69	855.25	860.75

B.2 CCIR system I Frequency Plan

The CCIR system I frequency plan, which is used in Great Britain, is similar to the CCIR system B/G used in Western Europe. In fact, the picture carrier frequencies from Channel 21 to channel 69 in Table B.2 are the same as E-21 to E-69 in Table B.1. However, the sound carrier frequency is 6-MHz above the picture carrier frequency, instead of 5.5-MHz as in Table B.1. See Section 2.1.3.

Table B.2 CCIR System I Cable TV Frequency Plan used in Great Britain.

Channel	Picture Carrier Frequency (MHz)	Sound Carrier Frequency (MHz)
21	471.25	477.25
22	479.25	485.25
23	487.25	493.25
24	495.25	501.25
25	503.25	509.25
26	511.25	517.25
27	519.25	525.25
28	527.25	533.25
29	535.25	541.25
30	543.25	549.25
31	551.25	557.25
32	559.25	565.25
33	567.25	573.25
34	575.25	581.25
35	583.25	589.25

Channel	Picture Carrier Frequency (MHz)	Sound Carrier Frequency (MHz)
36	591.25	597.25
37	599.25	605.25
38	607.25	613.25
39	615.25	621.25
40	623.25	629.25
41	631.25	637.25
42	639.25	645.25
43	647.25	653.25
44	655.25	661.25
45	663.25	669.25
46	671.25	677.25
47	679.25	685.25
48	687.25	693.25
49	695.25	701.25
50	703.25	709.25
51	711.25	717.25
52	719.25	725.25
53	727.25	733.25
54	735.25	741.25
55	743.25	749.25
56	751.25	757.25
57	759.25	765.25
58	767.25	773.25
59	775.25	781.25
60	783.25	789.25
61	791.25	797.25
62	799.25	805.25
63	807.25	813.25
64	815.25	821.25
65	823.25	829.25
66	831.25	837.25
67	839.25	845.25
68	847.25	853.25
69	855.25	861.25

B.3 CCIR System D Frequency Plan

The CCIR system D consists of 8-MHz wide channels. However, unlike its location in systems B/G and system I, the sound carrier frequency in system D is located 6.5-MHz above the picture carrier frequency.

Table B.3 CCIR System D Cable TV Frequency Plan used in China.

Channel	Picture Carrier Frequency (MHz)	Sound Carrier Frequency (MHz)
DS-1	49.75	56.25
DS-2	57.75	64.25
DS-3	65.75	72.25
DS-4	77.25	83.75
DS-5	85.25	91.75
DS-6	168.25	174.75
DS-7	176.25	182.75
DS-8	184.25	190.75
DS-9	192.25	198.75
DS-10	200.25	206.75
DS-11	208.25	214.75
DS-12	216.25	222.75
DS-13	471.25	477.75
DS-14	479.25	485.75
DS-15	487.25	493.75
DS-16	495.25	501.75
DS-17	503.25	509.75
DS-18	511.25	517.75
DS-19	519.25	525.75
DS-20	527.25	533.75
DS-21	535.25	541.75
DS-22	543.25	549.75
DS-23	551.25	557.75
DS-24	559.25	565.75
DS-25	607.25	613.75
DS-26	615.25	621.75
DS-27	623.25	629.75
DS-28	631.25	637.75
DS-29	639.25	645.75
DS-30	647.25	653.75

Channel	Picture Carrier Frequency (MHz)	Sound Carrier Frequency (MHz)
DS-31	655.25	661.75
DS-32	663.25	669.75
DS-33	671.25	677.75
DS-34	679.25	685.75
DS-35	687.25	693.75
DS-36	695.25	701.75
DS-37	703.25	709.75
DS-38	711.25	717.75
DS-39	719.25	725.75
DS-40	727.25	733.75
DS-41	735.25	741.75
DS-42	743.25	749.75
DS-43	751.25	757.75
DS-44	759.25	765.75
DS-45	767.25	773.75
DS-46	775.25	781.75
DS-47	783.25	789.75
DS-48	791.25	797.75
DS-49	799.25	805.75
DS-50	807.25	813.75
DS-51	815.25	821.75
DS-52	823.25	829.75
DS-53	831.25	837.75
DS-54	839.25	845.75
DS-55	847.25	853.75
DS-56	855.25	861.75

APPENDIX C

SATELLITE TRANSPONDER PARAMETERS FOR CABLE TV NETWORKS

The downstream information rate for a 6 MHz cable TV channel operating at a 64-QAM modulation is 26.97035-Mb/s according to the DOCSIS specifications. As pointed out previously in Chapter 7, many cable TV headends receive their downlink digital video feed from various satellite transponders operating at a QPSK modulation format with a matching information rate. The downlink C-Band satellite transponders are operating on vertically and horizontally polarized 36 MHz channels having center frequencies ranging from 3.72/3.74-GHz to 4.16/4.18-GHz with 40-MHz spacing between adjacent channels. The downlink K-Band satellite transponders are operating on right-hand and left-hand circularly polarized 24-MHz channels. Since these are bandwidth-limited systems, robust transmission is achieved using FEC code consisting of a R-S block code, which is interleaved and concatenated with convolutional code. Table C.1 summarizes the 36-MHz (wide) and 24-MHz (narrow) satellite transponder characteristics.

Table C.1 Common Satellite Transponder Parameters That Are Used by Cable TV Operators.

Parameter	Wide QPSK	Narrow QPSK
Transponder Bandwidth	36 MHz	24 MHz
Information Bit Rate	26.97035 Mb/s	26.97035 Mb/s
Total Channel Bit Rate	58.53140 Mb/s	39.02093 Mb/s
Symbol Rate	29.2657 MBaud	19.51047 MBaud
R-S Code	RS (204,188) T = 8	RS (204,188) T = 8
Trellis Code	R = 1/2	R = 3/4
Interleaver	I = 12, J = 19	I = 12, J = 19
Standard IF	70 MHz	70-MHz

According to Table C.1, the channel information rate of 26.97035 Mb/s is identical for both the wide and narrow QPSK channels. From a historical perspective, scientists at General Instrument have successfully selected the proper channel coding (R-S and Trellis codes) such that the 26.97035 Mb/s information rate would match that of ITU-T J.83 Annex B 64-QAM channel transmitted over the cable TV networks. Therefore, the information payload of the received satellite QPSK channel at the cable-TV headend can be used directly for a 64-QAM transmission with no need for bit stuffing. The corresponding symbol rate for a wide bandwidth QPSK channel is calculated as follows:

26.97-Mb/s x 204/188 x 2/1 = 58.53-Mb/s / 2bits/symbol = 29.266-MBaud

and for a narrow bandwidth QPSK channel

26.97-Mb/s x 204/188 x 4/3 = 39.02-Mb/s / 2bits/symbol = 19.51-MBaud

Unfortunately, there is no QPSK channel convolutional coding that can match the required information rate of 38.8107-Mb/s for an ITU-T J.83 Annex B 256-QAM channel.

ACRONYMS

ADC	Analog to Digital Converter
ADSL	Asymmetric Digital Subscriber Line
AGC	Automatic Gain Control
AM	Amplitude Modulation
AOTF	Acousto-Optic Tunable Filter
APC	Automatic Power Control
APE	Annealed Proton Exchange
API	Application Programming Interface
ARP	Address Resolution Protocol
ASE	Amplified Spontaneous Emission
ATC	Automatic Temperature Control
ATM	Asynchronous Transfer Mode
ATSC	Advanced Television System Committee
ATVEF	Advanced Television Enhancement Forum
AWG	Array Waveguide Grating
AWGN	Additive White Gaussian Noise
BBI	Balanced Bridge Interferometer
BCC	Binary Convolutional Code
BER	Bit Error Rate
BNU	Broadband Network Unit
BPF	Band Pass Filter
CCIR	Commite' Consulatif International Radiocommunications
CDF	Channel Definition Format
CDLI	Chroma Luma Delay Inequality
CIF	Common Intermediate Format
CM	Cable Modem
CMA	Constant Modulus Algorithm
CMTS	Cable Modem Termination System
CNR	Carrier to Noise Ratio
CPB	Constrained Parameters Bit stream
CSMA	Carrier Sensing Multiple-Access
CSMA/CD	Carrier Sensing Multiple-Access with Collision-Detection
CSO	Composite Second Order
CTB	Composite Triple Beat

DAC	Digital to Analog Converter
DASE	Digital television Application Software Environment
DAVIC	Digital Audio Visual Council
DBR	Double Rayleigh Backscattering
DBS	Direct Broadcast Satellite
DCT	Discrete Cosine Transform
DDFS	Direct Digital Frequency Synthesizer
DES	Data Encryption Standard
DFE	Decision Feedback Equalizer
DH	Double Heterostructure
DHCP	Dynamic Host Configuration Protocol
DLL	Data Link Layer
DOCSIS	Data-Over-Cable Service Interface Specifications
DSF	Dispersion Shifted Fiber
DVB	Digital Video Broadcast
DVS	Digital Video Subcommittee
DWDM	Dense Wavelength Division Multiplexing
ECM	Entitlement Control Message
EDF	Erbium-Doped Fiber
EDFA	Erbium-Doped Fiber Amplifier
EIA	Electronics Industry Association
EMM	Entitlement Management Message
EPG	Electronic Program Guide
EPM	External Phase Modulation
ESA	Excited state Absorption
ETSI	European Telecommunications Standardization Institute
EVM	Error Vector Magnitude
FBG	Fiber Bragg Grating
FCC	Federal Communications Commission
FDM	Frequency Division Multiplexing
FDMA	Frequency Division Multiple Access
FEC	Forward Error Correction
FFE	Feed-Forward Equalizer
FIR	Finite Impulse Response (filter)
FITL	Fiber In The Loop
FM	Frequency Modulation
FP	Fabry Perot
FPR	Free Propagation Range
FSR	Free Spectral Range
FTTC	Fiber To The Curb
FTTH	Fiber To The Home
FTTN	Fiber To The Node

FSS	Frequency Stacking Scheme
GDV	Group Delay Variation
GF	Galois Field
GOP	Group Of Pictures
GSA	Ground State Absorption
HCS	MAC Header Check Sequence
HDT	Host Digital Terminal
HDTV	High Definition Television
HFC	Hybrid Fiber Coax
HRC	Harmonically Related Carrier
HTML	Hyper Text Markup Language
HTTP	Hyper Text Transfer Protocol
IIN	Interference Intensity Noise
IIR	Infinite Impulse Response (filter)
IMAP	Internet Message Access Protocol
IPG	Interactive Program Guide
IPPV	Impulse Pay Per View
IRC	Incrementally Related Carrier
ISI	Inter-Symbol Interference
ITU	International Telecommunication Union
JBIG	Joint Bilevel Image experts Group
JPEG	Joint Photographic Experts Group
JVM	Java Virtual Machine
LLC	Logical Link Control
LMS	Least Mean Square
LOD	Level Of Detail
LPF	Low Pass Filter
LSB	Least Significant Bit
LTE	Linear Transversal Equalizer
MAC	Media Access Control
MER	Modulation Error Ratio
MGCP	Media Gateway Control Protocol
MHN	Mode Hopping Noise
MIPS	Million Instructions Per Second
MPEG	Moving Picture Expert Group
MPN	Mode Partition Noise
MQW	Multiple Quantum Well
MSB	Most Significant Bit
MSE	Mean Square Error
MZI	Mach-Zehnder Interferometer
NA	Numerical Aperture
NABTS	North American Basic Teletext Specification

NCO	Numerically Controlled Oscillator
NCTA	National Cable Television Association
NEP	Noise-Equivalent Photocurrent
NF	Noise Figure
NPR	Noise Power Ratio
NRZ	Non-Return to Zero
NTSC	National Television System Committee
NVOD	Near Video-On-Demand
OI	Optical Isolator
OOB	Out-Of-Band
OTN	Optical Transition Node
PAL	Phase Alteration Line
PBS	Polarization Beam Splitter
PCM	Pulse Code Modulation
PCR	Program Clock Reference
PDL	Polarization Dependent Loss
PDU	Packet Data Unit
PES	Packetized Elementary Stream
PGA	Programmable Gain Amplifier
PHB	Polarization Hole Burning
PID	Program (or packet) Identification
PLL	Phase Locked Loop
PMD	Physical Media Dependent
PMT	Program Management Table
POD	Point of Deployment
POP	Post-Office Protocol
PPV	Pay Per View
PSIP	Program and System Information Protocol
PSTN	Public Switched Telephone Network
PTS	Presentation Time Stamp
QAM	Quadrature Amplitude Modulation
QoS	Quality of Service
QPSK	Quadrature Phase Shift Keying
RAID	Redundant Array of Independent (or inexpensive) Disks
RAM	Random Access Memory
RGB	Red Green Blue
RGW	Residential Gateway
RIN	Relative Intensity Noise
RLS	Recursive Least Squares
ROM	Read Only Memory
RTOS	Real Time Operating System
SAW	Surface Acoustic Wave

SBS	Stimulated Brillouin Scattering
SCM	Sub-Carrier Multiplexing
SCTE	Society of Cable Television Engineers
SDRAM	Synchronous Dynamic Random Access Memory
SER	Symbol Error Rate
SFID	Service Flow ID
SHB	Spectral Hole Burning
SID	Service ID
SL	Strained Layer
SMF	Single Mode Fiber
S/MIME	Secure Multipurpose Internet Mail Extensions
SMTP	Simple Message Transfer Protocol
SNMP	Simple Network Management Protocol
SNR	Signal to Noise Ratio
SONET	Synchronous Optical Networks
SPM	Self Phase Modulation
SRRC	Square Root Raised Cosine
SRS	Stimulated Raman Scattering
SS7	Signaling System number 7
SSL	Secure Sockets Layer
STB	Set-Top Box
TCM	Trellis Coded Modulation
TDM	Time Division Multiplexing
TDMA	Time Division Multiple Access
TEC	Thermo Electric Cooler
TFTP	Trivial File Transfer Protocol
TGW	Trunking Gateway
TLV	Type Length Value
TVPAK	Television Platform Adaptation Kit
UCD	Upstream Channel Descriptor
UGS	Unsolicited Grant Service
UMA	Unified Memory Architecture
USB	Universal Serial Bus
URL	Uniform Resource Locator
UW	Unique Word
VBI	Vertical Blanking Interval
VBR	Variable Back Reflector
VCO	Voltage Controlled Oscillator
VITS	Vertical Interval Test Signals
VOD	Video-On-Demand
VoIP	Voice Over IP
VSB	Vestigal Side Band

WDM	Wavelength Division Multiplexing
XML	Extensible Markup Language
XMOD	Cross Modulation
XPM	Cross Phase Modulation
YAG	Yttrium Aluminum Garnet
YIG	Yttrium Iron Garnet

INDEX

A

AC-3, audio coding, 35
Absorption cross sections, 157–161
Acousto-optic tunable filters, *see* Filters
Adaptive equalizers, 208–214, *see also* Equalizers,
Address resolution protocol (ARP), 359
Advanced set-top box, 260–261
Advanced TV enhancement forum (ATVEF), 401, 407
Advanced Television System Committee (ATSC), 35, 382, 386, 389, 392
Aloha protocol, 380–381
Alpha blending, 254
AM hum modulation, 55, 273, 276
Amplified spontaneous emission (ASE), 157–161, 165–174, *see also* Noise
Amplifiers, optical, *see* Erbium-doped fiber amplifiers
Amplifiers, RF, 43–45, 90, 133–139, 142–144, 207–208, 232, 234
Amplitude ripple, QAM channel, 270–271
Analog to digital converter (ADC), 29 206–207, 354–356
Application program interface (API), 395–39, 400–401
Architecture:
  downstream HFC access, 4–6, 348–349
  downstream DWDM access, 349–351
  fiber-in-the-loop (FITL), 14–16
  digital set-top box software, 395–402

upstream DWDM access, 352–354
Array waveguide grating (AWG), 150–152
Asymmetric digital subscriber line (ADSL), 12–14
Aural carrier, 22, 24
Authorization module, 386
Automatic gain control (AGC), 46, 85–86, 119, 296
Automatic power control (APC), 85–86
Automatic slope control (ASC), 46
Automatic temperature control (ATC), 85–86

B

Backoff window, 382
Balanced-bridge interferometer (BBI) modulator, *see* Modulators
Bandwidth allocation, *see* Upstream,
Bipolar-junction transistors (BJT), 139, 143
Bit error rate (BER), of M-ary QAM signals, 223–226
Blind equalization, 212–213, *see also* Equalizers
Bragg grating, 152–153
Bragg wavelength, 153
Broadband network unit (BNU), 14–16
Burst frame structure, 371–372

Burst noise, 56, 271–273
Burst profiles, 365–367
Burst timing, 368–369
Bursty nonlinear distortions, 294–299

C

Call agent, 10–11
Cable modem (CM):
  downstream physical layer, 361–363
  downstream transmission convergence
    sublayer, 372–373
  interaction, with CMTS, 390–392
  media access control (MAC), see
    MAC layer
  random access methods, 380–382
  synchronization, 375–377
  upstream burst timing, 368–369
  upstream fragmentation, 388–390
  upstream physical layer, 351–352
  upstream quality of service, 384–388
  upstream spurious power outputs,
    357–358
  upstream transmitter, see Upstream
Cable modem termination system
  (CMTS), 360–362, 366–367, 374–379,
  390–392
Cable TV frequency plans, 38–42,
  423–428
Carrier-to-noise-ratio (CNR), of single
  and cascaded amplifiers, 47–48
Coaxial cable, 42–44
  return-path characteristics,
  RF tuner, see Tuner,
  RF amplifiers, 44–47, 90, 133–139
  taps, 47
Carrier sensing multiple access (CSMA)
  method, 380–381

Carrier recovery methods, in QAM
  receivers, 214–217
Cascode preamplifier configuration,
  143–144
Channel definition format (CDF), 411
Channel model, of HFC networks,
  53–55, 268–269
Characteristic temperature, of lasers, 71
Chroma-luma delay inequality (CLDI),
  25–27
Ciphertext, 247
Circuit model, large signal response,
  79–82
Clipping noise:
  BER of QAM channels due to, 283–287
  dynamic, 290–292
  double-level, 289–290
  reduction methods, 292–294
  statistical properties of, 281–282
Coding gain, of TCM, 196–197
Constant-modulus algorithm (CMA), 213
Constellation diagrams, 64/245-QAM,
  221–222, 265–266
Contention resolution protocols, 380–382
Composite second order (CSO), 49–50,
  265, 275, 294–299, see also Distortions
Composite triple beat (CTB), 49–50,
  265, 275, 294–299, see also Distortions
Constraint length, 193–196
Costas loop, 215–216
Cross phase modulation (XPM),
  322–323,
Coupled mode equations, 148, 150

D

Damping coefficient, 78
Data encryption standard (DES), 247
Data link layer (DLL), 359–361

Dense-wavelength-division multiplexing (DWDM), *see* Architecture

Decision-directed loops, 216–217

Decision-feedback equalizer (DFE), *see* Equalizers

Depletion region, p-i-n photodiode, 126–127

Differential detection scheme, 340–342

Differential gain (DG), NTSC signal, 25–27

Differential quantum efficiency, 72

Differential phase (DP), NTSC signal, 25, 27

Differential precoder, 200–201

Diffusion process,124–126

Digital encryption and decryption, 246–248

Digital Audio and Video Council (DAVIC), 183, 227, 380, 419–421

Digital to Analog converter (DAC), 206, 355

Digital video broadcast (DVB), 400–401, 404, 407

Digital video signals, *see* Video signals

Digitized return-path transport, 354–356

Direct broadcast satellite (DBS), 16–17

Discrete cosine transform (DCT), 30–31, 35

Dispersion shifted fiber (DSF), 311, 314, 318

Dissymmetrization scheme, 294

Distortions:
  chroma differential gain, 25–27
  chroma differential phase, 25–27
  composite second-order (CSO), 49–50, 265, 275, 294–299
  composite triple beat (CTB), 49–50, 265, 275, 294–299
  dispersion-induced, 309–311
  gain tilt, in EDFA, 346–348
  intermodulation, 49–50, 136–137

polarization-induced, 344–346

Double-heterostructure (DH), semiconductor lasers, 69–70

Double Rayleigh backscattering (DRB), 300–301

Downstream transmission parameters: ITU-T J.83 Annex A and B, 226–227

Dynamic service change request, 386

Dynamic host configuration protocol (DHCP), 360, 391

DTV application software environment (DASE), 400–401

E

Early-late timing recovery, 218–219

Einstein relation, 124

Electron mobility, 124–126

Electronic:
  commerce (E-commerce), 412–413
  home banking, 413–414
  mail (E-mail), 411–412
  predistortion techniques, *see* Linearization methods
  program guides (EPGs), 402–405
  wallets, 413

Emission cross sections, 157–161

Enhanced broadcast, 407

Entitlement control message (ECM), 239–240

Entitlement management message (EMM), 248–249, 404

Entropy coding, 31

Equalizers:
  adaptive, 208–214
  blind, 212–213
  decision feedback (DFE), 53, 212–214, 268–270
  feedforward, 53, 212–214, 268–270

fractionally spaced, 209
linear transversal (LTE), 209
zero-forcing, 210
Erbium doped fiber amplifier (EDFA):
  basic configurations, 163–165
  carrier-to-noise ratio (CNR),
    calculation,167–169
  dual-stage configurations, 164–165
  energy-level diagrams, 154–156
  gain-flattened, 176–178
  gain spectra, 159–161, 171–172
  noise figure (NF), measurement,
    169–170
  pump lasers, *see* Lasers
  requirements for cable TV networks,
    170–176
  three-level model, 156–160
Error vector modulation (EVM),
  219–223
Excited State Absorption, 154–156
External phase modulation (EPM), 319,
  342–344

F

Faraday effect, 87–88
Fiber:
  Bragg grating, 152–153
  Erbium doped, *see* Erbium doped fiber
  amplifiers (EDFA)
  In-the-loop (FITL), 15–17
Filters:
  acousto-optic tunable, 153–154,
    177–178
  bandpass, 214, 233–234, 283
  diplex, 46, 231–232, 261
  interpolation, 198
  Kalman, 212
  Nyquist, 202–203

raised cosine, 203, 222–223
SAW, 206–207
square-root raised cosine (SRRC),
  203–205, 208, 222–223
Flash memory, *see* Memory
Fractionally spaced equalizers, 202, *see
also* Equalizer,
Free propagation region (FPR), 152
Free spectral range (FSR), 152
Frequency chirp, of semiconductor lasers,
  81–82, 311, 309, 318, 346
Frequency Division Multiplexing (FDM),
  14
Frequency division multiple access
  (FDMA), 363
Frequency stacking scheme, 352–354

G

Gain:
  coefficient, optical, 64,
  compression coefficient, in lasers,
    75–79
  flattened EDFAs, 176–178
  optical coefficient, 66
  spectra, *see* Erbium doped fiber
  amplifier (EDFA)
  tilt, in EDFA, 346–348
Galois field (GF), 188–190, 238, 242,
  363
Generating polynomial, 188–190, 238,
  242, 363
Ground state absorption, 154
Group delay variations (GDV), 54–55,
  270–271

H

HBT, 139–140
Harmonically related carriers (HRC),
  37–38, 49, 423
Headends, master, primary and
  secondary, 4–6, 332, 348–349, 353
Hole mobility, 119–122
Hybrid fiber/coax (HFC) networks,
  architecture, 4–6, 348–349
  channel models, 55, 268–269
  two-way, overview, 6–9
Hypertext transfer protocol (HTTP), 395
  409–410, *see also* Protocol
Hypertext markup language (HTML),
  399–401

I

IEEE 802.14, model, 53–55, 268–269
Impulse pay per view (IPPV), 6–7,
  406–407
Incrementally related carriers (IRC),
  37–38, 423
Index-guided lasers, *see* Lasers
Interferometric noise, *see* Noise
Interference intensity noise, *see* Noise
Interleaving, downstream QAM channel,
  191–193
Intermodulation distortions, *see*
  Distortions
Internet-based applications, set-top box,
  409–414
Intersymbol interference (ISI), 51, 202,
  205, 209, 211, 213–214
IP telephony, 9–11
Isolators, optical, 87–88, 323–326, 344

ITU-T J.83 Annex A, downstream
parameters, 226–227
ITU-T J.83 Annex B, downstream
parameters, 226–227

J

Java, language, 399–401
Java virtual machine (JVM), 399–401

L

Large-signal circuit model, 79–81
Lasers:
  buried heterostructure (BH), 69–70
  directly modulated, transmitter,
    287–289, 303–304, 308–309, 311,
    316, 318
  see Laser Transmitter Design
  distributed-feedback (DFB), 72–73,
    85–87, 90–92
  dynamic characteristics, 74–82
  externally modulated, transmitter,
    116–120
  Fabry Perot (FP), 67–68, 92, 94
  frequency response, small-signal, 75–79
  gain-guided, 68–69
  index-guided, 69–70
  linewidth, 84–85
  multimode, 68
  multiple-quantum well (MQW), 70–71
  Nd-YAG, 116–118
  package and parasitics, 79–81
  pump, 161–162
  return-path, transmitters, 92–96,
    342–343

structures, 68–70
Laser transmitter design:
  directly modulated DFB, 85–90
  externally modulated DFB, 119–120
  externally modulated YAG, 116–118
Least mean square (LMS) algorithm,
  211–212
Lightwave multichannel trunking
  systems, 333–334
Linearization:
  circuits, 110–112
  feedforward, 107–110
  methods, overview, 106–115
  optical dual parallel, 112–114
  optical dual cascade predistortion,
    114–115
  predistortion, 108–110
Lithium Niobate modulator, basic
  operation, 100–104
Lumped circuit-element modulator, 101

M

Media access control (MAC) layer,
DOCSIS protocol:
  frame format, 374–375
  format and messages, 373–379
  management messages, 375–379
  operation, 383–384
  quality of service (QoS), 384–388
  ranging, 379, 391
  registration, 379, 391
  synchronization, 375–377
  upstream fragmentation, 388–390
Mean square error (MSE), 210–211
Memory, see also Set-top box:
  flash, 232, 258–260
  random access (RAM), 258–260
  read only (ROM), 258

synchronous dynamic random access
  (SDRAM), 232, 258-260
Middleware layer, 398–402
Mini-slot, 365–369, 371–374, 378
Modulation:
  AM hum, 55, 273, 276
  cross-phase (XPM), 322–323
  error vector (MER), 219–223
  external-phase (XPM), 319, 342–344
  index, optical, 87, 90, 227–280
  quadrature amplitude (QAM), modem,
    building blocks, 183–185
  self-phase (SPM), 316–319, 342–344
  trellis coded (TCM), 193–200
Modulators:
  balanced-bridge interferometer (BBI),
    102–105, 118
  Lithium Niobate modulator, basic
    operation, 100–104
  lumped circuit-element, 97–98
  Mach-Zehnder interferometer (MZI),
    102–105, 108, 112–114
  M-ary QAM modulator, upstream,
    239–243, 363–365, *see also* Upstream
    Burst Transmitter
  traveling-wave, 101–102
Multichannel multipoint distribution
services (MMDS), 18–19
Multipath reflections, 50–55, 267–270,
  276
Multiple reflections, optical, 299–309
Multiplexing:
  frequency-division (FDM), 354
  time-division (TDM), 352
  wavelength-division (WDM),
Multiple grant mode, 388
Multiple quantum well (MQW) lasers,
  73–74, 85, 92

N

Narrowcasting, 350
National Television Standard Committee
(NTSC) video format, 21–27, 244–245
Native middleware, 396, 398–402
Native applications, 402–405
Near video on demand (NVOD), 8
    407–409
Noise:
  additive white Gaussian (AWGN),
    223, 227, 262–267, 273
  amplified spontaneous emission (ASE),
    157–161, 165–175, 336
  burst, 57, 271–273
  clipping, 281–283, *see also* Clipping
    Noise
  equivalent photocurrent (NEP), 132
  funneling effect, 58–59
  impulse, 56, 271–273
  ingress, 56–60
  interference intensity (IIN), 300, 305
  interferometric, due to DBR, 300–301
  Langevin, 82–83
  mode-partition (MPN), 92–94
  mode-hopping (MHN), 92, 94
  phase, 254–256
  power ratio (NPR), 355–356
  relative-intensity (RIN), 82–84, 91–92,
    132, 289, 306–307
  shot, 129–130, 133–134
  signal-spontaneous beat, 162–164
  spontaneous-spontaneous beat, 162–164
  spurious, upstream, 370–371
  thermal, 130–134
Noise figure:
  optical amplifiers, 167–175
  RF amplifiers, 47–48
  saturated, 174–175

Non-blocking algorithm, 382
Nonlinear distortions, *see* Distortions
Nyquist bandwidth, 202

O

OpenCable, set-top box, 249–250, 273,
    275–276
Optical isolator, 87–88, 323–326, 344
Optical reflections, 299–309
Out-of-band (OOB):
  interleaver, 238–239
  mapping, 239
  R-S coding, 238
  parameters, receiver, 236
  randomizer, 237–238
  receiver, 232, 235–239, 261

P

p-persistent binary exponential backoff,
    382
Packet data unit (PDU), 240, 374–375,
    379, 383–384
Pay per view (PPV), 402, 406–407
Parental control, 404–405
Parity bytes, 188–190, 238, 363
Parity check matrix, 186–187
Peak optical gain, 66
Piggyback mode, cable modems, 389
p-i-n photodiode, 123–129
Physical media dependent (PMD)
sublayer, downstream, 361–363
Pockels effect, 100
Point of deployment (POD) module,
    249–250
Polarization-dependent distortions,

323–326, *see also* Distortions,
Polarization coupling, 300–301
Polarization hole burning (PHB), 170
Polygon mesh, 252
Power TV software, 398–399
Preamble, 240, 364–366
Preclipping methods, 292–294
Pre-equalization, adaptive, 367
Program identification (PID), 243, 245
Protocol:
  address resolution (ARP), 359
  data-over-cable interface specifications
  (DOCSIS), stack, 359–361
    dynamic host configuration (DHCP),
    360, 391
    hypertext transfer (HTTP), 395,
    409–410
  Internet message access (IMAP), 411
  network-based call signaling (NCS),
    11–12
  post-office (POP), 411
  program and system information (PSIP),
    409
  simple network management (SNMP),
    411
  simple message transfer (SMTP), 411
  trivial file transfer (TFTP), 360,
    391–392
Public key encryption, 247–248
Public switched telephone network
  (PSTN), 2, 10–11
Punctured convolutional coding,
  193–195, 198
Pump lasers, 161–162, *see also* Lasers
Push pull amplifiers, 90, 139–140

Q

Quadrature amplitude modulation

(QAM), *see* Modulation
QPSK, *see* Upstream Burst Transmitter
Quality of service (QoS), 384–388

R

RSA, cryptosystem, 247–248
Randering, 255–257
Random access methods, 366–369
Random access memory (RAM),
  249–250, *see also* Memory
Randomizer, 201–202, 242–243, 364
Ranging request, 379, 391, *see also*
  Cable Modem
Ranging response, 379, 391, *see also*
  Cable Modem
Rate equations, 75–76, 156–157
Rayleigh distribution, 282, 291
Rayleigh scattering, 299–300
Read-only-memory (ROM), *see* Memory
Registration:
  request, 379, 392
  response, 379
  acknowledge, 379
Receiver:
  basic configurations, 138–144
  front-end design, optical, 140–142
  M-ary QAM, operation, 206–208
  nonlinear behavior, p-i-n
   photodetectors, 135–138
  optical, configurations, 133–139
  out-of-band (OOB),8, 232, 235–239,
   248–250, 260–261, 419
  quantum efficiency, 128–129
  responsivity, 127–128
  transimpedance front-end design,
   141–143
Recursive least square (RLS) algorithm,
  212

Real time operating system (RTOS), 395–398

Reed Solomon (R-S) code, 183–185, 188–193, 227, 236, 238, 241–242, 267, 363–364, 419, 421

Resonance frequency, laser, 77–78

RMS modulation index, 91, 291–292

S

SECAM, signal format, 27–29

S-Aloha, 380

Saturated noise figure, 174–175

Secure sockets layer (SSL) protocol, 409–410

Semiconductor structures, *see* Lasers,

Semiconductor lasers, *see* Lasers

Self-phase modulation (SPM), 316–319, 342–344, *see also* Modulation

Service ID (SID), 374, 385, 387–389, 391

Service flow ID (SFID), 374

Set-top box:
  architecture, hardware, 231–233, 249–250, 260–261
  conditional access, 246–250
  CPU, 257–258
  graphics processor, 250–257
  memory, 249–250
  MPEG demultiplexer and decoder, 235–237
  on-screen video, 255–257
  renewable security, 249–251
  RF tuner, 231–235, 250, 260–261, 263–265
  system budget link, 273–274
  upstream transmitter, 239–243

Skin effect, 43

Simple message transfer protocol

(SMTP), 411

Signaling system number 7 (SS7), 11

Spectral-hole burning (SHB), 75–76

Spontaneous emission, 76, 154, 156–157

Square-root raised cosine (SRRC) filter, *see* Filters

Stimulated Brillouin scattering (SBS), 119, 312–316, 326

Stimulated emission, 75–76, 154, 156, 161

Stimulated Raman scattering (SRS), 319–323

Subcarrier multiplexed (SCM) lightwave systems, 91, 94, 149, 292, 294, 303–306 319–320

Stark levels, 155–156

Synchronization, upstream packets, *see* Cable Modem

Synchronization trellis group, for QAM, 199–200

T

Thermoelectric cooler (TEC), 86, 88–89

Time-Division Multiplexing (TDM), 6, 352, 363

Timing, burst, 355–357

Timing recovery methods, 217–219
  early-late, 218–219
  forward-acting, 217–218

Transimpedance receivers, 141–143, see *also* Receivers

Transmission convergence (TC) sublayer, DOCSIS protocol, 372–373

Traveling-wave modulator, 101–102

Trellis-coded modulation (TCM), 193–200

Trellis groups, for 64/256-QAM, 191–194

Trivial file transfer protocol (TFTP), *see* Protocol
Tuner, RF double-conversion, 231–235, 250, 260–261, 263–265
Turn-on delay, 290–291
TV platform adaptation kit (TVPAK), 397–398

U

Unsolicited grant service (UGS), 386–388
Unified memory architecture (UMA), 259
Unique word (UW), 240, 421
Uniform resource locator (URL), 411
Universal serial bus (USB), 232, 261
Upstream:
  channel descriptor (UCD), 364, 377–378, 391
  bandwidth allocation map, 365, 378, 383–384, 388
  channel change request, 379
  channel change response, 379
  fragmentation, 388–390
  channel parameters, 365–366
  service flow scheduling, 386–388
Upstream burst transmitter:
  block diagram, 239–242, 367
  forward-error-correction coding, 242–243, 363–365
  packet format, 240
  parameters, RF, 241, 419–421
  PHY layer, cable modem, 363–365
Upstream spurious output, cable modem, 370–371

V

Video on demand (VOD), 8, 402, 405 407–409
Video signals:
  DigiCipher I and II, 35–36
  NTSC system, 21–27
  MPEG-1 standard, 29–33, 37
  MPEG-2 standard, 33–34
  MPEG-4 standard, 35–36
  PAL system, 27–28
  SECAM system, 28–29
Visual carrier, 22, 24–25
Vertical blanking interval (VBI), 24, 245–246
Viterbi decoding, 195–197

W

Web browsers, set-top box, 409–411
Webcasting, 410–411
Window CE operating system, 397–398

Y

YAG laser transmitter, 116–118
Yttrium iron Garnet (YIG), 88
Y junction coupler, 102–103

Z

Z buffering technique, 253–254